Nutritional Neurosciences

Series Editor
Mohamed Essa, Sultan Qaboos University, Qaboos, Oman

This book series aims to publish volumes focusing on both basic and clinical research in the field of nutritional neuroscience with a focus on delineating the effect of nutrition on brain function and behavior. The books will examine the role of different nutrients, food agents and supplements (both macro and micro) on brain health, neurodevelopment, neurochemistry, and behaviour. The books will examine the influence of diet, including phytochemicals, antioxidants, dietary supplements, food additives, and other nutrients on the physiology and metabolism of neurons, neurotransmitters and their receptors, cognition, behavior, and hormonal regulations.

The books will also cover the influence of nutrients and dietary supplements on the management of neurological disorders. It details the mechanism of action of phytonutrients on signaling pathways linked with protein folding, aggregation, and neuroinflammation. The books published in the series will be useful for neuroscientists, nutritionists, neurologists, psychiatrists, and those interested in preventive medicine.

Wael Mohamed • Mohd Basri Mat-Nor
Editors

Nutrition and Stroke

Current State and Future Perspectives

Springer

Editors
Wael Mohamed
BMS Department
International Islamic University
Malaysia (IIUM)
Kuantan, Pahang, Malaysia

Mohd Basri Mat-Nor
Anesthesiology and Intensive Care
International Islamic University
Malaysia (IIUM)
Kuantan, Pahang, Malaysia

ISSN 2730-6712 ISSN 2730-6720 (electronic)
Nutritional Neurosciences
ISBN 978-981-97-6817-2 ISBN 978-981-97-6818-9 (eBook)
https://doi.org/10.1007/978-981-97-6818-9

This Springer imprint is published by the registered company Springer Nature Singapore Pte Ltd.
The registered company address is: 152 Beach Road, #21-01/04 Gateway East, Singapore 189721, Singapore

I express heartfelt gratitude and deep appreciation as I dedicate this work to my wife, Dr. Rehab Ismaeil. Your unwavering support and unconditional love have been constant pillars throughout my academic journey and beyond. I am truly grateful for the love and inspiration you bring to my life, and I acknowledge that none of this would have been possible without you.
Wael Mohamed
Pahang, Malaysia

Preface

> Exploring the intricacies of illness without books is like embarking on an uncertain voyage, while delving into books without the insights of patients is akin to remaining anchored to the shore.—Sir William Withey Gull (1816–1890)

Recently, the impact of nutrition and food intake has been highly investigated to study its impact on our brain function and development as it was shown that the diet we take will determine the outcome of certain brain disorders such as brain injury and stroke. Along with its effects on cardiovascular diseases and cancer development, nutrition and diet have been shown to be involved in preserving our mental cognitive function and behavior. Recent research have shown that the onset or worsening of certain strokes is associated with an imbalance in nutritional intake, as evidenced by the manifestation of stroke symptoms and indications.

These findings have been driven by revolutionary application of different "omics" fields and their application to studying the central nervous system (CNS) which broadened our understanding of fundamental neurobiological processes and has enabled the identification of proteins and pathways related to the complex molecular mechanisms underlying various diseases of the CNS. In fact, among these disciplines is the field of proteomics that evaluates the role of protein alterations in strokes, aiming to identify biomarkers of such disorders. Furthermore, the fields of metabolomics and microbiome assessment have emerged to study the role of the gut serotonin secretion and how it is implicated by "good" bacteria contributing to our sleep cycle, moods, and pain. Surveying the literature, we have noticed that there is a huge knowledge gap that discusses brain health and the role of nutrition in modulating their outcomes. We are not implying that changes to our daily diet may be an alternative substitute for brain health intervention such as medication; however, what we would like to highlight is the role of healthy diet and sound nutrition in alleviating certain strokes. Coming from a background of brain health research, the editors (Drs. Mohamed and Basri) have decided to collaborate with other colleagues with expertise in the areas of strokes and nutrition to address these knowledge gaps. Overall, this new book provides updated and novel concepts in the field of strokes and its relation to food intake. This new compilation will be of high interest among researchers and clinical scientists involved in neurology, nutrition, and biochemistry.

Finally, we thank all the authors for their significant efforts in contributing such excellent chapters for this new edition. We are also sincerely grateful to each author for their patience during the compilation and final editing of this book.

Kuantan, Pahang, Malaysia Wael Mohamed
Kuantan, Pahang, Malaysia Mohd Basri Mat-Nor

Acknowledgments

First, we would like to express a great appreciation to all the authors who contributed to this timely project. The high level of devotion and dedication between the authors and editors made writing this book an enjoyable journey. In addition, we also extend our gratefulness to the authors who are in the fields of medical psychiatry and neuropsychiatric research for delivering years of their experience and work in different areas of psychiatric disorders to deliver such an elegant piece of work. The herein-discussed topics are of great value in the areas of nutrition, psychiatry, neurological disorders, and neurodegeneration. Finally, we would like to thank the encouragement of many of our friends and colleagues for their unconditional love, encouragement, and inspiration throughout the endeavor of the project. Thank you.

Wael Mohamed
Mohd Basri Mat-Nor

About this Book

This book provides a thorough examination of the correlation between nutrition and stroke. It delves into the interplay between nutrition and brain functions, neurogenesis, and cognitive functions. Additionally, it investigates how various types of nutrition are related to stroke. Moreover, it discusses recent progress in recognizing the significant therapeutic role of different nutrients in addressing stroke. Finally, it offers an overview of nutrients as neuroprotective agents and outlines the fundamental principles of nutrigenomics.

Contents

Editors and Contributors

About the Editors

Wael Mohamed a Physician Neuroscientist, earned his Ph.D. from PSU, USA, and currently holds the position of Professor Madya at IIUM Medical School, Malaysia. With over 200 speaking engagements both domestically and internationally, Dr. Mohamed is a seasoned lecturer. He has authored more than 150 peer-reviewed papers in the field of neuroscience/psychiatry, boasting an h-index of 23. Additionally, he serves as an editor for several international Journals, having contributed to numerous journal special issues focusing on brain disorders. Furthermore, he is involved in editing several neuroscience books for various publishers. Dr. Mohamed has received research grants from esteemed national and international organizations including IBRO, ISN, MJF, STDF, FRGS, and INDO-ASEAN, totaling half a million US dollars in research funding. He is also the founder of the AfrAbia-PD-Genomic Consortium (AA-PD-GC).

Mohd Basri Mat-Nor is a Senior Lecturer at Department of Anaesthesiology and Intensive Care, Kulliyyah (School) of Medicine, International Islamic University Malaysia (IIUM). He is also a Senior Consultant in Intensive Care and Head of Intensive Care Service at IIUM Medical Centre. He is one of the pioneers in intensive care medicine, a specialist in the country, and has been elected as President of Malaysian Society of Intensive Care since 2017. He obtained his medical degree from Royal College of Surgeons in Ireland (RCSI) and National University of Ireland, Dublin, in 1990. He obtained M.Med (Anaesthesiology) from University of Malaya in 1998 and then pursued subspecialty training (fellowship) in intensive care medicine for four years in Australia. After serving Hospital Kuala Lumpur for 2 years as specialist, he joined IIUM in 2000. He is one of the members of writing committee for national ICU protocol and expert panel for national ICU antimicrobial guidelines. Currently, he is the Head of Research and Innovation (HORI) of IIUM Medical Faculty. His research areas include sepsis biomarkers, nutrition in critically ill, mechanical ventilation in acute respiratory failure, and antibiotics PK/PD.

Contributors

Fa'iza Abdullah Family Medicine Department, IIUM, Kuantan, Pahang, Malaysia

Taoheed Adedeji Abdulsalam Bioresources Development Centre (BIODEC), National Biotechnology Development Agency (NABDA), Ogbomoso, Nigeria

Wesam I. Abo Elenien Faculty of Pharmacy, Clinical Department, Alexandria, Egypt

Ebele Joan Ajagun Bioresources Development Centre (BIODEC), National Biotechnology Development Agency (NABDA), Ogbomoso, Nigeria

Saba Akram Nuclear Institute for Agriculture and Biology College, Pakistan Institute of Engineering and Applied Sciences (NIAB-C, PIEAS), Faisalabad, Pakistan

Mohamed Mahmoud Rabie Ali Beni-Seuf University, Beni Seuf, Egypt

Huda Aljumayi Department of Food Science and Nutrition, College of Sciences, Taif University, Taif, Saudi Arabia

Manal Almatrafi Department of Food Science and Nutrition, College of Sciences, Taif University, Taif, Saudi Arabia

Muhamad Azamin Anuar Department of Paediatrics, Kulliyyah of Medicine, International Islamic University Malaysia, Kuantan Campus, Kuantan, Pahang, Malaysia

Abraham Olufemi Asuku Bioresources Development Centre (BIODEC), National Biotechnology Development Agency (NABDA), Ogbomoso, Nigeria

Muhammad Asif Atiq Department of Clinical Pharmacology, Kabul University of Medical Sciences, Kabul, Afghanistan

Mahmoodullah Azimi Department of Clinical Pharmacology, Kabul University of Medical Sciences, Kabul, Afghanistan

Nur Athirah binti Azlan School of Graduate Studies (SGS), Management and Science University, Shah Alam, Selangor, Malaysia

Lobna Balouta Faculty of Exact Sciences and Natural and Life Sciences, University of Mohammed Khaider, Biskra, Algeria

Aroosa Faheem Department of Biochemistry, Quaid-i-Azam University, Islamabad, Pakistan

Farah Nadia Mohd Hanafiah Department of Anaesthesiology and Intensive Care, International Islamic University Malaysia, Kuala Lumpur, Malaysia

Azarinah Izaham Department of Anaesthesiology & Intensive Care, Faculty of Medicine, Universiti Kebangsaan Malaysia, Kuala Lumpur, Malaysia

Lukman Akanni Jayeoba Bioresources Development Centre (BIODEC), National Biotechnology Development Agency (NABDA), Ogbomoso, Nigeria

Manar Kahoul Faculty of Medicine, University of Constantine, Constantine, Algeria

Nazeha A. Khalil Nutrition and Food Sciences Department, Faculty of Home Economics, Menoufia University, Shibin El-Kom, Egypt

Rony Ibne Masud Department of Microbiology and Hygiene, Bangladesh Agricultural University, Mymensingh, Bangladesh

Wael Mohamed Basic Medical Science Department, Kulliyyah of Medicine, International Islamic University Malaysia (IIUM), Kuantan, Pahang, Malaysia

Clinical Pharmacology Department, Menoufia Medical School, Menoufia University, Menoufia, Egypt

Farwa Munir Faculty of Medicine, University of Sultan Zainal Abidin, Terengganu, Malaysia

Toheeb O. Oyerinde Laboratory for Experimental and Translational Neurobiology, University of Medical Sciences, Ondo, Nigeria

Muhammad Danial Che Ramli Faculty of Health and Life Science, Management and Science University, Shah Alam, Selangor, Malaysia

Iqbalmunauwir Ab-Rashid Department of Anaesthesiology and Intensive Care, Kulliyyah of Medicine, International Islamic University Malaysia, Kuala Lumpur, Malaysia

Hammed A. Sangodipe Department of Medicine, Ladoke Akintola University of Technology, Ogbomoso, Oyo State, Nigeria

Wasiu Sanusi Department of Biochemistry and Nutrition, Fountain University, Osogbo, Nigeria

Ina Ismiarti Shariffuddin Department of Anaesthesiology, Faculty of Medicine, Universiti Malaya, Kuala Lumpur, Malaysia

Kamrul Hassan Suman Ministry of Fisheries and Livestock, Dhaka, Bangladesh

Shadia Tasnim Department of Pathology, Bangladesh Agricultural University, Mymensingh, Bangladesh

Misya Afiqah binti Noor Tuah Faculty of Health and Life Science, Management and Science University, Shah Alam, Selangor, Malaysia

Al-Hassan Soliman Wadan Faculty of Dentistry, Sinai University, Arish Campus, Arish, North Sinai Peninsula, Egypt

Sinai University Research Center, Sinai University, Arish, Egypt

Ayesha Wadood Nuclear Institute for Agriculture and Biology College, Pakistan Institute of Engineering and Applied Sciences (NIAB-C, PIEAS), Faisalabad, Pakistan

Hamza Yassaad Department of Dentistry, Faculty of Medicine, University of Constantine, Constantine, Algeria

Part I

Understanding Stroke

Navigating the Landscape: Understanding Stroke Epidemiology

1

Ina Ismiarti Shariffuddin

Abstract

Stroke represents a significant global health challenge, ranking as the second-leading cause of mortality worldwide and the third-leading cause of death and disability combined. This chapter comprehensively overviews stroke epidemiology, risk factors, implications, and future perspectives. The incidence of strokes has surged dramatically over recent decades, particularly in low-income and middle-income countries (LMICs), posing profound implications for global health demographics. Understanding stroke epidemiology is crucial in addressing this public health burden, allowing for the design of targeted interventions and resource allocation. Stroke risk factors can be categorised as modifiable and non-modifiable, encompassing age, gender, genetics, lifestyle, medical conditions, and environment. Hypertension, diabetes mellitus, obesity, smoking, and air pollution emerge as key modifiable risk factors significantly influencing stroke occurrence. Stroke presents various subtypes, including ischemic stroke, haemorrhagic stroke, transient ischemic attacks (TIA), and cryptogenic strokes, each with distinct characteristics and implications. Effective stroke management involves both immediate treatment, such as thrombolysis and thrombectomy, and long-term rehabilitation and secondary prevention strategies. Global health disparities contribute to variations in stroke incidence and outcomes, with LMICs facing increasing stroke risk correlated with socioeconomic shifts. Access to quality medical care and awareness of stroke prevention play pivotal roles in reducing these disparities. Stroke prevention and therapeutic strategies rely on multidisciplinary collaboration and innovative approaches, such as artificial intelligence (AI) for precision medicine and advancements in stem cell technolo-

I. I. Shariffuddin (✉)
Department of Anaesthesiology, Faculty of Medicine, Universiti Malaya,
Kuala Lumpur, Malaysia
e-mail: ismiarti@ummc.edu.my

W. Mohamed, M. B. Mat-Nor (eds.), *Nutrition and Stroke*, Nutritional Neurosciences, https://doi.org/10.1007/978-981-97-6818-9_1

gies and genomics for regenerative therapy. Addressing challenges in stroke research, including data quality and accessibility to care, will be essential for improving stroke management globally. In conclusion, a comprehensive understanding of stroke epidemiology, risk factors, and implications is essential for effective prevention, treatment, and management, ultimately reducing the burden of stroke on individuals, healthcare systems, and societies worldwide.

Keywords

Stroke · Epidemiology · Risk factors · Disease management · Artificial intelligence

1.1 Introduction

Stroke is the second leading cause of mortality worldwide and is the third leading cause of death and disability combined, as measured by disability-adjusted life-years lost (DALYs) [1, 2]. In 2019, it accounted for 11.6% of all reported deaths, profoundly impacting global health demographics. Between 1990 and 2019, incidents of strokes have surged by 70%, with the prevalence of stroke increased by 85%, its mortality increased by 43%, and DALYs due to stroke increased by 32%. The increase in stroke burden is more significant in low-income and middle-income countries (LMICs) than in high-income countries (HICs) [3].

Stroke is a significant contributor to morbidity and mortality in Asia. However it has varying patterns across different Asian countries. The incidence rates range from 116 to 483 per 100,000 individuals annually [4]. High-income countries like Japan and Singapore exhibit lower mortality rates than low-income countries like Indonesia and Mongolia. In Japan, the mortality rate is 43.3 per 100,000 person-years, with a burden of 706.6 per 100,000 people. Similarly, Singapore reports a mortality rate of 47.9 per 100,000 person-years, coupled with a burden of 804.2 per 100,000 people. In contrast, Indonesia and Mongolia record a higher mortality rate of 193.3 per 100,000 person-years, with a burden of disease of 3382.2 per 100,000 people and a mortality rate of 222.6 per 100,000 person-years and a burden of disease of 4409.8 per 100,000 people respectively [5].

1.2 Definition

The term 'stroke' originates from the Greek word 'apoplexia', suggesting a severe and fatal blow [6]. In 1970, stroke was defined by the World Health Organisation, as 'rapidly developed clinical signs of focal (or global) disturbance of cerebral function, lasting more than 24 h or leading to death, with no apparent cause other than of vascular origin' [6]. However, this definition only concentrates on the clinical symptoms displayed by the patients. Stroke symptoms, ranging widely, often manifest as facial, arm, or leg weakness on one side, comprehension issues, or severe headaches. With the advancement of radio-imaging, the definition was updated by the American Stroke Association for the twenty-first century to a broader definition and it includes any objective evidence of permanent brain, spinal cord, or retinal cell death attributed to a vascular aetiology based on pathological or imaging evidence with or without the presence of clinical symptoms. Diagnosis entails assessing blood flow or utilizing brain imaging like CT, MRI, or Doppler sonography. The CNS infarction also includes infarction attributable to haemorrhage [7].

1.3 Risk Factors

The risk factors for strokes can be divided into modifiable and non-modifiable. Non-modifiable risk factors include age, gender, and genetics, while modifiable risk factors include lifestyle, diet and smoking, medical conditions, and environment. The modifiable factors were listed as the top leading factors for strokes [3].

1. Non-modifiable factors:
   (a) Age:

   Stroke is not confined to the elderly, as it can also affect children and young adults. While strokes in children are rare, with an incidence of 2.5–13 per 100,000 per year, boys exhibit higher prevalence and incidence than girls, with black children facing a greater risk than their Caucasian and Asian counterparts [8]. The mortality rate for paediatric strokes ranges from 10% to 25%, emphasizing the severity of the condition, with haemorrhagic strokes in children having a considerably higher mortality rate compared to ischemic strokes.

   The incidence of stroke increases with age, and it doubles after the age of 55 years old [9]. Recent research indicates a concerning trend of rising stroke incidence among younger demographics. From 2005, individuals aged 20–54, irrespective of race, exhibited significantly elevated incidence rates compared to previous periods. This bears significant public health implications, as strokes in younger individuals pose a higher potential for lifelong disability [10].

   (b) Gender:

   Stroke risk concerning gender varies with age. In younger years, women exhibit an equal or higher risk than men, likely influenced by pregnancy, the postpartum period, and hormonal factors like contraceptive use. At older ages, men's relative risk edges slightly higher. However, more strokes occur in women due to their longer lifespans. It was estimated that strokes are higher in females (6.4 million incident strokes) than in males (5.8 million incident strokes). Still, the age-standardized death rates were more significant in males than in females (96.4 per 100,000 vs 73.5 per 100,000) and DALY rates (2024.3 per 100,000 vs 1531.3 per 100,000) [11].

   (c) Genetics:

   Evidence from heritability studies on strokes concerning twins and family history suggests that genetic predisposition is an essential factor for stroke—the risk of an individual with a family history of stroke increases by

30% [12]. Genetic variability can influence stroke risk through diverse mechanisms. Rare single-gene disorders may lead to familial syndromes where stroke is the primary manifestation (e.g., CADASIL). Conversely, in conditions like sickle cell anaemia, stroke may manifest as part of a broader multisystem disorder. Additionally, genetic predispositions to conventional stroke risk factors, such as atrial fibrillation and diabetes, increased the likelihood of experiencing a stroke. Understanding these genetics is crucial for effective risk assessment and targeted intervention strategies in stroke prevention. By elucidating the diverse mechanisms through which genetic variability influences stroke susceptibility, healthcare professionals can tailor preventative measures and treatment approaches to individual patients [13].

2. Modifiable risk factors:

The incidence of stroke is related to several risk factors such as lifestyle, medical conditions and environment. It was shown that the five leading risk factors for stroke were high systolic blood pressure (SBP; contributing to 55.5% of total stroke DALYs), obesity, Diabetes Mellitus, smoking and ambient particulate matter pollution [11].

(a) Hypertension:

Hypertension plays a pivotal role in the occurrence of ischemic and haemorrhagic strokes. Hypertension is diagnosed when blood pressure exceeds 130/80 mmHg and it is the most important modifiable risk factor for stroke, with a linear, and continuous correlation with blood pressure and stroke incidence [14]. Thus, regular blood pressure screenings are essential for early detection and intervention in hypertension, as timely identification allows for proactive measures to be taken.

For individuals with hypertension, a comprehensive approach involving lifestyle adjustments and medical therapy is necessary to attain a normal blood pressure level and thus reduce the risk of stroke [15]. Lifestyle modifications encompass dietary changes, regular exercise, and stress management. Medical interventions, such as prescribed medications, complement these changes to ensure blood pressure control. Striving to maintain blood pressure within the recommended range not only addresses hypertension but also diminishes the risk of associated complications, particularly in the context of ischemic and haemorrhagic strokes.

(b) Diabetes Mellitus:

Diabetes Mellitus (DM) is an independent risk factor for stroke. DM doubles the risk of stroke in patients. Having diabetes for over 3 years amplifies the stroke risk by 74% [13]. While a glycated haemoglobin goal below 7% is recommended to prevent microvascular complications in Type II diabetes, intensive glucose control does not seem to diminish the risk of stroke. Alarmingly, pre-diabetic individuals also are at a higher risk of developing stroke [16]. Ultimately, stroke contributes to about 20% of deaths among individuals with DM. The use of combined behavioural modification and medical therapy in people with diabetes has been shown to reduce the risk of stroke [17]. This emphasizes the urgent need for proactive measures to manage and prevent diabetes, as well as to implement targeted interventions to reduce the incidence of stroke among diabetic and pre-diabetic populations.

(c) Obesity:

Obesity is an established independent risk factor for stroke, [18, 19] with a 18% increase in stroke risk for every five-point surge in Body Mass Index (BMI) [20]. The linear relationship shows that people who are overweight or obese have the highest risk of both haemorrhagic and ischemic strokes [21]. The risk of stroke is higher in young obese patients have, which also correlates with increased mortality. Hypertension emerges as the major mediator between higher BMI and stroke risk, showcasing a direct, linear connection to cardiovascular disease likelihood [20].

(d) Smoking:

Active smoking escalates stroke risk 2–4 times in comparison to lifelong non-smokers and individuals who have quit smoking more than 10 years, while second-hand smoke exposure increases the stroke risk by 25% [22]. An increase in stroke incidence occurred in a linear proportion with rising cigarette consumption, with each additional increment of 5 cigarettes per day raising the risk of stroke by 12% [23].

Numerous mechanisms have been proposed to link primary and environmental tobacco smoke exposure to increased stroke and heart disease risk. Smoking increases carboxyhaemoglobinaemia and platelet aggregability, elevated fibrinogen levels, reduced HDL-cholesterol, which leads to carotid atherosclerosis [24]. In addition, the direct toxic effects of compounds like 1,3-butadiene, found in environmental tobacco smoke, accelerate atherosclerosis as shown in animal models [25].

Reducing this risk of stroke in smokers is achievable through smoking cessation, combining counselling with medications like nicotine replacement, bupropion, or varenicline. This effective strategy not only addresses immediate health concerns but also emphasizes the crucial role of comprehensive interventions in combating the dangers associated with both active and passive smoking.

(e) Air pollution:

The increase in air pollution levels has been associated with 14% of all stroke deaths [1]. This is particularly prominent in countries with high mean air pollution levels, affecting predominantly low-income and middle-income nations. Particulate Matter, a complex blend of solid and liquid particles containing organic and inorganic substances, is closely linked to elevated mortality rates. Specifically, fine particles with a diameter of ≤ 2.5 μm (PM2.5), generated from fossil fuel combustion, can infiltrate the bronchioles, alveoli, and bloodstream, triggering systemic harm [26].

In a large systematic review and meta-analysis, it was found that approximately 2.2 million individuals with long-term exposure to PM2.5 levels were associated with stroke [27]. In fact, the risk of stroke increases by 31%, for every 10 $\mu g/m^3$ increase in PM2.5 levels [28]. The increased risk of strokes due to air pollution is observed the most in patients with stroke secondary to large-artery atherosclerosis and small-vessel occlusion.

1.3.1 Stroke Subtypes

Stroke, a cerebrovascular event, can be broadly categorized into ischaemic stroke, haemorrhagic stroke, transient ischaemic attacks (TIA) and cryptogenic, each with distinct characteristics and implications. The predominant type, ischaemic stroke, stands as the most frequent subtype, constituting approximately 62.4% of all reported stroke cases across the globe in 2019 [2]. Ischaemic occlusion initiates thrombotic and embolic conditions in the brain. Thrombosis arises from vessel narrowing due to atherosclerosis, leading to clot formation and thrombotic stroke. In embolic strokes, reduced blood flow due to embolism induces severe stress and premature cell death (necrosis). Necrosis triggers plasma membrane disruption, organelle swelling, and leakage of cellular contents into extracellular space, causing loss of neuronal function.

Another type of stroke is haemorrhagic stroke, which accounts for approximately 37.6% of all strokes [2] and has a high mortality rate [29]. This variant occurs due to bleeding within the brain, typically following the rupture of blood vessels or bleeding within brain tissue. The damage caused by a haemorrhagic stroke can be multi-faceted, ranging from the direct pressure exerted by the bleeding to subsequent complications such as oedema or an inadequate blood supply to affected brain regions. The TIA, which is caused by a severe, temporary clot and cryptogenic stroke (unknown cause of strokes), contributes to a smaller percentage of strokes overall.

The consequences of stroke extend far beyond its immediate impact, often resulting in neurological death and long-term adult disabilities. The aftermath of such a stroke imposes an immense burden, both in terms of health repercussions and economic strains, on affected individuals and healthcare systems.

Implications of understanding the epidemiology of stroke.

As there is a projected surge in both cases and incidence rates of stroke across diverse age groups between 2020 and 2030, understanding stroke epidemiology is paramount in addressing the significant public health burden posed by stroke [30]. Comprehensive knowledge of stroke incidence, prevalence, and distribution allows healthcare professionals, policymakers, and researchers to design targeted interventions and allocate resources effectively. Stroke epidemiology provides insights into the diverse risk factors contributing to strokes, ranging from modifiable lifestyle factors to non-modifiable demographic elements, enabling the development of tailored prevention strategies. Identifying trends in stroke occurrence among different populations and demographics helps address health disparities and ensure equitable access to care.

1.3.1.1 Burden of Disease

Stroke burden encompasses mortality, long-term outcomes, and associated costs across diverse dimensions. The global direct expenses of stroke, such as treatment and rehabilitation, and indirect costs to cover the income losses from patients who suffered are tremendous [31]. The escalating burden poses urgent challenges for governments and healthcare systems globally, demanding prompt and comprehensive interventions to manage the economic and societal impacts of stroke effectively. Good stroke epidemiology can commonly be improved by combining it with excellent clinical and diagnostic skills so that stroke can be diagnosed and treated promptly. Nevertheless, stroke management depends on treating its pathophysiology at a timely manner.

Post-stroke patient care addresses both immediate treatment and long-term disabilities.

(i) Immediate treatment:

The immediate intervention is to manage the risk factors that cause the patient to suffer stroke. This means addressing and treating all the modifiable risk factors, especially high systolic blood pressure. Following that, immediate interventions, such as intravenous thrombolysis and thrombectomy, form a critical aspect of stroke care [32]. The principal objective of stroke treatment is to reperfuse the brain by reinstating the blood flow to the brain to reduce neurological damage [29]. Interventions that can achieve this encompass intravenous or intra-arterial thrombolytics, fibrinolytics, craniotomy, or mechanical thrombectomy.

The timely administration of this treatment is paramount, emphasizing the critical importance of swift intervention to maximize benefits and mitigate potential complications in the aftermath of a stroke. However, these procedures are highly specialized and costly, limiting their availability to major hospitals in numerous countries. Many healthcare systems, particularly in resource-constrained settings, face challenges in providing widespread access to these advanced stroke interventions due to their specialized nature and associated costs.

(ii) Rehabilitation care:

The long-term disabilities resulting from strokes significantly impact individuals' quality of life while straining healthcare resources and social support systems. Post-stroke patients suffer from long-term disability and will need rehabilitation care and secondary stroke prevention [33]. Stroke rehabilitation requires a tailored approach, merging physical, occupational, speech, cognitive therapies, and psychological support to promote functional independence. The diverse nature and severity of impairments among patients preclude a one-size-fits-all approach, making stroke management challenging. This individualized strategy acknowledges the diverse needs in post-stroke recovery to enhance the overall functional well-being of stroke patients. All this treatment can lead to substantial societal and economic burdens to the families and eventually to the country's health and economy.

(iii) Secondary stroke prevention:

For secondary stroke prevention, a comprehensive approach encompasses cardiovascular risk reduction, stroke-specific measures based on subtype and cause, and behavioural interventions. These interventions target health and stroke literacy, lifestyle risk factors, and adherence to therapeutic strategies to diminish the likelihood of stroke recurrence.

1.3.2 Global Health Disparities

The incidence of stroke has increased rapidly in low-income and middle-income countries (LMICs), attributable to the shift in the epidemiology, demographic and socioeconomic of the population. Emerging evidence elucidates that the escalating stroke risk and mortality rates in developing nations are concomitant with the upward trajectory of socioeconomic status (SES). Conversely, in developed countries, there is an inverse relationship between increased SES and occurrences of stroke risk and mortality. Introducing new financial resources into historically deprived areas is thought to lead to a change in lifestyle. The population may have adopted unhealthy behavioural patterns, including increased consumption of meat and sugar-rich food and reduced physical activities such as walking as the primary mode of mobility over motorised modes of transportation [34]. On the other hand, in developed countries, improving health literacy has allowed the population to improve their diet and exercise.

The disparities also result from reduced accessibility to quality medical care, which can be attributed to patient, provider, and system factors. Poor awareness of stroke prevention and patient mistrust of the healthcare system can be some of the factors that cause delays in patients obtaining medical treatment. From the provider and system level, the lack of financial support may delay the hospital to provide timely medical intervention for emergency stroke treatment, for example, the prompt transfer of individuals exhibiting symptoms indicative of stroke or transient ischemic attack (TIA) to the Emergency Department (ED) has been shown to not only save lives but also to enhance overall outcomes for affected individuals [35]. Thus, the critical goal in reducing stroke health disparities across the globe is to help developing countries control increasing stroke risk and ensure that the population in these countries obtain the same health benefits as countries in the West.

1.3.2.1 The Future

Prevention and therapeutic strategies following a stroke are indispensable globally to mitigate its impact on public health and alleviate associated economic burdens. However, the current challenges of embarking on stroke research include data quality, surveillance systems, and accessibility to care. Presently, existing data in Asia is collected predominantly from hospital-based studies, demonstrating methodological inconsistencies. Unfortunately, there is a dearth of population-based research, which is most likely due to the poor accessibility to good data collection and well-designed population-based studies in regions with limited financial resources. As a result, most available data stems from hospital-based case series characterised by heterogeneous methodologies, age stratifications, and data analysis techniques, which considerably impedes the prospect of conducting meaningful inter-study comparisons.

Modifying risk factors within a population or individuals is anticipated for stroke prevention. Artificial intelligence(AI) approaches are predicted to allow the application of precision medicine for all individuals, especially those at risk of getting a stroke. With AI, health providers can predict stroke risk for a single patient, thus allowing appropriate medical therapy to be started early in these patients [36]. AI could also be used to predict the outcomes in patients with acute ischemic stroke. A study using a deep neural network model showed that AI could predict a 3-month modified Rankin Scale score better than the traditional Acute Stroke Registry and Analysis of Lausanne score (AUC, 0.888 versus 0.839; $P < 0.001$) [37].

The management of prevention and treatment of stroke lies in multidisciplinary collaboration. The rapid pre-hospital care of stroke, leading to rapid initiation of in-hospital management, such as intravenous thrombolysis and thrombectomy, followed by a good rehabilitation programme, has improved stroke outcomes tremendously [35]. Other recent advances in treating stroke include the development of stem cell technologies and genomics that facilitate regenerative therapy and allow the rebuilding of neural networks and repair of damaged neurons due to ischemic insult.

1.4 Conclusion

Stroke is a complex disease, and its management is challenging. The upward trend in stroke incidence represents a considerable challenge to global public health systems. This chapter highlighted the importance of understanding stroke epidemiology in order to address the growing burden of stroke globally, particularly in developing countries and LMICs. It is crucial to tackle the growing problem of stroke through various approaches covering prevention, treatment, and rehabilitation. This involves supporting policies encouraging healthy habits, detecting and managing risk factors early, and improving access to specialized stroke treatment. Additionally, we should focus on research efforts to understand stroke causes better and develop new treatments.

References

1. Verhoeven, J. I., Allach, Y., Vaartjes, I. C. H., Klijn, C. J. M., & de Leeuw, F. E. (2021). Ambient air pollution and the risk of ischaemic and haemorrhagic stroke. *The Lancet Planetary Health, 5*(8), e542–e552.
2. GBD 2019 Stroke Collaborators. (2021). Global, regional, and national burden of stroke and its risk factors, 1990–2019: A systematic analysis for the Global Burden of Disease Study 2019. *The Lancet Neurology, 20*(10), 795–820.

3. Owolabi, M. O., Thrift, A. G., Mahal, A., Ishida, M., Martins, S., Johnson, W. D., et al. (2022). Primary stroke prevention worldwide: Translating evidence into action. *The Lancet Public Health, 7*(1), e74–e85.
4. Turana, Y., Tengkawan, J., Chia, Y. C., Nathaniel, M., Wang, J. G., Sukonthasarn, A., et al. (2021). Hypertension and stroke in Asia: A comprehensive review from HOPE Asia. *The Journal of Clinical Hypertension (Greenwich), 23*(3), 513–521.
5. Venketasubramanian, N., Yoon, B. W., Pandian, J., & Navarro, J. C. (2017). Stroke epidemiology in South, East, and South-East Asia: A review. *Journal of Stroke, 19*(3), 286–294.
6. Coupland, A. P., Thapar, A., Qureshi, M. I., Jenkins, H., & Davies, A. H. (2017). The definition of stroke. *Journal of the Royal Society of Medicine, 110*(1), 9–12.
7. Sacco, R. L., Kasner, S. E., Broderick, J. P., Caplan, L. R., Connors, J. J., Culebras, A., et al. (2013). An updated definition of stroke for the 21st century: A statement for healthcare professionals from the American Heart Association/American Stroke Association. *Stroke, 44*(7), 2064–2089.
8. Chiang, K. L., & Cheng, C. Y. (2018). Epidemiology, risk factors and characteristics of pediatric stroke: A nationwide population-based study. *QJM: An International Journal of Medicine, 111*(7), 445–454.
9. Roger, V. L., Go, A. S., Lloyd-Jones, D. M., Benjamin, E. J., Berry, J. D., Borden, W. B., et al. (2012). Executive summary: Heart disease and stroke statistics—2012 update: A report from the American heart association. *Circulation, 125*(1), 188–197.
10. Kissela, B. M., Khoury, J. C., Alwell, K., Moomaw, C. J., Woo, D., Adeoye, O., et al. (2012). Age at stroke: Temporal trends in stroke incidence in a large, biracial population. *Neurology, 79*(17), 1781–1787.
11. Feigin, V. L., Brainin, M., Norrving, B., Martins, S., Sacco, R. L., Hacke, W., et al. (2022). World stroke organization (WSO): Global stroke fact sheet 2022. *International Journal of Stroke, 17*(1), 18–29.
12. Flossmann, E., Schulz, U. G., & Rothwell, P. M. (2004). Systematic review of methods and results of studies of the genetic epidemiology of ischemic stroke. *Stroke, 35*(1), 212–227.
13. Boehme, A. K., Esenwa, C., & Elkind, M. S. (2017). Stroke risk factors, genetics, and prevention. *Circulation Research, 120*(3), 472–495.
14. Stansbury, J. P., Jia, H., Williams, L. S., Vogel, W. B., & Duncan, P. W. (2005). Ethnic disparities in stroke: Epidemiology, acute care, and postacute outcomes. *Stroke, 36*(2), 374–386.
15. Schulz, U. G., Flossmann, E., & Rothwell, P. M. (2004). Heritability of ischemic stroke in relation to age, vascular risk factors, and subtypes of incident stroke in population-based studies. *Stroke, 35*(4), 819–824.
16. Sui, X., Lavie, C. J., Hooker, S. P., Lee, D. C., Colabianchi, N., Lee, C. D., & Blair, S. N. (2011). A prospective study of fasting plasma glucose and risk of stroke in asymptomatic men. *Mayo Clinic Proceedings, 86*(11), 1042–1049.
17. Gaede, P., Lund-Andersen, H., Parving, H. H., & Pedersen, O. (2008). Effect of a multifactorial intervention on mortality in type 2 diabetes. *New England Journal of Medicine, 358*(6), 580–591.
18. Strazzullo, P., D'Elia, L., Cairella, G., Garbagnati, F., Cappuccio, F. P., & Scalfi, L. (2010). Excess body weight and incidence of stroke: Meta-analysis of prospective studies with 2 million participants. *Stroke, 41*(5), e418–e426.
19. Kurth, T., Gaziano, J. M., Berger, K., Kase, C. S., Rexrode, K. M., Cook, N. R., et al. (2002). Body mass index and the risk of stroke in men. *Archives of Internal Medicine, 162*(22), 2557–2562.
20. Lu, Y., Hajifathalian, K., Ezzati, M., Woodward, M., Rimm, E. B., & Danaei, G. (2014). Metabolic mediators of the effects of body-mass index, overweight, and obesity on coronary heart disease and stroke: A pooled analysis of 97 prospective cohorts with 1·8 million participants. *Lancet, 383*(9921), 970–983.
21. Darke, S., Duflou, J., Kaye, S., Farrell, M., & Lappin, J. (2019). Body mass index and fatal stroke in young adults: A national study. *Journal of Forensic and Legal Medicine, 63*, 1–6.

22. Shah, R. S., & Cole, J. W. (2010). Smoking and stroke: The more you smoke the more you stroke. *Expert Review of Cardiovascular Therapy, 8*(7), 917–932.
23. Pan, B., Jin, X., Jun, L., Qiu, S., Zheng, Q., & Pan, M. (2019). The relationship between smoking and stroke: A meta-analysis. *Medicine (Baltimore), 98*(12), e14872.
24. Rehill, N., Beck, C. R., Yeo, K. R., & Yeo, W. W. (2006). The effect of chronic tobacco smoking on arterial stiffness. *British Journal of Clinical Pharmacology, 61*(6), 767–773.
25. Penn, A., & Snyder, C. A. (1996). 1,3 Butadiene, a vapor phase component of environmental tobacco smoke, accelerates arteriosclerotic plaque development. *Circulation, 93*(3), 552–557.
26. Ho, J. S. Y., Jou, E., Tan, B. Y. Q., & Sharma, V. K. (2022). Air pollution and ischaemic stroke. *Annals of Indian Academy of Neurology, 25*(Suppl 1), S26–S34.
27. Yuan, S., Wang, J., Jiang, Q., He, Z., Huang, Y., Li, Z., et al. (2019). Long-term exposure to PM(2.5) and stroke: A systematic review and meta-analysis of cohort studies. *Environmental Research, 177*, 108587.
28. Yang, X., Zhang, L., Chen, X., Liu, F., Shan, A., Liang, F., et al. (2021). Long-term exposure to ambient PM(2.5) and stroke mortality among urban residents in northern China. *Ecotoxicology and Environmental Safety, 213*, 112063.
29. Kuriakose, D., & Xiao, Z. (2020). Pathophysiology and treatment of stroke: Present status and future perspectives. *International Journal of Molecular Sciences, 21*(20), 7609.
30. Kupferman, J. C., Zafeiriou, D. I., Lande, M. B., Kirkham, F. J., & Pavlakis, S. G. (2017). Stroke and hypertension in children and adolescents. *Journal of Child Neurology, 32*(4), 408–417.
31. eClinicalMedicine. (2023). The rising global burden of stroke. *eClinicalMedicine, 59*, 102028.
32. Barreto, A. D. (2011). Intravenous thrombolytics for ischemic stroke. *Neurotherapeutics, 8*(3), 388–399.
33. Patel, A. T., Duncan, P. W., Lai, S. M., & Studenski, S. (2000). The relation between impairments and functional outcomes poststroke. *Archives of Physical Medicine and Rehabilitation, 81*(10), 1357–1363.
34. Morgenstern, L. B., & Kissela, B. M. (2015). Stroke disparities: Large global problem that must be addressed. *Stroke, 46*(12), 3560–3563.
35. Clarke, D. J., & Forster, A. (2015). Improving post-stroke recovery: The role of the multidisciplinary health care team. *Journal of Multidisciplinary Healthcare, 8*, 433–442.
36. Attia, Z. I., Noseworthy, P. A., Lopez-Jimenez, F., Asirvatham, S. J., Deshmukh, A. J., Gersh, B. J., et al. (2019). An artificial intelligence-enabled ECG algorithm for the identification of patients with atrial fibrillation during sinus rhythm: A retrospective analysis of outcome prediction. *The Lancet, 394*(10201), 861–867.
37. Heo, J., Yoon, J. G., Park, H., Kim, Y. D., Nam, H. S., & Heo, J. H. (2019). Machine learning-based model for prediction of outcomes in acute stroke. *Stroke, 50*(5), 1263–1265.

2 Understanding Pathophysiological Changes of Stroke: Genetic and Racial Factors for Stroke

Muhammad Danial Che Ramli, Nur Athirah binti Azlan, and Misya Afiqah binti Noor Tuah

Abstract

Stroke is the leading cause of disabilities and cognitive deficits, accounting for 5.2% of all mortalities worldwide. Transient or permanent occlusion of cerebral vessels leads to ischemic strokes, which constitutes most strokes. Ischemic strokes induce brain infarcts, along with cerebral tissue death and focal neuronal damage. Stroke remains a significant cause of mortality and morbidity worldwide, with its pathophysiology influenced by a complex interplay of genetic and racial factors. This chapter aims to elucidate the intricate mechanisms underlying stroke development and progression, focusing on the roles played by genetic variations and racial disparities. Through an extensive examination of current literature, we explore genetic polymorphisms associated with stroke susceptibility, severity, and outcomes, encompassing genes involved in vascular function, inflammation, thrombosis, and neuronal integrity. Furthermore, we delve into the impact of racial and ethnic backgrounds on stroke risk, prognosis, and response to treatment, highlighting disparities in incidence, prevalence, and healthcare access. By integrating insights from genetics and race, this review seeks to enhance our understanding of stroke pathophysiology, inform targeted prevention strategies, and promote equitable healthcare delivery for all populations.

M. D. C. Ramli (✉) · Misya Afiqah binti Noor Tuah
Faculty of Health and Life Sciences, Management and Science University, Shah Alam, Selangor, Malaysia
e-mail: muhddanial_cheramli@msu.edu.my

N. A. b. Azlan
School of Graduate Studies (SGS), Management and Science University, Shah Alam, Selangor, Malaysia

W. Mohamed, M. B. Mat-Nor (eds.), *Nutrition and Stroke*, Nutritional Neurosciences, https://doi.org/10.1007/978-981-97-6818-9_2

Keywords

Stroke · Ischemic stroke · Neurovascular unit · Cerberovascular disease · Neuroinflammation

2.1 Definition of Stroke

A stroke is a complex set of pathophysiological events brought on by hypoxia and ischemia. The pathophysiology of stroke needs thorough understanding of its biology to prevent mortality and disability. Reduce blood supply to the brain is the root of stroke and it divides into ischemic, haemorrhagic, thrombotic and embolic strokes. Among all, ischemic stroke is the most common with the higher statistic in the world compare to other types of strokes [1]. Studies on ischemic stroke and neuronal injury have shown that excitotoxicity effects, oxidative stress injury, inflammation and apoptosis all contribute to neuronal damage [2]. Numerous biochemical pathways and biomarkers play a role in repairing the damage to neurons brought on by ischemic stroke and cerebral infarction.

2.1.1 Ischemic Stroke

Ischemic stroke has a statistic of 85% of all stroke cases, the balance of cases is in other types of strokes. In this case, ischemic stroke is caused by the blockage of an artery in the main cerebral artery either temporarily or permanently that supplying to specific region of the brain, which restricts the oxygen supply and ultimately causes neuronal cell death. It also leads to the increasing in glutamate supply, the discrepancy between neuronal, glial, and endothelial cell energy requirements and availability. The circular of blood after occlusion are go out of way and not uniform. The insufficient of blood supply can lead to neuronal damage in the part of the brain known as ischemic core whereas inactive activity surround the core but cell death contain metabolically active cells known as penumbra [3, 4]. The area of penumbra has a normal oxygen and hypoperfusion consumption which needs the bloody to flow as the infarct area will continue to expand when the blood supply obstruction is not restored. As the case, the infarct region grows as neuronal and glial cell depolarization rises [2].

2.1.2 Excitotoxicity of Ischemic Stroke

Ischemic occlusion leads to insufficient blood flow to the brain whereas oxygen consumption needed 20% and the ATP must be generated sufficiently via the electron transport chain in the mitochondria to preserve and replenish ionic gradients. Failure of oxidative phosphorylation ischemia leads to the ATP been consumed in range of 2 minutes and gives effects to the imbalance of sodium ion (Na^+) and

Potassium ion (K^+) channels ATPase pump. Glucose and oxygen deficiency during ischemia leading to the depolarization of the plasma membrane and prompting glutamate release, the release of (K^+) into the extracellular space, and the entrance of sodium into the cells [5]. The dramatic will rice in intracellular calcium concentration due to failure in Ca^{2+} pump during ischemia. This results in cell death in the ischemia core by activating many death-signalling proteins, including lipases, DNases, and calcium-dependent proteases [2]. NMDARs and Na^+/Ca^{2+} channels are then stimulated by the latter [6]. Increased Ca^{2+} influx disturbs ionic balance, which causes Ca^{2+} overload in the cytosol and mitochondria. Numerous proteases, lipases, kinases, phosphatases, endonucleases, and free radicals are all stimulated by these modifications [7, 8] as well as physiologic processes that lead to cell death, include mitochondrial dysfunction, oxidative stress, and calpain activation.

2.1.3 The Getaway to Excitotoxicity of Glutamate Receptors

The vesicles fuse with the vesicular membrane in response to an increase in intracellular calcium concentration, releasing excitatory neurotransmitters like glutamate. An increase in glutamate-induced hyperstimulation raises the amount of Na and Cl ions in neurons as well as the input of calcium [9]. Besides, excitotoxicity may result from the activation of glutamate receptors of the N-methyl-D-aspartate (NMDA) and a-amino-3-hydroxy-5-methyl-4-isoxazolepropionic acid (AMPA) types, which cause membrane depolarization and make it easier for calcium ions to enter neurons. The vesicles fuse with the vesicular membrane in response to an increase in intracellular calcium concentration, releasing excitatory neurotransmitters like glutamate. NMDARs is actually responsible for neuronal damage after stroke NMDARs have a double-edged effect even though they are involved in excitotoxicity related to ischemic stroke. Recent study from Pajarillo et al., in [6] shows excitotoxicity during ischemic stroke and neuronal apoptosis is triggered by activation of NMDARs carrying the GluN2B subunit whereas GluN2A subunit has a protective impact on neurons. In addition, NMADRs are the most ionotropic glutamate receptors that are calcium-permeable. The combination of both NMDARs and AMPA receptors as blocking inhibits calcium entry and has neuroprotective effects in focal ischemia models.

Raising the concentration of calcium in the mitochondria causes excitotoxicity by causing the release of reactive oxygen species (ROS), depolarization of the mitochondria through opening the permeability transition pore [10], induction of calcium dysregulation, and ultimately, the induction of neuronal death. Thus, the brain can cause death and edema when the ROS increase due to excessive in Ca. Moreover, the mitochondrial NCX (mNCX) largely contributes to the delayed cytoplasmic calcium loading during excitotoxicity [11] and it occurs as a result of high calcium consumption that impairs mitochondrial performance. The enzyme and protease will release because of the free-radical formation. Through mitochondrial transition permeability pores, mitochondria release cytochrome C, which activates caspase and causes DNA cleavage and initiates apoptosis [12]. The protein in mitochondria that primarily controls cell death is called BCL-2. These cause the cell to lose its homeostasis, degrade proteins and enzymes, and eventually cause the cell death [13].

2.1.4 Oxidative Stress and Nitrative Stress

Upon cerebral ischemia, the body will produce more free radicals than antioxidant defense system and ATP production is declines together the mitochondrial respiratory chain is damaged as it can naturally able to overpowers the endogenous scavenge, which is accompanied by reactive oxygen and nitrogen species (ROS/RNS) [14, 15]. Oxidative stress caused by excessive ROS accumulate in tissue and it can help trigger the signaling pathways that cause apoptosis in cells. Furthermore, ROS are physiological mediators that control immune system activity, preserve redox homeostasis, and participate in numerous pathways [16]. A prior study from Margaill et al., found that there are four mechanisms that cause an excess of reactive oxygen species (ROS) to be produced: the respiratory chain of the mitochondria; NADPH oxidases; the reaction of arachidonic acid catalyzed by cyclooxygenase 2; and xanthine and hypoxanthine via xanthine oxidase. The primary source of ROS during the early stages of a stroke is mitochondria, then, the generation of ATP produces when there is lack of oxygen and glucose in the brain tissue. Directly, the concentration of calcium in neurons is increasing due to lack supply of ATP, and resulting in the enormous depolarization of the mitochondria and production of ROS [17, 18]. The activation of NADPH oxidase that known as "oxygen burst" produce ROS in other cells such vascular endothelial cells. Furthermore, several significant molecules present in cellular signaling pathways, such as protein kinases and ion channels, may be directly regulated by oxidative stress [19, 20].

Besides, its potent oxidation capacity disrupts the blood-brain barrier (BBB), causing irreversible harm to the metabolism of nucleic acids and carbohydrates. It also causes the peroxidation of lipids, proteins, and unsaturated fatty acids on the basement membrane [21, 22]. Indirect effects of oxidative stress also include the BBB permeability via controlling the distribution and expression of tight junction proteins and encouraging the production of these causes of inflammation metabolic process. In ischemic stroke, both core and penumbra have different oxidative level where severe oxidative stress in the ischemic penumbra triggers apoptosis by activating the NF-κB and MAPK signaling pathways, but in the ischemic core it can cause cell necrosis by breaking down essential cell components.

2.1.5 Inflammation

Inflammation is a critical step after ischemia and plays important role in the development of ischemic stroke, immune cells, cytokines and neuro-inflammatory response. Tumor necrosis factor-α (TNF-α), platelet activating factor (PAF), interleukin-1β (IL-1β), interleukin-6 (IL-6), and interferon (IFN-γ) are produced by activated astrocytes and microglia during cerebral ischemia [23, 24]. These inflammatory cytokines have two opposing effects as they enhance the inflammatory response and draw in immune cells. They also encourage signal cascade processes and boost the synthesis of stimulating signals like chemokines and free radicals. However, microglia cells activated in brain injury in which is resident macrophages in brain and they become highly active [25]. Microglia get activated during ischemia and develop into phagocytic cells, which emit a range of chemicals that are either cytotoxic or cytoprotective. Microglia produce neurotrophic factors such brain-derived neurotrophic factor (BDNF) and insulin-like growth factor I (IGF-I), which have neuroprotective effects on the brain. The initial purpose of microglia is to protect neurons, when there is over activation, it can lead to neuronal death [26].

In ischemia stroke, neutrophil and monocytes has been reported to be increase even though blood brain barrier (BBB) blocked blood inflammatory cells like neutrophils and monocytes from penetrating the brain, and the general consensus was that microglial cells were the sole ones causing brain inflammation [27]. Similar to microglial cells, astrocytes have the ability to distribute inflammatory mediators like cytokines, chemokines, and NO [28]. Others cytokines that involved in ischemic inflammation are including T lymphocytes, mononuclear phagocytes, natural killer (NK) cells, and poly morph nuclear leukocytes. During stroke, NK cells invade brain lesions caused by ischemia. Additionally, by boosting local inflammation and neuronal excitability and ultimately causing neuronal death, NK cells worsen the brain infarction [29, 30]. Lymphocytes and neutrophils migrate into the brain during ischemic stroke after the initial injury, followed by macrophages and natural killer cells [31]. Neutrophils also infiltrate the brain through the rupture of the BBB, which is promoted by cell adhesion molecules such as immunoglobulins, integrins, and selectins. Proteolytic enzymes and oxygen-free radicals are released when neutrophils enter an ischemic brain, causing tissue damage.

2.2 Pathophysiology of Hemorrhagic Stroke (Intracerebral Hemorrhage and Subarachnoid Hemorrhage)

Hemorrhagic stroke makes for 15% of all stroke cases. Approximately two thirds of stroke cases are classified as intracerebral hemorrhage (ICH) or subarachnoid hemorrhage (SAH) based on anatomic compartmentalization. A blood artery rupture that causes bleeding into the brain is the cause of a hemorrhagic stroke. ICH is known as a bleeding into the brain parenchyma, and SAH is bleeding into the subarachnoid space of the brain. This form of stroke is negative consequences stem from: (a) hypoxia brought on by a disturbed vascular supply; (b) the irritating effect of blood leaking on the brain parenchyma and vasculature; and (c) elevated intracranial pressure (ICP) from ongoing bleeding, which may further reduce cerebral blood flow [32]. The deadly kind of stroke is intracerebral hemorrhage (ICH), which has a 40% in-hospital death rate. 75% of patients are either dead or seriously incapacitated a year following the hemorrhage [33].

2.2.1 Intracerebral Hemorrhage (IHC)

As contrast to ischemic stroke, ICH is linked to increased rates of morbidity and death [34]. Primary and secondary ICH are separated. Primary ischemic heart disease (ICH) is caused by cerebral amyloid angiopathy (CAA) and hypertension, with the former accounting for roughly 65% of cases [35]. Secondary cerebral cavernous malformations, aneurysms, arteriovenous malformations, brain tumors, and coagulopathy can all induce haemorrhages [36].

2.2.1.1 Vessel Rupture

Pathophysiology of ICH happens in vessel rupture where it seems that a single vessel bursts to provide an initial hematoma, which is the first sign of hypertensive intracerebral hemorrhage. It is called chronic hypertension when vessel rupture happens which long before the event begins slowly deteriorates the vessel walls. The morphology of the cells of chronic hypertension will lead to smooth muscle cells proliferate, necrosis, and are eventually replaced by collagen, which is more brittle and structurally rigid than the typical components of the vessel wall [37]. The arterioles that affected with hematoma will typically dilate and form Charcot-Bouchart aneurysms. The size of the originally burst vessel, the features of the vessel wall opening through which blood can escape, and the vascular system's ability to control the bleeding are most likely to determine the size of the initial hematoma expansion [38].

Intracerebral hemorrhage (ICH) caused by cerebral amyloid angiopathy (CAA) the pathophysiological sequence differs from other types of hemorrhagic strokes. This condition involves the deposition of ß-amyloid peptide (Aβ) in the media of small- and medium-sized leptomeningeal and cortical vessels surrounding vascular smooth muscle cells (SMC). As the disease progresses, Aβ replaces the SMC, becoming present in all layers of the vessel wall. At this stage, microaneurysm formation, fibrinoid necrosis, and perivascular leakage become apparent. The weakened blood vessels, affected by these changes, can rupture, leading to cerebral microbleeds or lobar ICH. In simpler terms, vessel rupture in this scenario results in small bleeding events within the brain, contributing to the occurrence of hemorrhagic strokes [39]

2.2.1.2 Early Growth of Hematoma

The hematoma causes glia and neurons to malfunction. This leads to cellular edema, neurotransmitter release, oligaemia, and mitochondrial malfunction. Thrombin triggers edema and inflammation by activating microglia. Typically, imaging is not conducted during the hyperacute phase, which refers to the very early stages of the stroke. In this phase, one would anticipate a rapid expansion rate of the hematoma. This rapid growth is likely to cause sudden and pronounced stroke-like symptoms. The term "stroke-like symptoms" refers to the signs and effects commonly associated with a stroke, such as difficulty speaking, weakness on one side of the body, or other neurological deficits. As the hematoma grows, it puts pressure on the surrounding structures, leading to the manifestation of stroke symptoms. It is possible that the initial growth of the hematoma will slow down or cease when the force of the blood jet leaving from the main stream is countered by the counterpressure of the surrounding tissue, which grows in tandem with the hematoma formation. However, bleeding into the hematoma may still occur, and more enlargement may arise from nearby veins finally rupturing and giving way to the strain created by the initial hematoma [38].

2.2.1.3 Hematoma Expansion

The pooled blood forms a hematoma, which is a localized collection of blood within the brain. The hematoma exerts pressure on the brain tissue, causing further damage and potentially leading to increased intracranial pressure. Hematomas can grow at varying rates in different axial directions over time, and they frequently take on asymmetrical shapes [40]. While local variations in the impacted tissue may contribute to this, secondary bursts of nearby vessel clusters may also be the cause. According to a recent study, a higher expansion rate was predicted when a point on the hematoma surface was closer to the hematoma center [41]. This finding suggests that secondary bursts of local clusters of vessels may be responsible for secondary hemorrhage, which may be most effective near the initial bleeding site tissue. Thus, hematomas may tend to swell more globularly as they expand.

2.2.1.4 Blood Brain Barrier Dysfunction

The concentration of different blood degradation components of the blood clot and its impact on the blood brain barrier (BBB) determine the formation of perihematomal edema and the length of time it persists. In the animal model, the formation of the perihematomal edema can be observed in three main phases: (1) the very early phase, which involves the hydrostatic pressure and clot retraction during the first few hours; (2) the second phase, which continues on days 1 and 2 due to the activation of the coagulation cascade and thrombin production; and (3) the third phase, which occurs on day 3 or later and is characterized by blood lysis and hemoglobin's toxic effect on neuron tissue [42].

2.2.1.5 Secondary Blood Brain Barrier Dysfunction

The initial hemorrhage may cause physical disruption to the brain, increased intracranial pressure, and brain herniation, all of which can lead to brain injury and death. Nevertheless, ICH causes subsequent brain injury in the same way as ischemic stroke [43]. The majority of ICH models used in animal models include intracerebral collagenase injections or direct blood injections into the brain. The latter breaks down the basement membrane, which results in bleeding known as vascular disruption. Following blood injection intracerebrally, there is a delayed BBB hyperpermeability [36]. The secondary disruption with BBB in ICH may take part in neuroinflammation by promoting leukocyte migration into the brain, will aid in the development of brain edema (vasogenic edema), and will permit the entry of potentially hazardous chemicals into the brain.

2.2.2 Subarachnoid Hemorrhage (SAH)

Subarachnoid hemorrhage (SAH) is a medical condition characterized by bleeding that occurs between the arachnoid and pia mater, which are layers of the meninges, the protective membranes surrounding the brain and spinal cord in the central nervous system (CNS). SAH can be classified into two main types: traumatic and spontaneous [44]. Traumatic subarachnoid hemorrhage (tSAH) refers to bleeding into the subarachnoid space, which is the area between the arachnoid mater and the pia mater surrounding the brain, as a result of head trauma. This type of bleeding can occur due to various traumatic incidents such as falls, gunshot wounds, motor vehicle accidents, or any other event that causes significant trauma to the head [44]. Spontaneous subarachnoid hemorrhage (sSAH) refers to non-traumatic bleeding into the subarachnoid space, and it is most commonly caused by the rupture of a cerebral aneurysm. Cerebral aneurysms are weak areas in the walls of blood vessels within the brain that can balloon out and potentially rupture, leading to bleeding in the subarachnoid space [45]. An additional subcategory of SAH is an aneurysmal SAH (aSAH).

2.2.2.1 Aneurysm Rupture

Since aneurysmal subarachnoid hemorrhage (aSAH) is the primary cause of brain tissue inflammation, brain tissue nearly entirely depends on aerobic respiration to sustain its metabolism [46]. The most common cause of spontaneous SAH is the rupture of a cerebral aneurysm. Aneurysms are weakened areas in the walls of blood vessels within the brain that can balloon out over time. When an aneurysm ruptures, blood is released into the subarachnoid space at arterial pressure and results in early brain damage (EBI) and acute pathophysiological repercussions [47]. Poor early neurological status is indicative of EBI. Increased intracranial pressure, acute hydrocephalus, cerebral ischemia, tissue shifts and herniations, and cerebral edema are among the effects of aneurysm rupture that cause additional brain injury. Raising cytokine levels create a vicious loop whereby they recruit additional inflammatory cells and further damage brain tissue and the microcirculation. Ischemia can be made worse by seizures, rebleeding, and hydrocephalus. Rebleeding also increases the risk of subarachnoid blood. A more severe initial systemic response is frequently seen in aSAH compared to other cerebral injuries. This is typified by modifications that are hypothalamically driven, such as elevated sympathetic and parasympathetic activity that lead to problems with the heart and lungs [48].

2.2.2.2 Endothelial Dysfunction and Neuroinflammation

Loss of NO production, uncontrolled coagulation (by coagulation cascade malfunction), and increased permeability as a result of glycocalyx damage are the hallmarks of endothelial dysfunction in aSAH. The glycocalyx is a crucial layer that lines the inside of blood arteries because it acts as a vasculoprotective barrier between the endothelium that faces the lumen and the circulating blood by reducing BBB permeability [49].

2.2.3 Transient Ischemic Attack (TIA)

Annually, more than 240,000 individuals in the US undergo a transient ischemic attack, a significant precursor to future strokes. Within 90 days following such an attack, the risk of stroke can surge to 17.8%, with a majority of cases occurring within just 2 days of the initial event. Historically, a transient ischemic attack (TIA) was attributed to localized ischemia in the brain, spinal cord, or retina, excluding acute infarction. The associated neurological symptoms were deemed temporary and reversible, with minimal risk of brain cell damage. Following a transient ischemic attack (TIA), there is a notable risk of cerebrovascular events, cardiovascular issues, and mortality. The origins and mechanisms of both transient ischemic attack (TIA) and ischemic stroke (IS) share similarities, with both conditions involving localized hypoperfusion, oligemia, and disturbances in cerebral oxygenation and glucose metabolism [50].

2.2.3.1 Pathophysiology of TIA

As previously mentioned, transient ischemic attacks (TIAs) resemble ischemic strokes in that they result from a localized reduction in blood flow to the brain, leading to specific neurological symptoms. This diminished blood flow can be attributed to either an embolism in a cerebral supply artery (originating from the heart, large proximal vessels, or intracranial and extracranial arteries, typically affected by atherosclerosis) or the occlusion of small perforating arteries. Resolution of symptoms may occur spontaneously as the obstructing blood clot or embolus dissolves or moves distally, or through compensatory collateral circulation restoring blood flow to the ischemic brain region. Rarely, a transient ischemic attack (TIA) can occur due to localized hypoperfusion caused by significant arterial narrowing. These TIAs often manifest with distinct symptoms and are sometimes associated with an upright posture. The duration and pattern of symptom development or resolution may also be influenced by variables including blood pressure, blood oxygen levels, or blood viscosity.

2.2.3.2 Symptoms Prior to TIA

Transient ischemic attacks (TIAs) present with symptoms akin to those of ischemic stroke, including unilateral limb weakness, speech impairment, sensory issues, visual disturbances, and gait problems. Certain symptoms, like monocular blindness, are more common in TIAs than in strokes, and limb shaking is nearly always present in TIAs. However, symptoms such as limb ataxia or hemivisual field abnormalities are frequently disregarded in TIA cases. Transient neurological symptoms, although commonly indicative of cerebrovascular conditions like TIAs, can also arise from non-vascular disorders. Distinguishing between genuine TIAs and TIA mimics requires comprehensive history-taking, physical examination, and relevant imaging tests. Each TIA symptom varies in specificity and sensitivity, prompting efforts to evaluate patient outcomes based on specific clinical aspects. Advances in imaging technology have further enhanced our ability to accurately predict the prognosis of TIA patients.

2.3 Genetic Risk Factors

Globally, the occurrence of stroke has seen a marked rise, with more than 12 million new cases documented in 2019 alone. Predominantly ischemic in nature, strokes have diverse underlying causes. Despite the emphasis on lifestyle factors such as smoking, diabetes, hypertension, and high cholesterol levels, the substantial influence of hereditary factors is at times overlooked [52]. Early-onset strokes are primarily attributed to genetic factors, with certain individuals exhibiting a heightened genetic predisposition. Single-nucleotide polymorphisms (SNPs), which are genetic alterations, contribute to this susceptibility. However, the findings of genetic testing may be skewed due to the lack of genetic analysis in individuals with milder or later-onset strokes. This bias limits our understanding of the genetic components of stroke and may result in overlooking unique genetic profiles when assessing and managing stroke risk [53].

2.3.1 Family History

The presence of strokes among immediate family members, like parents, siblings, or grandparents, constitutes a family history of stroke, suggesting a potential genetic predisposition within the family. However, ongoing scientific discourse surrounds the precise impact of different family connections on stroke risk. While some studies show a clear link between family history and heightened stroke risk, others present conflicting data regarding the influence of specific familial relationships on stroke susceptibility [54]. This uncertainty highlights a significant research gap, underscoring the importance of further study to develop more precise preventive and therapeutic strategies.

The interplay between genetics, shared environmental factors, and a family history of stroke among first-degree relatives is widely acknowledged. The presence of a family history significantly heightens the risk of experiencing a stroke, as evidenced by numerous studies. However, the precise impact of specific familial relationships on stroke risk remains uncertain. According to a meta-analysis by Floßmann et al. [55] individuals with a positive family history had a remarkable 30% higher chance of experiencing a stroke compared to those without such a history. This notable increase underscores the importance of considering family history when assessing stroke risk. Furthermore, Tian et al. [56] found that individuals with one or more first-degree relatives who had suffered a stroke faced approximately double the risk of having a stroke themselves. These findings highlight the significant impact of familial predisposition on stroke susceptibility, emphasizing the need for tailored preventive interventions in individuals with a positive family history.

Individuals often adopt familial lifestyle patterns, alongside inheriting genetic predispositions to stroke risk factors from their parents, particularly in parent-child interactions. Factors like dietary habits, levels of physical activity, and exposure to second hand smoke, which greatly influence the risk of stroke, are commonly passed down from parents [57]. Seshadri et al. [58] discovered a somewhat stronger correlation between paternal stroke and offspring stroke compared to maternal stroke. Conversely, other studies suggest a more robust association between maternal stroke and offspring stroke [59], this finding was not supported by Seshadri et al.'s investigation. Discrepancies in the methods used to identify and document stroke events, such as variations in the depth of medical histories reported for strokes occurring in mothers versus fathers, may explain the discrepancies in these findings. Additionally, the utilization of survival analysis by Seshadri et al. diminishes the influence of gender-specific survival differences on the observed associations. The study's strengths include a community-based sample, validation of stroke statuses in both parents and offspring, and meticulous classification of stroke subtypes. However, limitations regarding subgroup analyses and generalizability arise due to factors such as the predominantly European population and inadequate adjustments for newer stroke risk variables. Consequently, there is a possibility that environmental and lifestyle factors may not be recognized for their impact on the accumulation of familial stroke risk.

Sibling relationships can also play a role in shaping the familial factors influencing stroke risk due to the sharing of genetic material and early environmental exposures. However, the impact of sibling connections on stroke risk can vary depending on factors such as birth order and shared lifestyle choices. As per Tian et al. [56], a person's susceptibility to stroke is influenced by shared genetic and environmental factors within their family. This investigation elucidated the unique dynamics within sibling connections by revealing that shared environmental influences on cardiovascular risk factors were notably more pronounced among siblings compared to parent-offspring relationships. Additionally, a sibling's stroke history exhibited a stronger correlation with both the occurrence and severity of stroke than did the parental history, suggesting potential differences in genetic and environmental factors among siblings. Despite the absence of a statistically significant difference in stroke risk between sibling and parental histories, the numerical findings indicated a stronger association with sibling history, underscoring the importance of considering shared environmental factors.

These results are consistent with recent findings from an extensive study carried out by Yu et al. [60], which challenged conventional assumptions about the comparative significance of different types of family history of stroke (FHS). Involving a substantial sample size of 655,552 individuals, this meta-analysis provides compelling evidence supporting the distinct risk associated with each form of FHS. Despite previous studies yielding conflicting results regarding the associations between various types of FHS and stroke risk, this research demonstrates that the risk associated with a sibling's history of stroke is comparable to that linked to a parent's or mother's history.

Extended family members, such as grandparents, can contribute to the clustering of stroke risk factors within families. Although the influence of extended family connections on stroke risk may not be as significant as that of immediate family members, the accumulation of risk factors across extended family networks can still substantially affect an individual's overall stroke risk profile.

In summary, the impact of family ties on stroke risk involves various behavioral, environmental, and genetic factors within parent-child, sibling, and extended family relationships. Understanding the distinct roles of each familial connection is crucial for tailoring preventive measures to mitigate genetic predispositions and promote optimal health outcomes for future generations. Emphasizing the importance of thorough family history documentation is essential for primary stroke prevention efforts. By meticulously collecting and analyzing family medical histories, healthcare professionals can gain valuable insights into individuals' susceptibilities to stroke risk factors. This proactive approach facilitates the identification of high-risk individuals and enables focused interventions such as lifestyle modifications and preventive medical care.

2.3.2 Specific Genetic Mutations

An individual's susceptibility to stroke, a multifaceted medical condition characterized by abnormal blood flow to the brain, is significantly influenced by genetic anomalies. These mutations can impact various biological processes related to brain function, clotting mechanisms, vascular health, and lipid metabolism, potentially contributing to the onset and progression of stroke. Identifying the specific genetic variations associated with stroke is crucial for comprehending the underlying mechanisms of stroke pathogenesis, assessing an individual's stroke risk, and devising tailored preventive and therapeutic strategies. Healthcare professionals can enhance the customization of treatments to address the root causes of stroke and mitigate its profound impact on individuals and communities by gaining insights into the genetic factors linked to stroke.

2.3.2.1 NOTCH3 Mutations in CADASIL

When investigating specific genetic variants linked to stroke, several notable cases underscore the direct genetic elements influencing susceptibility to stroke. One prominent mutation involves alterations in the NOTCH3 gene, notably identified in association with CADASIL (Cerebral Autosomal Dominant Arteriopathy with Subcortical Infarcts and Leukoencephalopathy), a hereditary condition characterized by damage to the brain's small blood vessels [61], this indicates that individuals who inherit only one copy of the faulty gene from a parent can still develop the condition.

Preserving the structural integrity of these blood vessels relies on the NOTCH3 gene. Mutations in the NOTCH3 gene on chromosome 19 can lead to the degeneration of vascular smooth muscle cells (VSMCs), characterized by the accumulation of abnormal protein deposits known as granular osmiophilic material (GOM) in the extracellular space surrounding VSMCs [62]. The buildup of these deposits disrupts regular blood flow and hampers vascular function, heightening the risk of experiencing a stroke. CADASIL often presents with migraine auras, ischemic strokes or transient ischemic attacks, psychiatric symptoms, and cognitive deterioration [63], appearing in middle age, that may progress to frank vascular dementia [61]. Diagnosis of CADASIL is often based on clinical symptoms, brain imaging findings such as white matter lesions [64], and genetic testing to confirm the presence of NOTCH3 gene mutations. Hence, understanding CADASIL and its genetic basis is essential for accurate diagnosis and management of affected individuals.

2.3.2.2 RNF213 Mutations in Moyamoya Disease

A rare cerebrovascular condition referred to as "moyamoya disease" is defined by the gradual constriction or narrowing of the internal carotid arteries on both sides, resulting in the emergence of delicate collateral vessels termed "moyamoya vessels" [65]. These anomalous blood vessels are linked to ischemic strokes, hemorrhagic strokes, and recurrent transient ischemic attacks, heightening the risk of stroke by decreasing blood flow to the brain. Mutations in the RNF213 gene on chromosome 17q25.3 have been associated with about 15% of familial moyamoya cases. While different modes of transmission have been observed, many families exhibit autosomal dominant inheritance with incomplete penetrance [66].

The maintenance of vascular cells is significantly influenced by the RNF213 gene, which regulates inflammatory responses, lipid metabolism, oxygen consumption, and the cell cycle [67]. Alterations in the RNF213 gene interfere with the typical development of blood vessels, leading to the progressive obstruction of the brain's major branches and the terminal internal carotid arteries (ICA). As a compensatory mechanism, collateral capillary vessels emerge to supply blood to the affected areas. This phenomenon indicates an abnormal increase in the formation of new blood vessels, which is a hallmark characteristic of moyamoya disease [68]. The predominant clinical features include recurrent episodes of transient ischemic attacks (TIAs) or strokes, along with headaches and cognitive deficits. Children with moyamoya disease are more prone to ischemic strokes and TIAs, which can be provoked by crying, physical activity, fever, coughing, or hyperventilation. Conversely, older individuals are more predisposed to intraventricular and intraparenchymal hemorrhage [66]. In imaging studies Cerebral angiography and magnetic resonance angiography (MRA) depict a distinct "puff of smoke" pattern when moyamoya arteries are present [68]. Therapeutic interventions for moyamoya disease encompass revascularization surgeries, including direct or indirect bypass procedures. These treatments aim to mitigate the risk of recurrent strokes by enhancing blood circulation to the affected regions of the brain [69].

2.3.2.3 GLA Mutations in Fabry Disease

Anderson-Fabry disease, also called Fabry disease, is a rare X-linked lysosomal storage disorder caused by a mutation in the alpha-galactosidase A (GLA) gene, located on the long arm of the X chromosome. This mutation results in a deficiency of the alpha-galactosidase A (α-Gal A) enzyme [70]. In young males, stroke is nearly 12 times more common, while in young females, it occurs about 10 times more often than in the general population, highlighting the widespread occurrence of cerebrovascular disease [71]. Other CNS manifestations include vertigo, hearing impairment, tinnitus, and cognitive disturbances [66]. This condition typically emerges during infancy or adolescence and is characterized by discomfort in the hands and feet, reduced sweating, gastrointestinal disturbances, and skin abnormalities such as reddened areas near the navel and cloudy patches on the eye lenses. It often results in the presence of protein and albumin in the urine, which can eventually lead to kidney failure. Additionally, nervous system complications commonly include white matter lesions or vascular issues like stroke, transient ischemic attack, and dilation of posterior circulation arteries. These abnormalities manifest as heightened signals in the white matter and an enlarged basilar artery diameter on MRI scans [72].

The α-Gal A enzyme, coded by the GLA gene, is crucial for glycolipid metabolism. Its primary function is to break down the fatty substance globotriaosylceramide (Gb3). This degradation process takes place within lysosomes, cellular organelles responsible for breaking down a variety of substances. When the GLA gene functions correctly, α-Gal A assists in the hydrolysis of Gb3 into simpler compounds that can be readily digested and eliminated by the body [73]. When there's a shortage of α-Gal A, globotriaosylceramide (Gb3) builds up in lysosomes across various cell types, resulting in cell damage and dysfunction. Initially, the condition affects microvasculature before progressing to involve medium and large arteries, leading to thickening of the intima-media layers and arterial remodeling. This dysfunction extends to multiple organs, including the heart, kidneys, blood vessels, and nervous system, among others [74].

Dried blood spot (DBS) samples are commonly employed to measure α-Gal A activity. For males, a diagnosis can be confirmed if α-Gal A activity levels in DBS, plasma, and leukocytes are either undetectable or significantly reduced. However, for female heterozygotes, genetic testing is necessary since random X-chromosome inactivation can result in normal activity levels. Thus, diagnosing females requires an investigation into mutations in the GLA gene [74]. Moreover, magnetic resonance imaging (MRI) serves as the optimal imaging modality in Fabry disease for identifying cerebrovascular complications. Renowned for its precision in assessing the extent and type of brain damage in Fabry patients with focal neurological symptoms, MRI is the preferred choice. Advanced techniques like diffusion tensor imaging, functional MRI, MRI spectroscopy, and positron emission tomography have also been employed to further investigate the underlying causes of brain injury [72].

As previously indicated, abnormalities in the posterior cerebral circulation and white matter hyperintensities in the periventricular and subcortical regions are common features of Fabry disease. While basilar artery dilation and elongation are not consistently observed in Fabry disease, they can serve as an indication. Another notable feature, referred to as the pulvinar sign, is more prevalent in male patients with significant cardiac and renal complications. This marker of advanced disease progression manifests as heightened signal intensity in the pulvinar region of the brain [72].

As for its treatment, enzyme replacement therapy using recombinant GLA is an important treatment option to consider. This approach has demonstrated effectiveness in reducing left ventricular mass, alleviating discomfort, and improving overall quality of life. Typically, treatment begins promptly after diagnosis and involves regular intravenous infusions every two weeks for as long as necessary [66]. Despite the benefits observed with recombinant GLA-based enzyme replacement therapy, it hasn't led to a reduction in stroke incidence among patients. To address this risk in Fabry disease, antiplatelet medications like ticlopidine, clopidogrel, or aspirin are commonly prescribed to mitigate the occurrence of stroke and transient ischemic episodes. Effective management of blood pressure and adequate hydration are also essential for maintaining optimal cerebral blood flow [71].

In summary, genetic testing plays a crucial role in identifying individuals with specific genetic mutations that increase their susceptibility to stroke. By pinpointing these genetic predispositions, healthcare providers can tailor preventive and treatment strategies to each patient, reducing the risk of stroke and improving outcomes. However, genetic variability and the intricate interplay between genes and environmental factors present challenges for genetic testing in stroke patients. To advance our knowledge of the genetic factors involved in stroke and develop more effective prevention and treatment approaches, future research efforts should focus on overcoming these challenges.

2.3.3 Heritability of Stroke

The heritability of stroke assessed by genome-wide association studies is estimated to be 38%. Contributing factors include monogenic diseases predisposing a person to stroke (such as Fabry disease), single-nucleotide gene polymorphisms (SNPs) such as blood type gene ABO or forkhead transcription factor (FOXF2), and polygenic risk factors (such as hypertension), as well as environmental and lifestyle factors which are often shared by family members. Instead, the guidelines of the American Heart Academy/American Stroke Association suggest obtaining "family history of stroke (FHS)" as a way of identifying individuals at increased risk. The association of FHS with stroke risk is well established and has been studied systematically; however, to our knowledge its link to risk factors and stroke outcomes has not been updated systematically since 2004.

It is stated in a cross-sectional study by that positive family history of stroke is associated with increased prevalence of hypertension and transient ischemic attack (TIA), as well as decreased prevalence of obesity in the general population and increase prevalence of atrial fibrillation (AF) in women. Given that hypertension coaggregate strongly in families, it might be one of the main culprits which accounts for the heritability of stroke. TIA on the other hand was not confirmed by other authors other than the study itself. Other authors, often used a combined endpoint of stroke/TIA and studied patients with first-ever stroke, who constitute a high-risk population. Moreover, AF is one of the major etiologies of stroke, with female sex increasing stroke risk by 1.3fold. According to the study, the association between

AF and the family history of stroke has been already shown in the literature but with regards to general populations.

In the past, scientists employed methods like twin and family history studies to figure out how much of a role genetics plays in causing strokes. These analyses, which are often referred to as linkage analysis, were effective in establishing a relationship between monogenic genetic illnesses such as CADASIL and stroke. On the other hand, polygenetic studies were less successful in comprehending the combined effects of numerous genes [53].

2.3.3.1 Strategies for Unraveling Stroke Genetics

The most common method used in the past to determine the genetic contribution to stroke vulnerability was twin and familial history exams, or linkage analysis [55]. The candidate gene approach is another often used strategy for identifying genetic variables associated with stroke. This method entails locating genetic variants within a candidate gene which a gene that has already been linked to a particular disorder [75]. The newest approach, called genome-wide association studies (GWAS), is becoming more and more popular because of its efficacy and potential. With the use of these investigations, more than a million polymorphisms throughout the entire genome can be genotyped simultaneously, enabling thorough genetic research.

2.3.3.2 Genetic Epidemiology Insights into Ischemic Stroke

Significant hereditary factors on stroke have been found in a number of epidemiological studies involving twins, afflicted sibling pairs, and family history information. Due to their superior ability to distinguish genetic impacts from those resulting from a common environment, twin studies give the strongest evidence for the involvement of genes in complex disorders. Our meta-analysis revealed a minor genetic impact on the risk of fatal stroke, indicating an increasing significance of hereditary factors in stroke risk among younger patients. However, this initial study included only eight stroke-concordant twin pairs. Over a ten-year follow-up period, this difference decreased to 12.8% and 8.0%.

A family history of stroke was identified as a moderate risk factor for stroke in both case-control and cohort studies. However, case-control studies tended to indicate a higher risk compared to prospective cohort studies. Nevertheless, the results of the most comprehensive case-control studies closely mirrored those of prospective cohort studies. The age of afflicted relatives below a threshold was used in several studies to identify a positive family history of stroke, and this definition produced larger associations than the combined estimate [76].

In addition, varied types of strokes may have varied heritability, and the makeup of the cases is anticipated to affect the results in accordance with this. For instance, subarachnoid hemorrhage, featured in specific studies, might possess a genetic makeup distinct from other stroke types [77]. In a nutshell, certain kinds of stroke may be more strongly influenced by genetics. However, just two studies looked at stroke phenotype in detail, and many more did not differentiate between hemorrhagic and ischemic stroke in the proband.

2.3.3.3 Genetic Variants Associated with Stroke Heritability

Stroke is a multifactorial illness with polygenic features and intricate processes of interaction, impacted by both hereditary and non-genetic variables [78]. Numerous genes associated with stroke have been found through genetic study, especially those related to the development and maintenance of the peripheral nervous system, which are important in the inheritance of both ischemic stroke and intracerebral hemorrhage [79]. Single nucleotide polymorphisms (SNPs) with a minor allele frequency (MAF) ranging from 0.6% to 1.8% associated to stroke were found using a thorough meta-analysis of genome-wide association studies spanning several ancestries and encompassing over eight million SNPs. Remarkably, among distinct populations, specific connections were found for several SNPs. The majority of the genetic areas linked to stroke shared a common genetic connection with other vascular characteristics, according to the meta-analysis. These locations demonstrated robust associations with cardiac mechanisms that go beyond those commonly linked to cardio-embolism, offering fresh perspectives on the pathogenesis of stroke [80].

Study from Weng et al., in [81] and Jeanne et al., in [82] shows the variants identified in certain regions of the HTRA1, COL4A1, and COL4A2 genes have the potential to influence the hereditary transmission of characteristics linked to stroke in a way consistent with Mendelian inheritance patterns. Variants associated with a higher stroke risk include nucleotide substitutions in prothrombin (G20210A), a polymorphism in nitric oxide synthase (NOS3) at position E298D, a mutation in factor V Leiden (R506Q), and polymorphisms in Methylenetetrahydrofolate reductase (A222V). Additionally, a polymorphism in intron 16 of the ACE gene independently increases the risk of spontaneous intracerebral hemorrhagic stroke [83]. Linkage study identifies the genetic markers rs11833579 and rs12425791, which are located inside or close to the NINJ2 gene on chromosome 12p13, as hereditary factors linked to the familial clustering of stroke [84]. The protein-coding gene loci rs2280887 and rs4376531 of ARHGEF10 have been linked to ischemic stroke and atherosclerosis. Therefore, the stroke risk associated with these loci offers an approximation of the hereditary risk and underscores the significance of stroke genetics in identifying novel pharmaceutical targets [85].

2.3.4 Genetic Biomarkers

According to Biomarkers Definitions Working Group [86], a biomarker is defined as "a characteristic that is objectively measured and evaluated as an indicator of pathogenic processes, normal biological processes, or pharmacologic responses to a therapeutic intervention." This term covers a broad spectrum of indicators. It is commonly acknowledged that early-onset strokes are significantly influenced by hereditary factors [53]. Research on the prevention and identification of biomarkers that could enhance treatment response is currently recommended [87]. Despite extensive research efforts over the years, stroke biomarkers suitable for therapeutic applications are still absent, and the molecular pathways underlying stroke remain largely unknown [88]. To ensure the reliability and consistency of results, standardization or harmonization of techniques is necessary before translating biomarkers into clinical practice [89].

2.3.4.1 Candidate Genes

Studying the genetics of stroke has been particularly challenging due to its diverse nature, making it more complex to unravel compared to other common vascular and neurological diseases [89]. The candidate gene approach aids in identifying the genetic basis of stroke by examining differences in genes known to be associated with specific stroke pathologies. However, there are several limitations to this strategy. Firstly, accurate characterization of the stroke under study is crucial, as insufficient or inaccurate descriptions of stroke characteristics may lead to the selection of the wrong gene for investigation. Additionally, focusing on genes already linked to stroke may overlook entirely new genetic factors contributing to the condition. Moreover, challenges such as limited sample sizes and potential publication biases can make it difficult to accurately identify and replicate stroke phenotypes, further complicating data interpretation [53].

Even so, using this approach has revealed that mutations in the collagen IV alpha (COL4A1) and alpha 2 (COL4A2) genes are associated with intracerebral haemorrhage (ICH). These genes encode heterotrimers, vital components of nearly all basement membranes in the body. Initially, mutations in COL4A1 were linked to porencephaly type 1, which subsequent research associated with both sporadic and recurrent ICH, as well as lacunar stroke, through familial inheritance studies. A recent investigation also found that patients with certain APOE gene haplotypes, specifically haplotypes 2 and 4, have a higher risk of developing lobar ICH. Furthermore, the same study revealed that the APOE*E2 mutation was correlated with larger ICH volumes, resulting in poorer functional outcomes and higher mortality rates [90].

2.3.4.2 Genome-Wide Association Studies (GWAS)

Genome-wide association studies (GWAS) on human diseases have effectively identified numerous genetic variations associated with susceptibility to illness. When these genetic connections for disease traits align with intermediate molecular features such as gene expression and metabolomics, it provides robust evidence supporting hypotheses about the genes and pathways through which these variations influence disease [91]. GWAS has proven to be a valuable and widely adopted approach for identifying genetic factors linked to stroke risk. Unlike the candidate gene method, which focuses on specific genes and single nucleotide polymorphisms (SNPs), GWAS examines the entire genome. This broader approach allows for the identification of significant SNPs within known genes and the discovery of novel associations between chromosomal regions and diseases. However, one limitation of GWAS is their reduced ability to detect rare variants due to their lower statistical power [53].

According to Montaner et al. [89], a total of 42 loci that are robustly associated with stroke have been identified covering various types and subtypes, predominantly in European populations. While two initially significant loci couldn't be replicated in subsequent larger studies, several others exhibited robust associations with subtypes like ischemic stroke and intracerebral hemorrhage (ICH). Additional correlations at lower significance levels suggest a shared genetic pathway between monogenic and multifactorial stroke. Chauhan and Debette [52] highlighted a noteworthy finding from GWAS studies, revealing shared genetic variations between complex traits and stroke, shedding light on the underlying disease mechanisms. They emphasized significant overlap in the genetic risk profiles of coronary artery disease and ischemic stroke, particularly the large artery subtype. This shared genetic basis enhances our understanding of the common molecular pathways involved in these conditions. Deciphering the statistical associations between specific risk loci and phenotypes, like the PHACTR1 gene crucial to vascular biology, necessitates an understanding of these shared processes.

2.4 Racial and Ethnic Disparities in Stroke

2.4.1 Epidemiology of Stroke by Race

Differences in age-specific mortality rates from ischemic stroke, intracerebral hemorrhage (ICH), and subarachnoid hemorrhage (SAH) among young patients (ages 25–44) by race/ethnic groups highlight significant variations in health outcomes within this demographic. These differences could stem from various factors including genetic predispositions, socioeconomic disparities, access to healthcare, cultural differences, and lifestyle choices. Interestingly, the disparities in mortality rates based on race/ethnicity are more pronounced among younger age groups. For instance, in older patients aged over 65, the relative risk of death from ischemic stroke in black individuals compared to white individuals was 1.17. However, among younger patients aged 25–44, the relative risk was significantly higher at 3.5 for black individuals compared to white individuals [92].

These findings are corroborated by the Northern Manhattan Study (NOMAS), which is a prospective, population-based study examining stroke incidence. Age-specific death rates from ischemic stroke, ICH, and SAH varied by race/ethnic group among young patients that ages 25–44 as shown in Table 2.1. The thirty-day case-fatality risks following ischemic stroke among patients aged 20–44 were notably higher among Blacks (25%) and Hispanics (4%) compared to Whites (0%).

Table 2.1 Age-specific death rates from ischemic stroke, ICH, and SAH varied by race/ethnic group among young patients (ages 25–44) from [92]

	Ischemic stroke	Intracerebral hemorrhage	Subarachnoid hemorrhage
Whites	0.8	1	1.3
Black	2.8	5.2	2.9
American Indian	1.4	1.6	1.8
Asian/PI	0.6	1.4	1.0
Hispanics	0.8	1.5	1.3

Similarly, for intracerebral hemorrhage (ICH), the case-fatality risks were substantially higher with 67% for Blacks, 32% for Hispanics, and 0% for Whites.

For stroke incidence and race ethnicity, the NOMAS investigators studied stroke incidence in an urban community and discovered that Black and Hispanic adults had higher rates of stroke across all age groups compared to Whites. The annual age-adjusted incidence rates were 223 per 100,000 for Black individuals, 93 for Whites, and 196 for Hispanics. Even among younger age groups, significant differences persisted. For individuals aged 35–44, the incidence rates were markedly different: 6 per 100,000 per year for Whites, 54 for Blacks, and 53 for Hispanics. These findings highlight notable racial and ethnic disparities in stroke incidence, emphasizing the need for targeted interventions to address these differences [93].

In Northern Manhattan cohort, of the 924 first stroke cases identified over a span of 4 years, 74 occurred in young adults. Researchers then calculated the age-adjusted stroke incidence rates among young adults from different racial backgrounds, including Blacks, Whites, and Hispanics. The incidence rates of all stroke were 25 per 100,000 per year in Blacks, 26 in Hispanics, and 10 in Whites. This translates to an increased age-adjusted relative risk of all stroke of 2.4 (95% CI: 0.8–6.7) in Blacks compared to Whites and 2.5 (95% CI: 1.1–5.8) in Hispanics compared to Whites as shown in Table 2.2.

The incidence rates of ischemic stroke were 7 in White people, 11 in Hispanic people, and 11 per 100,000 Black people annually [93]. The incidence rate of intracerebral hemorrhage by NOMAS cohort are Men of Hispanic descent had a higher frequency of ICH (44/100,000/year), followed by Black men (37/100,000/year) and White men (15/100,000/year) where in women, black women had the highest rates of ICH (35/100,000/year) among all women [95].

Table 2.2 Age-adjusted stroke incidence rates (cases/100,000 persons/year) and stroke subtype percentages in Northern Manhattan adults 20–44 years old according to [94]

Study	All stroke	Infarct	ICH	SAH
Northern Manhattan	23	10	7	6
Blacks	25	11	8	6
Hispanics	26	11	8	6
Whites	10	7	0	3
Women	21	11	3	7
Men	25	9	11	4

To compare, the similarity conclusion stated in study from Hanchate et al., in [96], the findings from numerous well-known epidemiological studies conducted over the past two decades consistently demonstrate that the incidence rate of stroke is higher among Black and Hispanic populations compared to White populations. These studies provide robust evidence highlighting the disproportionate burden of stroke within these racial and ethnic groups. The results of a recent study, which utilized brain imaging, revealed fewer occurrences of cardio-embolic strokes and strokes resulting from large vessel atherosclerosis among Black individuals compared to White individuals [97]. This study's findings align with our own research, as Black participants tended to be younger and had a higher prevalence of hypertension and diabetes, while also demonstrating a lower prevalence of atrial fibrillation. These patterns lend support to the hypothesis that the etiologic subtype of small vessel disease may play a significant role in the observed racial disparities in stroke incidence and subtype distribution [98, 99].

2.4.2 Risk Factor in Different Racial Groups

The reduction of one-third of the burden of non-communicable diseases, including stroke, by 2030 is a notable Sustainable Development Goal of the United Nations.1. Nonetheless, during the previous 10 years, the incidence of stroke in young adults has increased by up to 50% globally [100]. Stroke still affects Blacks and Hispanics at disproportionately higher rates than it does White people. Even after taking socio-economic level variations into consideration, mortality differences have persisted to be considerable [101].

The research examined the risk factors specified in the 2014 guidelines, from the American Stroke Association including diabetes, hypertension, smoking history, patent foramen ovale (PFO) dyslipidemia, obesity and atrial fibrillation (AF). Cohorts that used definitions for a risk factor which couldn't be reclassified based on the studys criteria were excluded from relevant sub analyses. The table provides an overview of risk factors categorized by characteristics among patients with risk factors such as hypertension, diabetes, dyslipidemia, smoking, coronary artery disease, atrial fibrillation, PFO, obesity, migraine, excessive alcohol consumption and drug use. It is organized by sex, age group, ethnic and racial subgroups, as geographical income and country level income. Globally smoking was found to be the risk factor affecting 49.2% of participants followed by hypertension at 36.6% and dyslipidemia at 31.7%.

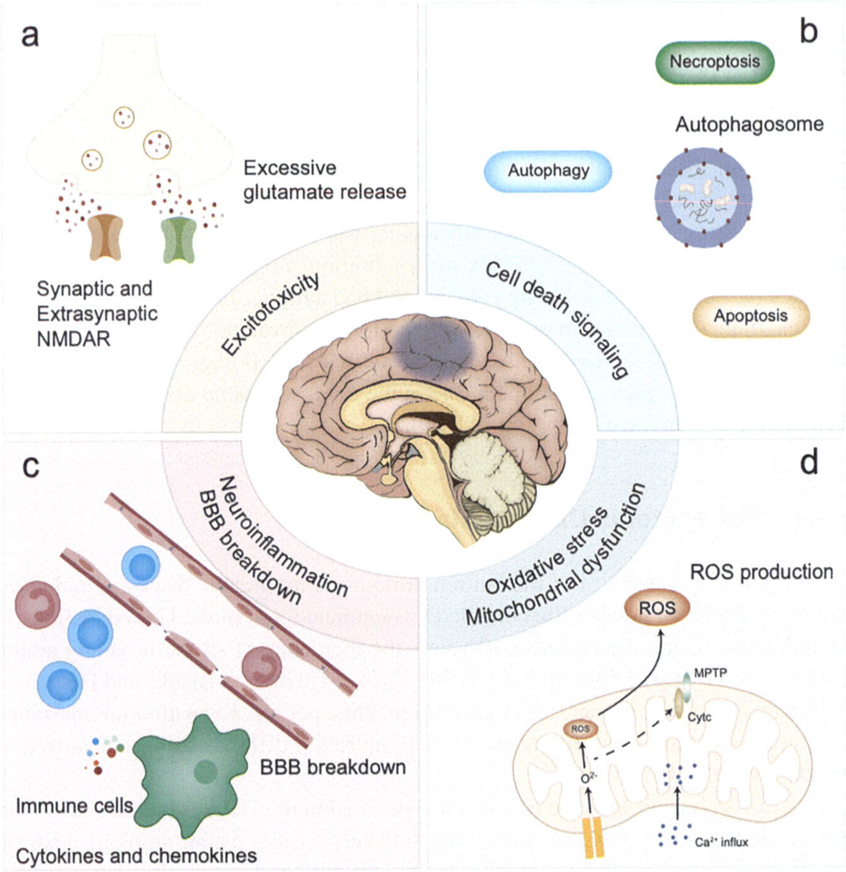

Fig. 2.1 The figure show the explanation of pathway of the stroke [51]

A significant number of individuals, under the age of 30, 49.6% displayed at one vascular risk factor. Men showed rates of risk factors such as high blood pressure diabetes, high cholesterol and smoking compared to women. These risk factors became more common as age increased (all; p trend ≤ 0.001; refer to Fig. 2.1). Conversely women were more prone to experiencing migraines (28.1 vs. 14.0%; $p \leq 0.001$) and patent foramen ovale (PFO) (18.1 vs. 12.7%; $p < 0.001$) than men, among patients (both; p trend ≤ 0.001).

In addition to having the highest rates of diabetes (20.9%, Asian: 20.9%) and hypertension (52.1%), Black and Asian people were most likely to have two or more risk factors. Conversely, the highest prevalence of PFO (24.9%) and dyslipidemia (40.4%) was seen in White people. In comparison to patients from Europe,

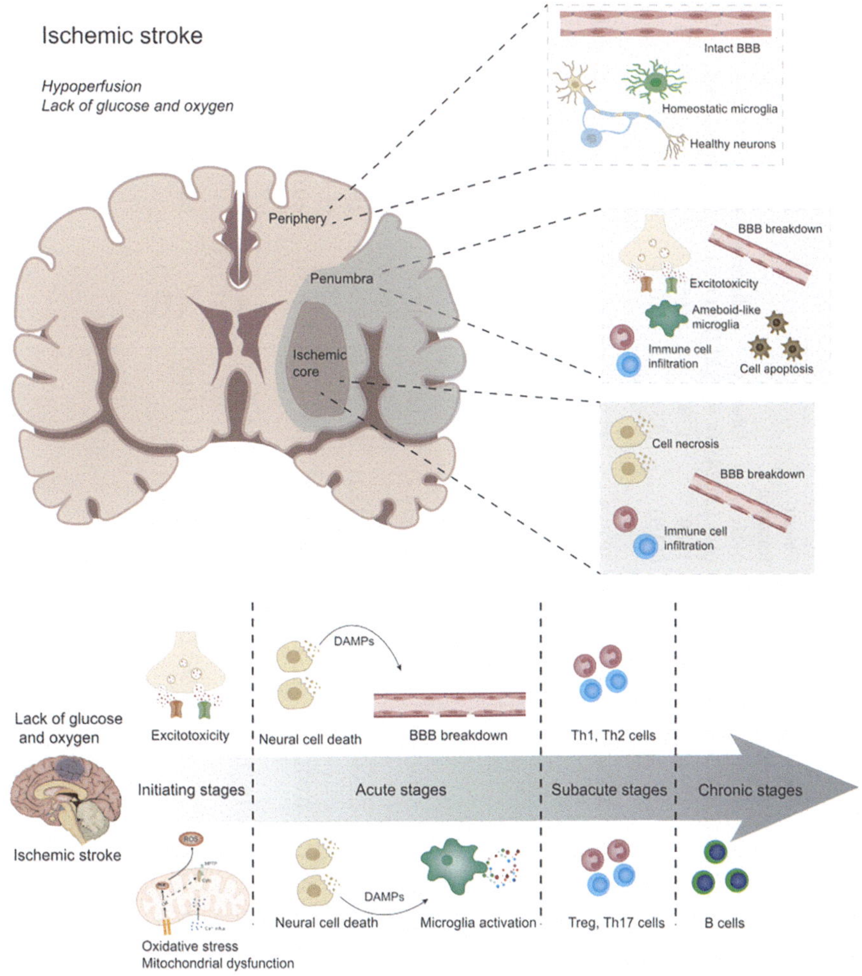

Fig. 2.2 The figure show the explanation of pathway of the stroke [51]

individuals from Asia, North America, and Oceania showed higher frequencies of most vascular risk factors, particularly obesity (Fig. 2.2). Additionally, compared to low- and middle-income countries (LMICs), the majority of vascular risk factors were considerably more common in high-income countries (HICs); dyslipidemia was one such factor, with a *p*-value of 0.001 for dyslipidemia (33.9 vs. 21.7%).

Table 2.3 shows the Risk Factors Stratified by Demographic Characteristics from a journal American Academy of Neurology [94].

Table 2.3 Risk factors stratified by demographic characteristics

	Patients with available data on risk factor										
	Hypertension (n = 17,409)	Diabetes (n = 17,631)	Dyslipidemia (n = 17,364)	Ever smoking (n = 17,328)	Coronary artery disease (n = 16,159)	Atrial fibrillation (n = 17,258)	Patent foramen ovale (n = 12,024)	Obesity (n = 13,217)	Migraine (n = 5138)	Excessive alcohol use (n = 9630)	Illicit drug use (n = 5891)
Patients with risk factor	6368 (36.6)	2432 (13.8)	5498 (31.7)	8525 (49.2)	759 (4.7)	691 (4.0)	1789 (14.9)	1796 (13.6)	1056 (20.6)	1030 (10.5)	254 (4.3)
Mean age (SD) of patients with risk factor, y	43.6 (5.5)	44.3 (5.3)	42.8 (6.1)	41.7 (7.0)	43.0 (6.3)	43.5 (6.3)	38.0 (8.1)	41.9 (6.4)	37.8 (7.7)	40.6 (7.3)	38.5 (8.0)
Sex											
Men	4294 (40.8)	1657 (15.5)	3668 (34.9)	6198 (59.0)	506 (5.1)	432 (4.1)	922 (12.7)	1114 (13.4)	383 (14.0)	896 (16.7)	174 (5.2)
Women	2074 (30.1)	75 (11.1)	1830 (26.7)	2327 (34.1)	253 (4.0)	259 (3.8)	867 (18.1)	682 (14.0)	673 (28.1)	187 (4.4)	80 (3.1)
p	<0.001	<0.001	<0.001	<0.001	0.002	0.33	<0.001	0.35	<0.001	<0.001	<0.001
Age group, years											
18–25	64 (6.9)	24. (2.6)	97 (10.5)	325 (35.3)	18 (2.1)	13 (1.4)	181 (27.2)	38 (6.1)	94 (27.2)	52 (7.8)	26 (6.9)
26–30	137 (12.3)	35 (3.1)	183 (16.3)	435 (39.0)	23 (2.2)	26 (2.3)	164 (20.1)	81 (10.5)	107 (25.2)	68 (8.3)	21 (4.3)

31–35)	367 (20.6)	112 (6.1)	415 (23.1)	743 (41.4)	51 (3.1)	38 (2.1)	258 (19.8)	180 (14.6)	162 (24.6)	131 (10.9)	26 (3.8)
36–40	914 (31.3)	307 (10.3)	904 (30.9)	1396 (47.5)	101 (3.7)	95 (3.3)	372 (17.8)	299 (14.4)	245 (24.0)	185 (10.1)	52 (5.0)
41–45	1955 (40.1)	693 (14.0)	1671 (34.2)	2534 (51.9)	226 (4.9)	177 (3.7)	496 (14.2)	545 (15.7)	306 (18.5)	313 (11.3)	78 (5.2)
46–50	2931 (50.5)	1261 (21.8)	2228 (39.0)	3092 (54.4)	340 (6.4)	342 (6.0)	318 (8.7)	653 (13.0)	142 (13.7)	334 (14.3)	51 (2.9)
$P_{\%}$	<0.001	<0.001	<0.001	<0.001	<0.001	<0.001	<0.001	<0.001	<0.001	<0.001	0.006
Ethnic and racial subgroups											
White	1709 (30.2)	497 (8.7)	2254 (40.4)	2559 (45.8)	174 (3.5)	184 (3.2)	1045 (24.9)	566 (19.2)	841 (20.9)	505 (9.4)	138 (6.3)
Black	264 (52.1)	105 (20.7)	128 (25.6)	262 (52.0)	55 (11.1)	14 (2.8)	6 (1.4)	221 (44.6)	0/42	15 (4.1)	48 (10.6)
Hispanic	397 (23.1)	190 (11.1)	247 (14.4)	840 (49.2)	52 (3.0)	74 (4.3)	171 (13.9)	157 (9.1)	17 (19.3)	248 (16.0)	38 (2.2)
Asian	3061 (46.1)	1426 (20.9)	1948 (28.8)	3611 (52.9)	335 (4.9)	324 (4.8)	239 (4.8)	634 (9.5)	48 (18.3)	215 (17.8)	15 (1.3)
p	<0.001	<0.001	<0.001	<0.001	<0.001	<0.001	<0.0001	<0.001	<0.001	<0.001	<0.001
Geographic location											
Oceania	235 (36.3)	106 (16.2)	277 (46.6)	293 (46.1)	48 (7.3)	43 (6.5)	110 (19.4)	153 (27.5)	0/207	79 (14.0)	59 (11.9)
Asia	3038 (46.2)	1423 (21.0)	1937 (28.8)	3588 (53.1)	330 (4.9)	323 (4.8)	231 (4.7)	637 (9.6)	45 (19.0)	206 (17.9)	17 (1.4)
Africa	20 (22.0)	11 (12.1)	16 (18.0)	29 (32.6)	1 (1.1)	0/89	0/70	8 (10.8)	0/0	13 (14.4)	2 (2.2)
Southern Europe	777 (26.2)	213 (7.1)	880 (29.5)	1166 (39.1)	35 (1.3)	92 (3.1)	742 (27.2)	59 (10.9)	558 (23.1)	255 (8.6)	13 (4.9)

(continued)

Table 2.3 (continued)

	Patients with available data on risk factor										
	Hypertension (n = 17,409)	Diabetes (n = 17,631)	Dyslipidemia (n = 17,364)	Ever smoking (n = 17,328)	Coronary artery disease (n = 16,159)	Atrial fibrillation (n = 17,258)	Patent foramen ovale (n = 12,024)	Obesity (n = 13,217)	Migraine (n = 5138)	Excessive alcohol use (n = 9630)	Illicit drug use (n = 5891)
Central Europe	669 (31.1)	138 (6.4)	849 (41.9)	1071 (50.7)	92 (6.1)	45 (2.4)	324 (36.2)	99 (23.2)	159 (22.7)	97 (8.4)	6 (3.9)
Northern Europe	775 (36.5)	181 (8.5)	1012 (48.1)	933 (49.3)	72 (4.6)	85 (4.1)	179 (32.0)	292 (13.9)	252 (18.2)	148 (12.0)	28 (2.8)
North America	427 (42.5)	163 (16.2)	266 (26.6)	578 (57.4)	128 (12.8)	26 (2.6)	25 (2.8)	387 (38.5)	25 (21.2)	20 (2.6)	82 (10.5)
Central and South America	427 (23.0)	197 (10.6)	261 (14.1)	867 (47.0)	53 (2.9)	77 (4.2)	178 (13.0)	161 (8.7)	17 (20.2)	265 (15.7)	47 (2.5)
P	<0.001	<0.001	<0.001	<0.001	<0.001	<0.001	<0.001	<0.001	<0.001	<0.001	<0.001
Country level of income											
High-income countries	5345 (37.2)	2054 (14.3)	4806 (33.9)	7137 (50.6)	588 (4.5)	539 (3.8)	1594 (15.4)	1494 (14.5)	987 (20.3)	739 (10.4)	190 (5.7)
Low and middle-income countries	1023 (33.5)	378 (11.6)	692 (21.7)	1388 (42.9)	171 (5.8)	152 (4.7)	195 (11.8)	302 (10.3)	69 (24.9)	344 (13.7)	64 (2.5)
P	<0.001	<0.001	<0.001	<0.001	<0.001	0.02	<0.001	<0.001	0.08	<0.001	<0.001

Data are n (%), mean (SD), or *p* value. Percentages are proportion of patients with the certain risk factor calculated based on the total patients with available data of the risk factor

Black and Asian individuals exhibited the highest prevalence of having two or more risk factors, including hypertension (Black: 52.1%, Asian: 47.1%) and diabetes (Black: 20.7%, Asian: 20.9%). Meanwhile, White individuals had the highest prevalence of dyslipidemia (40.4%) and patent foramen ovale (PFO) (24.9%). Furthermore, patients from Asia, North America, and Oceania demonstrated higher frequencies of most vascular risk factors, particularly obesity, compared to patients from Europe as shown in Fig. 2.3.

Regional, ethnic, and racial differences in the prevalence of vascular risk factors align with the reported disparities in stroke incidence among young adults. For instance, North America, Asia, and Oceania exhibit higher rates of most vascular risk factors compared to Europe [102] and South America, potentially explaining their elevated stroke incidence in young age groups. Notably, hypertension, diabetes mellitus, and smoking are more prevalent among Black and Asian individuals compared to White individuals, which are known risk factors for atherosclerosis of both larger and smaller vessels [103]. The increased cardiovascular burden seen in these ethnic groups corresponds to the higher prevalence of large vessel atherosclerosis or small vessel disease, such as stroke, especially in Black and Asian participants.

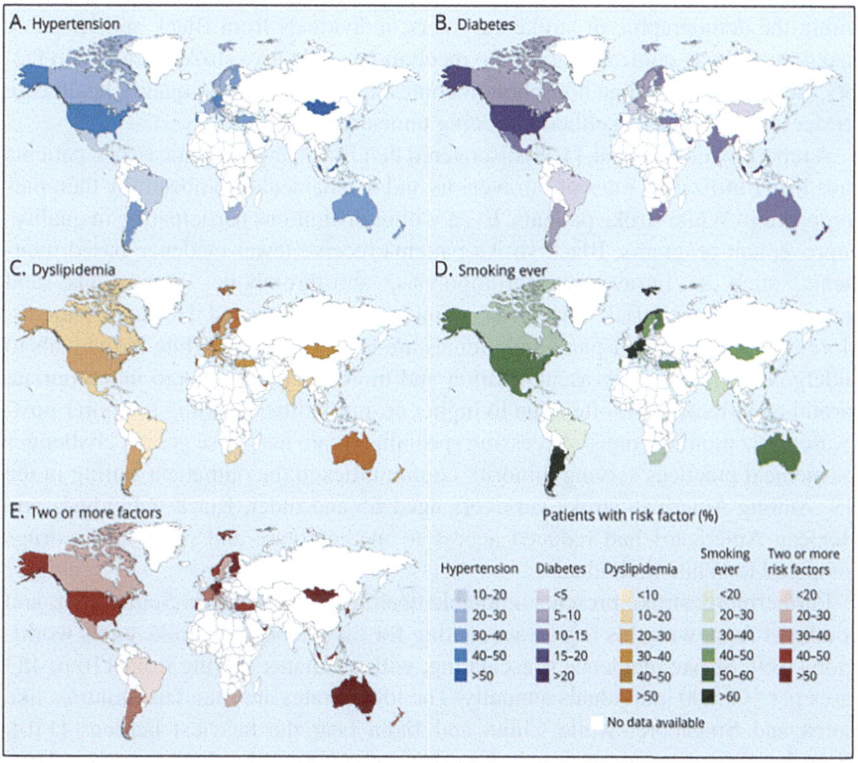

Fig. 2.3 The global maps showing the distribution of vascular risk factors by region in young stroke patients [94]

These disparities in risk factors are influenced by various factors including differences in healthcare access, socioeconomic determinants, and lifestyle choices, as shown by Kissela et al. [104] and O'Donnell et al. [103]. Genetic predispositions may also play a role, as demonstrated by the higher occurrence of Moyamoya disease in Asian populations, according to Putaala et al. [105] and Ekker et al. [106]. Further research is required to gain a better understanding of the relationship between vascular risk factors and the susceptibilities of different ethnic and racial subgroups.

2.4.3 Healthcare Access and Quality Disparities

The disparities in stroke rates across racial and ethnic groups underscore the pressing need for action from both public health and medical sectors, given the serious health and financial implications. Minority individuals who have survived strokes often encounter challenges accessing adequate secondary preventive measures, including limited access to medications and healthcare providers. In the United States, African American and Hispanic communities experience significantly higher rates of stroke incidence and prevalence compared to White populations. Moreover, within the demographic of stroke survivors, individuals from Black and Hispanic backgrounds may confront barriers to receiving preventative stroke care due to factors such as lower median household incomes, reduced access to quality healthcare services, and a higher likelihood of being uninsured [107].

A report by Levine et al. [108] discovered that Black and Hispanic stroke patients have lower utilization rates of intravenous and mechanical thrombectomy therapies compared to White stroke patients. Even within institutions participating in quality-improvement programs, Black stroke patients receive fewer evidence-based treatments, such as intravenous thrombolysis, anti-thrombotic medications, and anticoagulants for atrial fibrillation, compared to White and Hispanic patients. Moreover, Black and Hispanic individuals are less likely than White individuals to undergo carotid artery revascularization and more likely to undergo inappropriate carotid endarterectomy, often due to higher co-morbidities, leading to poorer post-revascularization outcomes. Accessing specialized care may pose greater challenges for medical practices serving minority communities in the outpatient setting in the US. Among American stroke survivors aged 65 and older, Black individuals and Mexican Americans had reduced access to medical care and prescription drugs compared to White individuals.

Furthermore, stroke presents a notable healthcare challenge in South, East, and Southeast Asia, with this region accounting for the majority of stroke cases worldwide [109]. Stroke incidence is escalating, with estimates varying from 116 to 483 cases per 100,000 individuals annually. The lowest rates are found in countries like Korea and Singapore, while China and India bear the heaviest burdens [110]. Globally, Asia, particularly South Asia, bears the highest burden of stroke-related mortality. While stroke prevalence tends to be lower in rural areas of South Asia compared to urban centers, further research is needed to delve into this disparity. In

China, stroke incidence varies across regions, with higher rates observed in the northern and western areas, where obesity and hypertension prevalence is also elevated. Challenges in reducing stroke-related mortality and disability persist due to insufficient awareness and the scarcity of specialized stroke care facilities [111].

A systematic review and meta-analysis by Gajurel et al. [110], also reported that intravenous thrombolysis (IVT), a proven treatment for acute ischemic stroke (AIS), is underutilized in Asia, with only 11.3% of high-income countries (HIC) and 8.1% of low- and middle-income countries (LMICs) administering thrombolysis. The greater utilization in HICs can be attributed to factors like abundant human resources and advanced infrastructure. However, the cost and limited availability of IVT in LMICs pose barriers to access, especially in the absence of comprehensive insurance systems. Additionally, inadequate policy and governance for stroke management, delays in prehospital and inpatient care, and suboptimal performance contribute to the disparities in IVT administration.

2.4.4 Cultural and Environmental Factors

Kleinman [112] proposed the concept of "explanatory models," which suggests that cultural perspectives and beliefs significantly influence individuals' attitudes toward seeking healthcare. These models are tailored to specific communities to address their concerns about health and treatment. These models assist individuals in managing their symptoms within the context of their social and cultural surroundings. Varied racial and ethnic groups demonstrate unique cultural influences by presenting divergent perspectives on stroke. For instance, African American elders commonly associate emotional and religious elements—such as anxiety and emotional repression—with stroke risk, turning to prayer as a preventive measure. This perspective may stem from the nation's extensive history of institutionalized racism, which has widened gaps in opportunities and living standards. Conversely, due to the pervasive influence of Catholicism and the cultural significance of religiosidad, Latino communities prioritize spirituality and faith in addressing health challenges [113].

The Chinese and Korean American communities prioritize emotional regulation and dietary moderation as strategies for lowering the risk of stroke, drawing from traditional medicine and Confucian philosophy. Moreover, elderly Korean Americans specifically highlight the perils of suppressing emotions, expressing concerns about hwa-byung, a condition associated with feelings of sadness and anger. Given the significant influence of these beliefs on health behaviors and risk perceptions, culturally sensitive approaches to stroke prevention and treatment are imperative [113]. Indeed, in Malaysia, a diverse country in Southeast Asia, cultural influences play a pivotal role in determining healthcare-seeking behaviors. Despite the presence of advanced tertiary care facilities, many stroke patients still opt for traditional and alternative medicine as their primary choice of treatment. Discovered that people's perceptions of what causes disrupted blood flow were shaped by their lifestyle choices, social interactions, and experiences with healthcare professionals. Similar beliefs were identified in Aceh, Indonesia, where stroke was attributed to various factors including dietary habits, past medical history, and supernatural influences

Furthermore, apart from cultural beliefs, cultural norms such as dietary habits and lifestyle factors may also contribute to stroke risk among racial and ethnic groups. For instance, a cross-sectional study conducted by Babatunde et al. [114], revealed that Black communities tend to consume more ready-to-eat foods compared to Hispanic groups. These dietary preferences, which typically include high levels of cholesterol and salt and low fiber content, increase the risk of stroke and other cardiovascular diseases within these populations. Moreover, researchers have identified several dietary disparities between the white and black populations. These differences include lower consumption of dietary cholesterol and higher intake of potassium, fiber, fruits, and vegetables among the white population [115].

Asian dietary customs vary widely across the continent, reflecting a rich tapestry of cuisines and culinary traditions. Traditional Asian diets often feature minimal consumption of meat and dairy products, with a heavy emphasis on grains, vegetables, legumes, and seafood. Fermented foods, such as miso in Japan, kimchi in Korea, and various pickled vegetables, are commonly integrated into meals, providing beneficial probiotics for gut health. However, the forces of globalization and rapid urbanization have led to shifts in dietary patterns, notably an increase in the consumption of processed and fast foods. These dietary changes have contributed to the growing prevalence of chronic diseases, including obesity, diabetes, and cardiovascular disorders, across the region.

A study by Zhang et al. [116] identified three distinct dietary patterns among Chinese adults: modern, meat-based, and southern. Their research suggests a noticeable shift towards Westernized eating habits in China from 1991 to 2015, with the southern pattern scoring lower compared to the modern and meat-based patterns. The southern dietary pattern, characterized by a high consumption of pork, vegetables, and rice, reflects the traditional eating practices observed in southern regions of China. Interestingly, this pattern demonstrated greater consistency over time compared to the others. Across all demographic groups, there was a consistent rise in scores for the modern dietary pattern, characterized mainly by snacks such as fruits, dairy products, and processed foods. This upward trend was primarily driven by changes in food availability resulting from modernization and convenience. However, the consumption of packaged foods, particularly those high in added sugars, unhealthy fats, and sodium, is associated with an increased risk of stroke. Additionally, there has been a significant increase in meat consumption, particularly pork. Nevertheless, excessive meat intake can have adverse health effects, especially in urban areas where consumption often exceeds recommended levels.

Furthermore, tobacco smoking represents a significant public health challenge in East and Southeast Asia, playing a substantial role in stroke occurrence. Despite being aware of the associated health risks, many Asians, especially males, continue to smoke. However, some Asian nations, such as Malaysia, have made gradual strides in efforts to decrease smoking rates [111]. Therefore, these variables together with cultural attitudes add to the complicated picture of stroke inequalities across various racial and ethnic groups.

2.5 Interaction Between Genetic and Racial Factors

2.5.1 Genetic Variations Among Racial Groups

Genes that have been found to be associated with an outcome even those with minor effects that fall short of the threshold for genome-wide significance are all taken into consideration when creating polygenic risk scores (PRS). A cumulative genetic influence on a given outcome is estimated by these scores, which are generated from genetic risk scores (GRS) [117]. Previously, genetic studies and detailed genetic analyses have identified 51 specific locations in the genome associated with different types and subtypes of stroke. In the largest combined analysis involving over 500,000 participants from the MEGASTROKE Consortium, 32 of these locations were confirmed to have significant associations with stroke across different populations. Previous studies in populations of European descent have also identified ten strong genetic associations with stroke, particularly with specific subtypes such as large artery stroke, cardioembolic stroke, small vessel stroke, and intracerebral hemorrhage. Multiple stroke subtypes or overall stroke risk showed associations with some loci [118]. In 2018, Malik et al. conducted a multiancestry genome-wide association meta-analysis involving 521,612 individuals, which included 67,162 cases and 454,450 controls. This research led to the identification of 22 previously undiscovered stroke risk loci, expanding the total number of known loci to 32.

Out of the 22 recently identified genetic sites, 18 were identified through analyses that considered multiple ancestral backgrounds, meeting specific criteria such as having fixed-effects with a p-value of less than 5.0×10^{-8} or a MANTRA log10(Bayes factor (BF)) greater than 6. The remaining four sites were revealed through analyses specific to individuals with European ancestry, with fixed-effects p-values less than 5.0×10^{-8}, as illustrated in Fig. 2.1. The Manhattan plot in Fig. 2.1 displays genetic sites associated with stroke or its subtypes, distinguishing between newly discovered (red) and previously known (black) associations. The threshold of statistical significance log10 BF > 6.0 is shown by the dashed line. With the exception of two of the new loci, which had a minor allele frequency between 0.01 and 0.05 and had substantial effect sizes odds ratios [ORs] of 2.33 and OR = 1.95, the 20 novel loci had common variants MAF 0.16–0.48 and ORs 1.05–1.20 across Europeans and East Asians, the two largest ancestral subgroups, the risk-allele frequencies and ORs of all 32 loci showed highly significant correlations between these populations and six loci had population-specific associations,

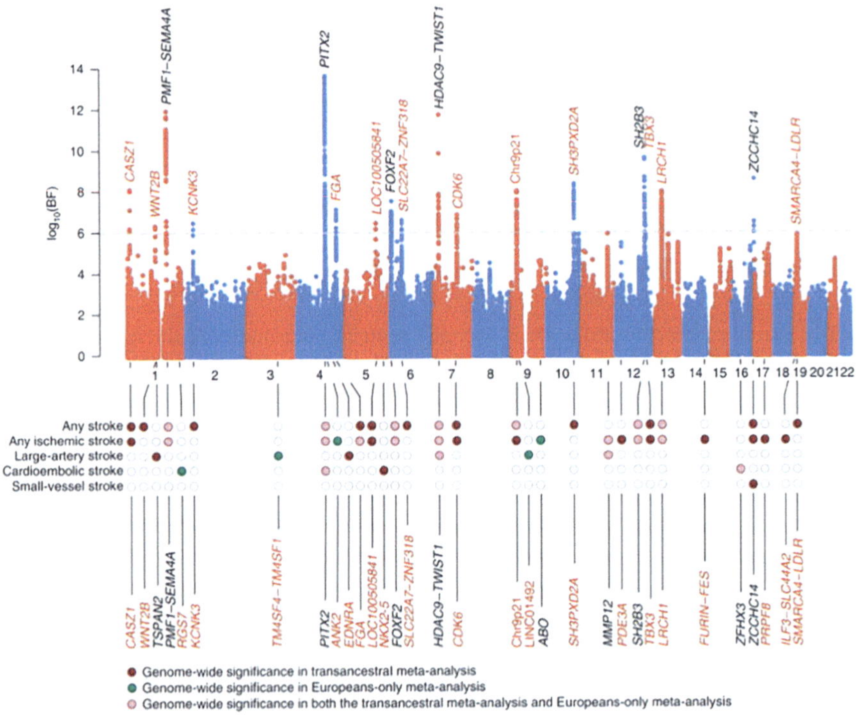

2.5.1.1 Connections to Subtypes of Etiological Stroke

For any stroke, genome-wide significance achieved of 18 loci, 12 of which were novel, while 20 loci, 12 of which unknown, were identified for AIS. Regarding LAS, we identified six loci, three of which were novel, while for CES, we reported four loci, two of which novel. ICA1L–WDR12, a novel locus discovered via gene-based tests, exhibited genome-wide significance solely for SVS. In a study from Malik et al., certain novel loci displayed an association limited only to a single stroke subtype.

2.5.1.2 Single Variant Association Analysis

In the study by Hu et al., during the discovery analysis in the TOPMed, it was considered about 40 million genetic variants with an MAC of over 20 in the single variant association analysis. This number is more than five times larger than the number of variants analyzed in a similar analysis conducted by the MEGASTROKE Consortium after imputation using the 1000 Genome reference panel [80]. All 5 novel loci detected in the single variant association analysis in TOPMed were created to contain the lead variants which were either monomorphic in European ancestry populations (7q22 and AUTS2) or the lead variants had a low minor allele frequencies in Europeans MAF < 1.0% for 13q33 and RAP1GAP2_, MAF = 0.2% for TEX13C. Certainly, RAP1GAP2 is the gene for which product is protein RAP1 GTPase-activating protein 2. The RAP1 GTPase-activating protein 2 protein is inhibitive in specific lateral processes and well-known to functionality as regulator in platelets.

GTPase-activating protein (GAP) is the function of the protein encoded by RAP1GAP2. GAPs play a role in controlling the actions of small GTPases, like Rap1. Rap1 is a tiny protein that binds guanine nucleotides and is involved in signaling and regulation within cells. Rap1 in platelets is activated by RAP1GAP2. Rap1 interacts with other proteins, such as Rab27 and synaptotagmin-like protein 1, when it is activated. These interactions are a part of a multifaceted process that controls platelet production of dense granules. At this particular genomic locus, CUL7 was prioritized with the help of GTEx expression data. The E3 ubiquitin-protein ligase complex includes the protein that CUL7 encodes. Studies in mouse models have highlighted its significant role in vascular morphogenesis [119] and its potential for enhancing cardiac function following myocardial infarction [120].

2.5.1.3 Genetic Contributors of Incident Stroke in African Americans

The Antihypertensive and Lipid-Lowering Treatment to Prevent Heart Attack Trial (ALLHAT) was accompanied by an adjunct pharmacogenetic investigation using GenHAT, according to additional research from Armstrong et al. in [121]. REGARDS, or "Reasons for Geographic and Racial Differences in Stroke," is a continuous cohort study that evaluates stroke risk throughout the continental United States.

The top 21 variations ($p < 1.00E-07$) for nine different genes are shown in Table 2.2. Ten variations were found to be statistically significant ($p < 5.00E-08$) in the inverse variance-weighted meta-analysis. On chromosome 18, the intergenic region between LINC01443 and LOC644669 included the most significant variant, rs117880209 ($p = 6.45E-09$). The direction of the effect was consistent across both GenHAT and REGARDS, and at this variation, the odds ratio for incident stroke was 2.98 (95% CI 2.06–4.31) for the C (v. T) allele. Ten more variations met or exceeded $p < 1.00E-06$ in this intergenic area.

Table 2.4 shows the Top variations ($p < 1.00E-07$) from an inverse variance-weighted meta- analysis linked to incident stroke.

Furthermore, in COL12A1 on chromosome 6, 13 intronic variants with $p < 1.00E-6$ were discovered. The rs144162260 intronic variation of the gene reached statistical significance ($p = 1.25E-08$; OR = 4.32; 95% CI = 2.61–7.15). Three other variants exceeded genome-wide significance: one intronic variant of PTPRR (rs568505299; $p = 2.64E-08$, OR = 3.95, 95% CI = 2.44–6.41) and two intergenic variants between TMTC1 and IP O 8 (top variant: rs536017124; $p = 1.32E-08$; OR = 4.79; 95% CI = 2.79–8.22). Please consult Table 2.4. Two intronic variants of NTM (top variant: rs185159493, $p = 1.37E-07$), three intronic variants of SNTG1 (top variant: rs117962542, $p = 5.91E-08$), and eleven intergenic variants between LINC02472 and PCDH7 (top variant: rs74599173, $p = 8.09E-08$) are additional biologically significant variants [121].

Table 2.4 The Top variations ($p < 1.00E-07$) from an inverse variance-weighted meta- analysis linked to incident stroke

rsID	CHR	BP (hg38)	A1/A2	EAF	OR	95% CI	[a]*P*	[b]Direction	Location	Gene(s)
rs117880209	18	14,996,132	C/T	0.012	2.98	2.06, 4.31	6.4SE-09	– –	Intergenic	*LINC01443; LOC64466S*
rs144162260	6	75,188,146	G/T	0.005	4.32	2.61, 7.15	1.25E-08	– –	Intronic	*COL12A1*
rs117791256	18	15,006,517	C/G	0.011	2.95	2.03, 4.28	1.31E-08	++	Intergenic	*LINC01443; LOC644669*
rs536017124	12	30,539,571	A/C	0.004	4.79	2.79, 8.22	1.32E-08	++	Intergenic	*TMTC1; IPO8*
rs142422295	18	15,015,740	C/T	0.011	2.93	2.02, 4.25	1.51E-08	– –	Intergenic	*LINC01443; LOC644669*
rs145341988	18	15,015,569	T/C	0.011	2.93	2.02, 4.25	1.51E-08	++	Intergenic	*LINC01443; LOC644669*
rs140550089	18	15,003,860	C/T	0.011	2.93	2.02, 4.25	1.51E-08	– –	Intergenic	*UNC01443; LOC644669*
rs192840029	12	30,513,818	A/G	0.004	4.67	2.72, 8.00	2.17E-08	++	Intergenic	*TMTC1; IPO8*
rs568505299	12	70,890,293	T/C	0.005	3.95	2.44, 6.41	2.64E-08	++	Intronic	*PTPRR*
rs58633304	18	14,988,000	C/A	0.013	2.76	1.93, 3.96	2.79E-08	– –	Intergenic	*LINC01443; LOC644669*
rs186234470	15	48,470,292	T/C	0.011	2.77	1.92, 4.00	5.20E-08	++	Intronic	*FBN1*
rs117962542	8	50,357,386	A/G	0.003	5.59	3.00, 10.41	5.91E-08	++	Intronic	*SNTG1*
rs116671900	5	1,833,726	T/C	0.002	7.60	3.64, 15.87	6.59E-08	++	Intergenic	*NDUFS6; LINC02116*
rs145817478	6	75,173,029	G/A	0.004	4.27	2.52, 7.23	6.70E-08	– –	Intronic	*COL12A1*
rs74599173	4	30,537,733	A/G	0.004	4.48	2.59, 7.74	8.09E-08	++	Intergenic	*LINC02472; PCDH7*
rs77853510	4	30,553,234	A/G	0.004	4.48	2.59, 7.74	8.09E-08	++	Intergenic	*LINC02472; PCDH7*
rs118141576	8	50,326,925	A/T	0.003	5.09	2.81,9.22	8.18E-08	++	Intronic	*SNTG1*
rs541454296	12	70,806,203	C/T	0.003	5.52	2.95, 10.31	8.72E-08	– –	Intronic	*PTPRR*
rs143089250	6	28,694,541	C/T	0.003	5.79	3.04, 11.04	9.02E-08	– –	Intergenic	*LINC00533; LINC01623*
rs117306,905	18	15,004,526	T/C	0.013	2.67	1.86, 3.84	9.50E-08	++	Intergenic	*LINC01443; LOC644669*
rs75989184	4	30,534,014	G/C	0.004	4.42	2.56, 7.63	9.93E-08	– –	Intergenic	*LINC02472; PCDH7*

Abbreviations: *rsID* reference SNP cluster ID, *CHR* chromosome number, *BP* base position from hg38, *A1* effect allele, *A2* allele 2, *EAF* effect allele frequency, *OR* odds ratio, *CI* confidence interval

[a]Genome-wide statistical significance alter multiple testing correction, $p < 5.00e-08$

[b]Direction order: GenHAT, REGARDS

2.5.2 Gene-environments Interactions

Gene-environment interactions elucidate the combined impact of genetic and environmental factors on an individual's stroke risk. These interactions can modify the effects of genetic variants linked to stroke risk, impacting processes like lipid metabolism, inflammation, and blood pressure regulation depending on environmental conditions. Various factors, including genetic predispositions, diet, exercise habits, smoking, air quality, and stress levels, collectively influence an individual's susceptibility to stroke and its progression.

Feng et al. [122] conducted a study investigating the impact of gene-environment interactions on stroke risk. They found that individuals with a high genetic predisposition, characterized by the combination of VKORC1 rs2359612C and Chr.9p21.3 rs10757274G variants, experienced a significant increase in stroke risk when exposed to hypertension or engaged in less than once-weekly exercise. These genetic variants play critical roles in processes such as inflammation, lipid metabolism, endothelial dysfunction, and vascular remodeling, all of which contribute to atherosclerosis progression. VKORC1, specifically, is involved in vitamin K recycling, coagulation cascade initiation, and vascular integrity maintenance. However, the study did not uncover any meaningful associations between VKORC1 and Chr.9p21.3 variants and lifestyle factors such as alcohol consumption, smoking, or obesity.

In contrast, Chen et al. [123], conducted a population-based case-control study to explore the potential link between two CRP single nucleotide polymorphisms (SNPs), namely rs1800947 and rs3093059, and the occurrence of ischemic stroke in the Chinese Han population. Their study also aimed to investigate any interactions between these genetic variations and environmental risk factors. The findings revealed a significant association between the rs1800947 GC genotype and the risk of ischemic stroke, particularly when considering small-vessel disease and its subtype. Furthermore, individuals with a history of drinking or smoking, combined with the rs1800947 GC genotype, exhibited heightened susceptibility to ischemic stroke.

Similarly, Cheng et al. [124] explored the correlation between genetic factors and behaviors such as smoking and alcohol consumption in determining the risk of ischemic stroke (IS). They compared the prevalence of these behaviors among IS patients and controls. Their study identified significant associations between alcohol consumption, smoking, and two genetic variations (rs12615793 and rs11896604). Among current smokers, individuals with the rs12615793-GA+AA genotype had the highest risk of IS compared to those with the rs12615793-GG genotype who had never smoked. Likewise, individuals carrying the rs11896604-CG+GG genotype and identified as current drinkers showed the highest risk of IS compared to those without this genotype who had the rs11896604-CC genotype. Recent research has echoed these findings regarding the relationship between alcohol consumption and the risk of stroke [125].

In summary, the risk of stroke is significantly impacted by gene-environment interactions. Studies examining how genetic variations interact with environmental factors have shed light on how lifestyle choices such as alcohol consumption, smoking, and physical activity can modify an individual's risk of stroke.

2.5.3 Epigenetics and Racial Disparities

The intricate regulation of gene expression and function in every cell of the body is orchestrated by sophisticated chemical processes termed epigenetic mechanisms. These mechanisms undergo dynamic adjustments throughout life to govern not just normal development, adult homeostasis, and aging, but also responses to external stimuli such as exposure to chemicals, physical activity, and dietary intake [126]. Recognized as a vital regulatory mechanism for tissue repair, epigenetics has gained prominence in both stroke genesis and post-stroke recovery. Three primary epigenetic factors—DNA methylation, post-translational modifications of histones, and non-coding RNAs—are indispensable for an organism's functioning [127].

Epigenetic mechanisms involved in stroke recovery encompass the suppression of histone deacetylase and methyltransferase functions. These processes initiate cellular responses that modify neuronal activity, support tissue healing, and facilitate the restructuring of damaged cells. The pathophysiology of intricate conditions like stroke, influenced by a multitude of genetic and environmental elements, correlates with alterations in overall DNA methylation patterns over time in response to environmental triggers. Consequently, epigenetics holds potential for identifying biomarkers that could enhance stroke diagnosis and prognosis [128].

DNA methylation involves the covalent modification of the fifth cytosine site, resulting in the formation of 5-methylcytosine (5mC). This process facilitates the binding of functionally active proteins to methylated genomic regions, which then attract regulatory complexes containing epigenetic factors. These complexes coordinate both reversible and long-term gene silencing processes. The levels of DNA methylation can be influenced by various factors, including inflammation and folate levels. Additionally, gender, ethnicity, lifestyle choices, and dietary habits can all affect DNA methylation levels [128]. Qureshi and Mehler [126] research highlighted the intricate involvement of DNA methylation levels in cerebral ischemia. Their findings indicated that DNA methylation levels increase in ischemic brain tissue following middle cerebral artery occlusion (MCAO) in mice, potentially contributing to cell death. Additionally, investigations into DNA methylation of Cyclin-dependent kinase inhibitor 2B (CDKN2B) have suggested that it might elevate the risk of arterial calcification in individuals who have experienced ischemic stroke [129]. Similarly, an Epigenome-Wide Association Study delving into DNA methylation and the risk of ischemic stroke (IS) discovered a distinct DNA methylation pattern that is causally linked to IS risk. [130]. Additionally, X chromosome inactivation and genomic imprinting, both governed by DNA methylation, are linked to other hereditary factors related to stroke [126]. For instance, in Fabry disease, an X-linked disorder stemming from a deficiency of the lysosomal enzyme α-galactosidase A, the accumulation of glycosphingolipids in the vascular endothelium can lead to stroke [70].

Histones, which are fundamental proteins, are situated within the nucleus of eukaryotic cells. They wrap around approximately 146 base pairs (bp) of DNA, forming a compact structure known as a nucleosome. The interaction between DNA and histone molecules is influenced by the presence of electrical charges on both molecules [131]. Changes to histones can regulate the accessibility of genomic programs to transcription factors, impacting processes such as gene repression and transcriptional activation. These modifications, which occur at various genomic sites, include phosphorylation, ubiquitination, acetylation, methylation, and adenosine diphosphate ribosylation, and they influence chromosomal structure and function [128]. Histone modification, in contrast to DNA methylation, occurs specifically on the amino-terminal tail extending from the histone subunit. Additionally, it is characterized by its transient and reversible nature [131]. In physiological conditions, histone acetylases (HATs) and histone deacetylases (HDACs) ensure the delicate balance between histone acetylation and deacetylation. However, conditions such as stroke often disrupt this equilibrium [128].

For example, the enzyme histone deacetylase (HDAC) plays a crucial role in modulating the lysine acetylation (H3KAc) of histone 3. Inhibiting HDAC results in the upregulation of anti-inflammatory genes like STAT3 and IL-10, while pro-inflammatory genes such as TNF-α, iNOS, STAT1, and IL-6 are downregulated in activated microglia, both in vivo and in vitro. This elevation in anti-inflammatory genes underscores the neuroprotective potential of HDAC inhibitors by promoting neuronal survival, reducing the extent of brain infarcts, and suppressing microglial activation (M1). Additionally, by regulating excitotoxicity, oxidative stress, endoplasmic reticulum stress (ER-stress), apoptosis, inflammation, and blood-brain barrier (BBB) breakdown, other HDAC inhibitors like valproic acid (VPA), sodium butyrate (SB), trichostatin-A (TSA), and sodium 4-phenylbutyrate (4-PBA) also demonstrate neuroprotective effects [131].

RNA modifications, akin to alterations in DNA, have emerged as crucial regulators of gene expression. The majority of the human genome comprises non-protein-coding DNA sequences, serving as the primary origin of this supplementary collection of functional RNAs. Among these RNA subclasses are transfer RNAs, ribosomal RNAs, microRNAs, small nucleolar RNAs, and long non-coding RNAs [126]. During transcription, primary miRNA transcripts serve as precursors for microRNAs (miRNAs), a extensively studied subgroup of non-coding RNAs (ncRNAs). Following transcriptional processing, these transcripts mature into sequences approximately twenty-two nucleotides in length. These miRNA sequences then engage in specific interactions with mRNA molecules, regulating numerous target genes. By binding adaptively to both the coding and regulatory domains of associated mRNAs, miRNAs effectively suppress translation. Alongside miRNAs, long non-coding RNAs (lncRNAs), which exceed 200 nucleotides in length, play pivotal roles in managing chromatin remodeling and transcriptional alterations, crucial processes in ischemic stroke (IS) [128].

References

1. Almutairi, F. M., Ullah, A., Althobaiti, Y. S., Irfan, H. M., Shareef, U., Usman, H., & Ahmed, S. (2022). A review on therapeutic potential of natural phytocompounds for stroke. *Biomedicines, 10*(10), 2566. https://doi.org/10.3390/biomedicines10102566
2. Khoshnam, S. E., Winlow, W., Farzaneh, M., Farbood, Y., & Moghaddam, H. F. (2017). Pathogenic mechanisms following ischemic stroke. *Neurological Sciences, 38*(7), 1167–1186. https://doi.org/10.1007/s10072-017-2938-1
3. Jung, S., Gilgen, M., Slotboom, J., El-Koussy, M., Zubler, C., Kiefer, C., Luedi, R., Mono, M.-L., Heldner, M. R., Weck, A., Mordasini, P., Schroth, G., Mattle, H. P., Arnold, M., Gralla, J., & Fischer, U. (2013). Factors that determine penumbral tissue loss in acute ischaemic stroke. *Brain, 136*(12), 3554–3560. https://doi.org/10.1093/brain/awt246
4. Xing, C., Arai, K., Lo, E. H., & Hommel, M. (2012). Pathophysiologic cascades in ischemic stroke. *International Journal of Stroke, 7*(5), 378–385. https://doi.org/10.1111/j.1747-4949.2012.00839.x
5. Datta, A., Sarmah, D., Mounica, L., Kaur, H., Kesharwani, R., Verma, G., Veeresh, P., Kotian, V., Kalia, K., Borah, A., Wang, X., Dave, K. R., Yavagal, D. R., & Bhattacharya, P. (2020). Cell death pathways in ischemic stroke and targeted pharmacotherapy. *Translational Stroke Research, 11*, 1185–1202. https://doi.org/10.1007/s12975-020-00806-z
6. Pajarillo, E., Rizor, A., Lee, J., Aschner, M., & Lee, E. (2019). The role of astrocytic glutamate transporters GLT-1 and GLAST in neurological disorders: Potential targets for neurotherapeutics. *Neuropharmacology, 161*, 107559. https://doi.org/10.1016/j.neuropharm.2019.03.002
7. Mázala, D. A. G., Grange, R. W., & Chin, E. R. (2015). The role of proteases in excitation-contraction coupling failure in muscular dystrophy. *American Journal of Physiology-Cell Physiology, 308*(1), C33–C40. https://doi.org/10.1152/ajpcell.00267.2013
8. Weber, J. T. (2012). Altered calcium signaling following traumatic brain injury. *Frontiers in Pharmacology, 3*, 60. https://doi.org/10.3389/fphar.2012.00060
9. Lai, T. W., Zhang, S., & Wang, Y. T. (2014). Excitotoxicity and stroke: Identifying novel targets for neuroprotection. *Progress in Neurobiology, 115*, 157–188. https://doi.org/10.1016/j.pneurobio.2013.11.006
10. Abramov, A. Y., & Duchen, M. R. (2008). Mechanisms underlying the loss of mitochondrial membrane potential in glutamate excitotoxicity. *Biochimica et Biophysica Acta (BBA) -Bioenergetics, 1777*(7-8), 953–964. https://doi.org/10.1016/j.bbabio.2008.04.017
11. Pérez-Pinzón, M. A., Mumford, P. L., Carranza, V., & Sick, T. J. (1998). calcium influx from the extracellular space promotes NADH hyperoxidation and electrical dysfunction after anoxia in hippocampal slices. *Journal of Cerebral Blood Flow & Metabolism, 18*(2), 215–221. https://doi.org/10.1097/00004647-199802000-00013
12. Dirnagl, U., Iadecola, C., & Moskowitz, M. A. (1999). Pathobiology of ischaemic stroke: An integrated view. *Trends in Neurosciences, 22*(9), 391–397. https://doi.org/10.1016/s0166-2236(99)01401-0
13. Sekerdag, E., Solaroglu, I., & Gursoy-Ozdemir, Y. (2018). Cell death mechanisms in stroke and novel molecular and cellular treatment options. *Current Neuropharmacology, 16*(9), 1396–1415. https://doi.org/10.2174/1570159x16666180302115544
14. Beckman, K. B., & Ames, B. N. (1998). Mitochondrial aging: Open questions[a]. *Annals of the New York Academy of Sciences, 854*(1 TOWARDS PROLO), 118–127. https://doi.org/10.1111/j.1749-6632.1998.tb09897.x
15. Achzet, L. M., Davison, C. J., Shea, M., Sturgeon, I., & Jackson, D. A. (2021). Oxidative Stress underlies the ischemia/reperfusion-induced internalization and degradation of AMPA receptors. *International Journal of Molecular Sciences, 22*(2), 717. https://doi.org/10.3390/ijms22020717
16. Valko, M., Leibfritz, D., Moncol, J., Cronin, M. T. D., Mazur, M., & Telser, J. (2007). Free radicals and antioxidants in normal physiological functions and human disease. *The*

International Journal of Biochemistry & Cell Biology, 39(1), 44–84. https://doi.org/10.1016/j.biocel.2006.07.001
17. Kalogeris, T., Bao, Y., & Korthuis, R. J. (2014). Mitochondrial reactive oxygen species: A double edged sword in ischemia/reperfusion vs preconditioning. *Redox Biology, 2*, 702–714. https://doi.org/10.1016/j.redox.2014.05.006
18. Shenoda, B. (2015). The role of Na^{+}/Ca^{2+} exchanger subtypes in neuronal ischemic injury. *Translational Stroke Research, 6*(3), 181–190. https://doi.org/10.1007/s12975-015-0395-9
19. Veit, F., Pak, O., Brandes, R. P., & Weissmann, N. (2015). Hypoxia-dependent reactive oxygen species signaling in the pulmonary circulation: Focus on ion channels. *Antioxidants & Redox Signaling, 22*(6), 537–552. https://doi.org/10.1089/ars.2014.6234
20. Wang, F., Reece, E. A., & Yang, P. (2015). Advances in revealing the molecular targets downstream of oxidative stress–induced proapoptotic kinase signaling in diabetic embryopathy. *American Journal of Obstetrics and Gynecology, 213*(2), 125–134. https://doi.org/10.1016/j.ajog.2015.01.016
21. Sun, M.-S., Jin, H., Sun, X., Huang, S., Zhang, F.-L., Guo, Z.-N., & Yang, Y. (2018). free radical damage in ischemia-reperfusion injury: An obstacle in acute ischemic stroke after revascularization therapy. *Oxidative Medicine and Cellular Longevity, 2018*, 1–17. https://doi.org/10.1155/2018/3804979
22. Sun, S., Gao, N., Hu, X., Luo, H., Peng, J., & Xia, Y. (2019). SOD3 overexpression alleviates cerebral ischemia-reperfusion injury in rats. *Molecular Genetics & Genomic Medicine, 7*(10), e00831. https://doi.org/10.1002/mgg3.831
23. Qin, C., Zhou, L.-Q., Ma, X.-T., Hu, Z.-W., Yang, S., Chen, M., Bosco, D. B., Wu, L.-J., & Tian, D.-S. (2019). Dual functions of microglia in ischemic stroke. *Neuroscience Bulletin, 35*(5), 921–933. https://doi.org/10.1007/s12264-019-00388-3
24. Xu, S., Lu, J., Shao, A., Zhang, J. H., & Zhang, J. (2020). Glial cells: Role of the immune response in ischemic stroke. *Frontiers in Immunology, 11*, 294. https://doi.org/10.3389/fimmu.2020.00294
25. Hoehn, B. D., Palmer, T. D., & Steinberg, G. K. (2005). Neurogenesis in rats after focal cerebral ischemia is enhanced by indomethacin. *Stroke, 36*(12), 2718–2724. https://doi.org/10.1161/01.str.0000190020.30282.cc
26. Fernandes, A., Miller-Fleming, L., & Pais, T. F. (2014). Microglia and inflammation: Conspiracy, controversy or control? *Cellular and Molecular Life Sciences, 71*(20), 3969–3985. https://doi.org/10.1007/s00018-014-1670-8
27. Kim, B., Jeong, H.-K., Kim, J., Lee, S. Y., Jou, I., & Joe, E. (2011). Uridine 5′-diphosphate induces chemokine expression in microglia and astrocytes through activation of the P2Y6 receptor. *Journal of Immunology, 186*(6), 3701–3709. https://doi.org/10.4049/jimmunol.1000212
28. Swanson, R., Ying, W., & Kauppinen, T. (2004). astrocyte influences on ischemic neuronal death. *Current Molecular Medicine, 4*(2), 193–205. https://doi.org/10.2174/1566524043479185
29. Gan, Y., Liu, Q., Wu, W., Yin, J.-X., Bai, X.-F., Shen, R., Wang, Y., Chen, J., La Cava, A., Poursine-Laurent, J., Yokoyama, W., & Shi, F.-D. (2014). Ischemic neurons recruit natural killer cells that accelerate brain infarction. *Proceedings of the National Academy of Sciences of the United States of America, 111*(7), 2704–2709. https://doi.org/10.1073/pnas.1315943111
30. Fu, Y., Liu, Q., Anrather, J., & Shi, F.-D. (2015). Immune interventions in stroke. *Nature Reviews Neurology, 11*(9), 524–535. https://doi.org/10.1038/nrneurol.2015.144
31. Hammond, M. D., Taylor, R. A., Mullen, M. T., Ai, Y., Aguila, H. L., Mack, M., Kasner, S. E., McCullough, L. D., & Sansing, L. H. (2014). CCR2+Ly6Chi Inflammatory monocyte recruitment exacerbates acute disability following intracerebral hemorrhage. *Journal of Neuroscience, 34*(11), 3901–3909. https://doi.org/10.1523/jneurosci.4070-13.2014
32. Escudero Augusto, D., Marqués Álvarez, L., & Taboada Costa, F. (2008). Actualización en hemorragia cerebral espontánea. *Medicina Intensiva, 32*(6), 282–295. https://doi.org/10.1016/s0210-5691(08)70956-2

33. van Asch, C. J., Luitse, M. J., Rinkel, G. J., van der Tweel, I., Algra, A., & Klijn, C. J. (2010). Incidence, case fatality, and functional outcome of intracerebral haemorrhage over time, according to age, sex, and ethnic origin: A systematic review and meta-analysis. *The Lancet. Neurology, 9*(2), 167–176. https://doi.org/10.1016/S1474-4422(09)70340-0
34. Adeoye, O., & Broderick, J. P. (2010). Advances in the management of intracerebral hemorrhage. *Nature Reviews Neurology, 6*(11), 593–601. https://doi.org/10.1038/nrneurol.2010.146
35. Aguilar, M. I., & Brott, T. G. (2011). *Update in intracerebral hemorrhage. The Neurohospitalist, 1*(3), 148–159. https://doi.org/10.1177/1941875211409050
36. Keep, R. F., Zhou, N., Xiang, J., Andjelkovic, A. V., Hua, Y., & Xi, G. (2014). Vascular disruption and blood–brain barrier dysfunction in intracerebral hemorrhage. *Fluids and Barriers of the CNS, 11*, 18. https://doi.org/10.1186/2045-8118-11-18
37. Sutherland, G. R., & Auer, R. N. (2006). Primary intracerebral hemorrhage. *Journal of Clinical Neuroscience, 13*(5), 511–517. https://doi.org/10.1016/j.jocn.2004.12.012
38. Schlunk, F., & Greenberg, S. M. (2015). The pathophysiology of intracerebral hemorrhage formation and expansion. *Translational Stroke Research, 6*(4), 257–263. https://doi.org/10.1007/s12975-015-0410-1
39. Biffi, A., & Greenberg, S. M. (2011). Cerebral amyloid angiopathy: A systematic review. *Journal of Clinical Neurology, 7*(1), 1. https://doi.org/10.3988/jcn.2011.7.1.1
40. Edlow, B. L., Bove, R., Viswanathan, A., Greenberg, S. M., & Silverman, S. (2012). The pattern and pace of hyperacute hemorrhage expansion. *Neurocritical Care, 17*(2), 250–254. https://doi.org/10.1007/s12028-012-9738-5
41. Boulouis, G., Dumas, A., Betensky, R. A., Brouwers, H. B., Fotiadis, P., Vashkevich, A., Ayres, A., Schwab, K., Romero, J. M., Smith, E. E., Viswanathan, A., Goldstein, J. N., Rosand, J., Gurol, M. E., & Greenberg, S. M. (2014). Anatomic pattern of intracerebral hemorrhage expansion. *Stroke, 45*(4), 1154–1156. https://doi.org/10.1161/strokeaha.114.004844
42. Xi, G., Keep, R. F., & Hoff, J. T. (2002). Pathophysiology of brain edema formation. *Neurosurgery Clinics of North America, 13*(3), 371–383. https://doi.org/10.1016/s1042-3680(02)00007-4
43. Keep, R. F., Hua, Y., & Xi, G. (2012). Intracerebral haemorrhage: Mechanisms of injury and therapeutic targets. *The Lancet Neurology, 11*(8), 720–731. https://doi.org/10.1016/s1474-4422(12)70104-7
44. Griswold, D. P., Fernandez, L., & Rubiano, A. M. (2021). Diagnosis and management of traumatic subarachnoid hemorrhage: Protocol for a scoping review. *JMIR Research Protocols, 10*(10), e26709. https://doi.org/10.2196/26709
45. Macdonald, R. L., & Schweizer, T. A. (2017). Spontaneous subarachnoid haemorrhage. *The Lancet, 389*(10069), 655–666. https://doi.org/10.1016/s0140-6736(16)30668-7
46. Bauernfeind, A. L., Barks, S. K., Duka, T., Grossman, L. I., Hof, P. R., & Sherwood, C. C. (2013). Aerobic glycolysis in the primate brain: reconsidering the implications for growth and maintenance. *Brain Structure and Function, 219*(4), 1149–1167. https://doi.org/10.1007/s00429-013-0662-z
47. Macdonald, R. L., Diringer, M. N., & Citerio, G. (2014). Understanding the disease: Aneurysmal subarachnoid hemorrhage. *Intensive Care Medicine, 40*(12), 1940–1943. https://doi.org/10.1007/s00134-014-3483-5
48. Lee, V. H., Oh, J. K., Mulvagh, S. L., & Wijdicks, E. F. M. (2006). Mechanisms in neurogenic stress cardiomyopathy after aneurysmal subarachnoid hemorrhage. *Neurocritical Care, 5*(3), 243–249. https://doi.org/10.1385/ncc:5:3:243
49. Cosgun, Z. C., Fels, B., & Kusche-Vihrog, K. (2020). Nanomechanics of the endothelial glycocalyx. *American Journal of Pathology, 190*(4), 732–741. https://doi.org/10.1016/j.ajpath.2019.07.021
50. Duca, A., & Jagoda, A. (2016). Transient Ischemic Attacks. *Emergency Medicine Clinics of North America, 34*(4), 811–835. https://doi.org/10.1016/j.emc.2016.06.007
51. Qin, C., Yang, S., Chu, Y. H., Zhang, H., Pang, X. W., Chen, L., & Wang, W. (2022). Signaling pathways involved in ischemic stroke: molecular mechanisms and therapeutic interventions. *Signal transduction and targeted therapy, 7*(1), 215

52. Chauhan, G., & Debette, S. (2016). Genetic risk factors for ischemic and hemorrhagic stroke. *Current Cardiology Reports, 18*(12), 124. https://doi.org/10.1007/s11886-016-0804-z
53. Ekkert, A., Šliachtenko, A., Grigaitė, J., Burnytė, B., Utkus, A., & Jatužis, D. (2021). Ischemic stroke genetics: What is new and how to apply it in clinical practice? *Genes, 13*(1), 48. https://doi.org/10.3390/genes13010048
54. Claeys, J., Gurvich, O., & Hadidi, N. N. (2020). Association between family history of stroke and stroke risk: A community survey. *Western Journal of Nursing Research, 42*(12), 1174–1181. https://doi.org/10.1177/0193945920957935
55. Floßmann, E., Schulz, U. G. R., & Rothwell, P. M. (2004). Systematic review of methods and results of studies of the genetic epidemiology of ischemic stroke. *Stroke, 35*(1), 212–227. https://doi.org/10.1161/01.str.0000107187.84390.aa
56. Tian, T., Jin, G., Yu, C., Lv, J., Guo, Y., Bian, Z., Yang, L., Chen, Y., Shen, H., Chen, Z., Hu, Z., & Li, L. (2017). Family history and stroke risk in China: Evidence from a large cohort study. *Journal of Stroke, 19*(2), 188–195. https://doi.org/10.5853/jos.2016.01270
57. Eguchi, E., Iso, H., Wada, Y., Kikuchi, S., Watanabe, Y., & Tamakoshi, A. (2012). Parental history and lifestyle behaviors in relation to mortality from stroke among Japanese men and women: The Japan collaborative cohort study. *Journal of Epidemiology, 22*(4), 331–339. https://doi.org/10.2188/jea.je20110163
58. Seshadri, S., Beiser, A., Pikula, A., Himali, J. J., Kelly-Hayes, M., Debette, S., DeStefano, A. L., Romero, J. R., Kase, C. S., & Wolf, P. A. (2010). Parental occurrence of stroke and risk of stroke in their children. *Circulation, 121*(11), 1304–1312. https://doi.org/10.1161/circulationaha.109.854240
59. Touzé, E., & Rothwell, P. M. (2007). Heritability of ischaemic stroke in women compared with men: A genetic epidemiological study. *The Lancet Neurology, 6*(2), 125–133. https://doi.org/10.1016/S1474-4422(06)70683-4
60. Yu, S., Su, Z., Miao, J., Yu, Y., Zhang, S., Wu, J., Zheng, H., Zhang, X., Zhong, S., Li, H., Qiao, L., Liang, J., Chu, M., & Tian, T. (2019). Different types of family history of stroke and stroke risk: Results based on 655,552 individuals. *Journal of Stroke and Cerebrovascular Diseases, 28*(3), 587–594. https://doi.org/10.1016/j.jstrokecerebrovasdis.2018.10.038
61. Wang, M. (2018). CADASIL. In *Handbook of clinical neurology* (Vol. 148, pp. 733–743). https://doi.org/10.1016/b978-0-444-64076-5.00047-8
62. Manini, A., & Pantoni, L. (2021). CADASIL from bench to bedside: Disease models and novel therapeutic approaches. *Molecular Neurobiology, 58*, 2558–2573. https://doi.org/10.1007/s12035-021-02282-4
63. Ni, W., Zhang, Y., Zhang, L., Xie, J., Li, H., & Wu, Z. (2022). Genetic spectrum of *NOTCH3* and clinical phenotype of CADASIL patients in different populations. *CNS Neuroscience & Therapeutics, 28*(11), 1779–1789. https://doi.org/10.1111/cns.13917
64. Mizuno, T., Mizuta, I., Watanabe-Hosomi, A., Mukai, M., & Koizumi, T. (2020). Clinical and genetic aspects of CADASIL. *Frontiers in Aging Neuroscience, 12*, 91. https://doi.org/10.3389/fnagi.2020.00091
65. Bang, O. Y., Chung, J.-W., Kim, D. H., Won, H.-H., Yeon, J. Y., Ki, C.-S., Shin, H. J., Kim, J.-S., Hong, S. C., Kim, D.-K., & Koizumi, A. (2019). Moyamoya disease and spectrums of RNF213 vasculopathy. *Translational Stroke Research, 11*(4), 580–589. https://doi.org/10.1007/s12975-019-00743-6
66. Barrett, K. M., & Meschia, J. F. (2014). Genetic stroke syndromes. *CONTINUUM: Lifelong Learning in Neurology, 20*, 399–411. https://doi.org/10.1212/01.con.0000446109.20539.68
67. Mineharu, Y., & Miyamoto, S. (2021). RNF213 and GUCY1A3 in Moyamoya disease: Key regulators of metabolism, inflammation, and vascular stability. *Frontiers in Neurology, 12*, 687088. https://doi.org/10.3389/fneur.2021.687088
68. Mertens, R. B., Graupera, M., Gerhardt, H., Bersano, A., Tournier-Lasserve, E., Mensah, M. A., Mundlos, S., & Vajkoczy, P. (2021). The genetic basis of moyamoya disease. *Translational Stroke Research, 13*(1), 25–45. https://doi.org/10.1007/s12975-021-00940-2
69. Bersano, A., Khan, N., Fuentes, B., Acerbi, F., Canavero, I., Tournier-Lasserve, E., Vajcoczy, P., Zedde, M. L., Hussain, S., Lémeret, S., Kraemer, M., & Herve, D. (2023). European

Stroke Organisation (ESO) guidelines on Moyamoya angiopathy: Endorsed by Vascular European Reference Network (VASCERN). *European Stroke Journal, 8*, 23969873221144O. https://doi.org/10.1177/23969873221144089
70. Zhao, Y., Zhu, Y., Li, F., Sun, Y., Ma, W., Wu, Y., Zhang, W., Wang, Z., Yuan, Y., & Huang, Y. (2023). Brain MRI correlations with disease burden and biomarkers in Fabry disease. *Journal of Neurology, 270*(10), 4939–4948. https://doi.org/10.1007/s00415-023-11826-8
71. Mishra, V., Banerjee, A. K., Gandhi, A. B., Kaleem, I., Alexander, J., Hisbulla, M., Kannichamy, V., Subas, S. V., & Hamid, P. (2020). Stroke and Fabry disease: A review of literature. *Cureus, 12*, e12083. https://doi.org/10.7759/cureus.12083
72. Ortíz, J. F., Solís, M. B., Ali, S. S., Khurana, M., Moncayo, J. A., Kothari, N. Y., Alzamora, M., Eissa-Garces, A., Patel, G., Monteros, G. A., Sen, M., & Quiñonez, J. (2022). Pulvinar sign, stroke and their relationship with Fabry disease: A systematic review and metanalysis. *Neurology International, 14*(2), 497–505. https://doi.org/10.3390/neurolint14020041
73. Li, X., Ren, X., Zhang, Y., Ding, L., Huo, M., & Li, Q. (2022). Fabry disease: Mechanism and therapeutics strategies. *Frontiers in Pharmacology, 13*, 1025740. https://doi.org/10.3389/fphar.2022.1025740
74. Choi, Y., Ok, T., & Lee, K.-Y. (2023). Fabry disease screening in young patients with acute ischemic stroke in Korea. *Cardiovascular Prevention and Pharmacotherapy, 5*(2), 54–60. https://doi.org/10.36011/cpp.2023.5.e5
75. Zhu, M., & Zhao, S. (2007). Candidate gene identification approach: Progress and challenges. *International Journal of Biological Sciences, 3*(7), 420–427. https://doi.org/10.7150/ijbs.3.420
76. Choi, J. C., Lee, J. S., Kang, S.-Y., Kang, J.-H., & Bae, J.-M. (2009). Family history and risk for ischemic stroke: Sibling history is more strongly correlated with the disease than parental history. *Journal of the Neurological Sciences, 284*(1–2), 29–32. https://doi.org/10.1016/j.jns.2009.03.015
77. Kubota, M., Yamaura, A., Ono, J., Itani, T., Tachi, N., Ueda, K., Nagata, I., & Sugimoto, S. (1997). Is family history an independent risk factor for stroke? *Journal of Neurology, Neurosurgery & Psychiatry, 62*(1), 66–70. https://doi.org/10.1136/jnnp.62.1.66
78. Dominiczak, A. F., & McBride, M. W. (2003). Genetics of common polygenic stroke. *Nature Genetics, 35*(2), 116–117. https://doi.org/10.1038/ng1003-116
79. Pickrell, J. K., Berisa, T., Liu, J. Z., Ségurel, L., Tung, J. Y., & Hinds, D. A. (2016). Detection and interpretation of shared genetic influences on 42 human traits. *Nature Genetics, 48*(7), 709–717. https://doi.org/10.1038/ng.3570
80. Malik, R., Chauhan, G., Traylor, M., Sargurupremraj, M., Okada, Y., Mishra, A., Rutten-Jacobs, L., Giese, A.-K., van der Laan, S. W., Gretarsdottir, S., Anderson, C. D., Chong, M., Adams, H. H. H., Ago, T., Almgren, P., Amouyel, P., Ay, H., Bartz, T. M., Benavente, O. R., & Bevan, S. (2018). Multiancestry genome-wide association study of 520,000 subjects identifies 32 loci associated with stroke and stroke subtypes. *Nature Genetics, 50*(4), 524–537. https://doi.org/10.1038/s41588-018-0058-3
81. Weng, Y.-C., Sonni, A., Labelle-Dumais, C., de Leau, M., Kauffman, W. B., Jeanne, M., Biffi, A., Greenberg, S. M., Rosand, J., & Gould, D. B. (2012). COL4A1 mutations in patients with sporadic late-onset intracerebral hemorrhage. *Annals of Neurology, 71*(4), 470–477. https://doi.org/10.1002/ana.22682
82. Jeanne, M., Labelle-Dumais, C., Jorgensen, J., Kauffman, W. B., Mancini, G. M., Favor, J., Valant, V., Greenberg, S. M., Rosand, J., & Gould, D. B. (2012). COL4A2 mutations impair COL4A1 and COL4A2 secretion and cause hemorrhagic stroke. *The American Journal of Human Genetics, 90*(1), 91–101. https://doi.org/10.1016/j.ajhg.2011.11.022
83. Matsushita, T., Umeno, J., Hirakawa, Y., Yonemoto, K., Ashikawa, K., Amitani, H., Ninomiya, T., Hata, J., Doi, Y., Kitazono, T., Iida, M., Nakamura, Y., Kiyohara, Y., & Kubo, M. (2010). Association study of the polymorphisms on chromosome 12p13 with atherothrombotic stroke in the Japanese population. *Journal of Human Genetics, 55*(7), 473–476. https://doi.org/10.1038/jhg.2010.45

84. Hsieh, Y.-C., Seshadri, S., Chung, W.-T., Hsieh, F.-I., Hsu, Y.-H., Lin, H.-J., Tseng, H.-P., Lien, L.-M., Bai, C.-H., Hu, C.-J., Jeng, J.-S., Tang, S.-C., Chen, C.-I., Yu, C.-C., & Chiou, H.-Y. (2012). Association between genetic variant on chromosome 12p13 and stroke survival and recurrence: A one year prospective study in Taiwan. *Journal of Biomedical Science, 19*(1), 1–8. https://doi.org/10.1186/1423-0127-19-1
85. Pourasgari, M., & Mohamadkhani, A. (2020). Heritability for stroke: Essential for taking family history. *Caspian Journal of Internal Medicine, 11*(3), 237–243. https://doi.org/10.22088/cjim.11.3.237
86. Biomarkers Definitions Working Group. (2001). Biomarkers and surrogate endpoints: Preferred definitions and conceptual framework. *Clinical Pharmacology & Therapeutics, 69*(3), 89–95. https://doi.org/10.1067/mcp.2001.113989
87. Ruscu, M. A., Burdusel, D., Cercel, A.-M., Aldea, M., Hermann, D. M., Cadenas, I. F., Doeppner, T. R., Surugiu, R., & Popa-Wagner, A. (2023). Genetic plasma biomarkers associated with ischemic stroke. *Exploration of Neuroscience, 2*, 27–47. https://doi.org/10.37349/en.2023.00010
88. Li, W., Shao, C., Zhou, H., Du, H., Chen, H., Wan, H., & He, Y. (2022). Multi-omics research strategies in ischemic stroke: A multidimensional perspective. *Ageing Research Reviews, 81*, 101730. https://doi.org/10.1016/j.arr.2022.101730
89. Montaner, J., Ramiro, L., Simats, A., Tiedt, S., Makris, K., Jickling, G. C., Debette, S., Sanchez, J.-C., & Bustamante, A. (2020). Multilevel omics for the discovery of biomarkers and therapeutic targets for stroke. *Nature Reviews Neurology, 16*(5), 247–264. https://doi.org/10.1038/s41582-020-0350-6
90. Giralt-Steinhauer, E., Jiménez-Conde, J., Tárraga, C. S., Mola, M., Rodríguez-Campello, A., Cuadrado-Godia, E., Ois, A., Fernández-Cádenas, I., Carrera, C., Montaner, J., Díaz Navarro, R. M., Vives-Bauzá, C., & Roquer, J. (2014). Aproximación al conocimiento de las bases genéticas del ictus. Consorcio español de genética del ictus. *Neurología, 29*(9), 560–566. https://doi.org/10.1016/j.nrl.2013.04.005
91. Suhre, K., McCarthy, M. I., & Schwenk, J. M. (2020). Genetics meets proteomics: Perspectives for large population-based studies. *Nature Reviews Genetics, 22*(1), 19–37. https://doi.org/10.1038/s41576-020-0268-2
92. Chong, J. Y., & Sacco, R. L. (2005). Epidemiology of stroke in young adults: Race/ethnic differences. *Journal of Thrombosis and Thrombolysis, 20*(2), 77–83. https://doi.org/10.1007/s11239-005-3201-9
93. Marini, C., & Carolei, A. (2003). Epidemiology of stroke in the young. *Stroke, 34*(5), e13–e14. https://doi.org/10.1161/01.str.0000065429.56621.6e
94. Jacob, M. A., Ekker, M. S., Allach, Y., Cai, M., Aarnio, K., Arauz, A., Arnold, M., Bae, H.-J., Bandeo, L., Barboza, M. A., Bolognese, M., Bonardo, P., Brouns, R., Chuluun, B., Chuluunbatar, E., Cordonnier, C., Dagvajantsan, B., Debette, S., Don, A., & Enzinger, C. (2021). Global differences in risk factors, etiology, and outcome of ischemic stroke in young adults—A worldwide meta-analysis. *Neurology, 98*(6), e573–e588. https://doi.org/10.1212/wnl.0000000000013195
95. Howard, V. J. (2013). Reasons underlying racial differences in stroke incidence and mortality. *Stroke, 44*(6, Supplement 1), S126–S128. https://doi.org/10.1161/strokeaha.111.000691
96. Hanchate, A. D., Schwamm, L. H., Huang, W., & Hylek, E. M. (2013). Comparison of ischemic stroke outcomes and patient and hospital characteristics by race/ethnicity and socioeconomic status. *Stroke, 44*(2), 469–476. https://doi.org/10.1161/strokeaha.112.669341
97. Wolf, P. A., & Kannel, W. B. (2007). Preventing stroke. *Circulation, 116*(19), 2099–2100. https://doi.org/10.1161/circulationaha.107.736942
98. Cockroft, K. M. (2006). Unruptured cerebral arteriovenous malformations. *Stroke, 37*(5), 1148–1149. https://doi.org/10.1161/01.str.0000220514.28847.ac
99. Morgenstern, L. B., & Kissela, B. M. (2015). Stroke disparities. *Stroke, 46*(12), 3560–3563. https://doi.org/10.1161/strokeaha.115.009533
100. Bennett, J. E., Stevens, G. A., Mathers, C. D., Bonita, R., Rehm, J., Kruk, M. E., Riley, L. M., Dain, K., Kengne, A. P., Chalkidou, K., Beagley, J., Kishore, S. P., Chen, W., Saxena, S.,

Bettcher, D. W., Grove, J. T., Beaglehole, R., & Ezzati, M. (2018). NCD countdown 2030: Worldwide trends in non-communicable disease mortality and progress towards sustainable development goal target 3.4. *The Lancet, 392*(10152), 1072–1088. https://doi.org/10.1016/s0140-6736(18)31992-5

101. Modan, B., & Wagener, D. K. (1992). Some epidemiological aspects of stroke: Mortality/morbidity trends, age, sex, race, socioeconomic status. *Stroke, 23*(9), 1230–1236. https://doi.org/10.1161/01.str.23.9.1230
102. Marini, C., Russo, T., & Felzani, G. (2011). Incidence of stroke in young adults: A review. *Stroke Research and Treatment, 2011*, 1–5. https://doi.org/10.4061/2011/535672
103. O'Donnell, M. J., Chin, S. L., Rangarajan, S., Xavier, D., Liu, L., Zhang, H., Rao-Melacini, P., Zhang, X., Pais, P., Agapay, S., Lopez-Jaramillo, P., Damasceno, A., Langhorne, P., McQueen, M. J., Rosengren, A., Dehghan, M., Hankey, G. J., Dans, A. L., Elsayed, A., & Avezum, A. (2016). Global and regional effects of potentially modifiable risk factors associated with acute stroke in 32 countries (INTERSTROKE): A case-control study. *Lancet (London, England), 388*(10046), 761–775. https://doi.org/10.1016/S0140-6736(16)30506-2
104. Kissela, B. M., Khoury, J. C., Alwell, K., Moomaw, C. J., Woo, D., Adeoye, O., Flaherty, M. L., Khatri, P., Ferioli, S., De Los Rios La Rosa, F., Broderick, J. P., & Kleindorfer, D. O. (2012). Age at stroke: Temporal trends in stroke incidence in a large, biracial population. *Neurology, 79*(17), 1781–1787. https://doi.org/10.1212/wnl.0b013e318270401d
105. Putaala, J., Metso, A. J., Metso, T. M., Konkola, N., Kraemer, Y., Haapaniemi, E., Kaste, M., & Tatlisumak, T. (2009). Analysis of 1008 consecutive patients aged 15 to 49 with first-ever ischemic stroke. *Stroke, 40*(4), 1195–1203. https://doi.org/10.1161/strokeaha.108.529883
106. Ekker, M. S., Verhoeven, J. I., Vaartjes, I., van Nieuwenhuizen, K. M., Klijn, C. J. M., & de Leeuw, F.-E. (2019). Stroke incidence in young adults according to age, subtype, sex, and time trends. *Neurology, 92*(21), e2444–e2454. https://doi.org/10.1212/wnl.0000000000007533
107. Levine, D. A., Neidecker, M. V., Kiefe, C. I., Karve, S., Williams, L. S., & Allison, J. J. (2010). Racial/ethnic disparities in access to physician care and medications among US stroke survivors. *Neurology, 76*(1), 53–61. https://doi.org/10.1212/wnl.0b013e318203e952
108. Levine, D. A., Duncan, P. W., Nguyen-Huynh, M. N., & Ogedegbe, O. G. (2020). Interventions targeting racial/ethnic disparities in stroke prevention and treatment. *Stroke, 51*(11), 3425–3432. https://doi.org/10.1161/strokeaha.120.030427
109. Venketasubramanian, N., Yoon, B. W., Pandian, J., & Navarro, J. C. (2017). Stroke epidemiology in South, East, and South-East Asia: A review. *Journal of Stroke, 19*(3), 286–294. https://doi.org/10.5853/jos.2017.00234
110. Gajurel, B. P., Nepal, G., Jaiswal, V., Ang, S. P., Nain, P., Shama, N., Ruchika, F. N. U., Bohara, S., Kharel, S., Yadav, J. K., Medina, J. R. T., & Shrestha, A. B. (2023). Utilization rates of intravenous thrombolysis for acute ischemic stroke in Asian countries: A systematic review and meta-analysis. *Medicine, 102*(42), e35560. https://doi.org/10.1097/MD.0000000000035560
111. Ravindranath, V., Dang, H. V., Goya, R. G., Mansour, H., Nimgaonkar, V. L., Russell, V. A., & Xin, Y. (2015). Regional research priorities in brain and nervous system disorders. *Nature, 527*(7578), S198–S206. https://doi.org/10.1038/nature16036
112. Kleinman, A. (1978). Culture, illness, and care. *Annals of Internal Medicine, 88*(2), 251–258. https://doi.org/10.7326/0003-4819-88-2-251
113. Chang, E., Choi, S., Kwon, I., Araiza, D., Moore, M., Trejo, L., & Sarkisian, C. (2018). Characterizing beliefs about stroke and walking for exercise among seniors from four racial/ethnic minority communities. *Journal of Cross-Cultural Gerontology, 33*(4), 387–410. https://doi.org/10.1007/s10823-018-9356-6
114. Babatunde, O. T., Briley, P. M., White, B. M., Fang, X., & Ellis, C. (2018). Dietary practices among stroke-survivors—Racial/ethnic differences. *Journal of Stroke and Cerebrovascular Diseases, 27*(11), 2926–2931. https://doi.org/10.1016/j.jstrokecerebrovasdis.2018.06.026
115. Newby, P. K., Noel, S. E., Grant, R., Judd, S., Shikany, J. M., & Ard, J. (2012). Race and region have independent and synergistic effects on dietary intakes in black and white women. *Nutrition Journal, 11*(1), 1–12. https://doi.org/10.1186/1475-2891-11-25

116. Zhang, J., Wang, Z., Du, W., Huang, F., Jiang, H., Bai, J., Zhang, X., Zhang, B., & Wang, H. (2021). Twenty-five-year trends in dietary patterns among Chinese adults from 1991 to 2015. *Nutrients, 13*(4), 1327. https://doi.org/10.3390/nu13041327
117. Malik, R., Bevan, S., Nalls, M. A., Holliday, E. G., Devan, W. J., Cheng, Y.-C., Ibrahim-Verbaas, C. A., Verhaaren, B. F. J., Bis, J. C., Joon, A. Y., de Stefano, A. L., Fornage, M., Psaty, B. M., Arfan Ikram, M., Launer, L. J., van Duijn, C. M., Sharma, P., Mitchell, B. D., Rosand, J., & Meschia, J. F. (2014). Multilocus genetic risk score associates with ischemic stroke in case–control and prospective cohort studies. *Stroke, 45*(2), 394–402. https://doi.org/10.1161/strokeaha.113.002938
118. Chauhan, G., Arnold, C. R., Chu, A. Y., Fornage, M., Reyahi, A., Bis, J. C., Havulinna, A. S., Sargurupremraj, M., Smith, A. V., Adams, H. H. H., Choi, S. H., Pulit, S. L., Trompet, S., Garcia, M. E., Manichaikul, A., Teumer, A., Gustafsson, S., Bartz, T. M., Bellenguez, C., & Vidal, J. S. (2016). Identification of additional risk loci for stroke and small vessel disease: A meta-analysis of genome-wide association studies. *The Lancet Neurology, 15*(7), 695–707. https://doi.org/10.1016/s1474-4422(16)00102-2
119. Arai, T., Kasper, J. S., Skaar, J. R., Ali, S., Takahashi, C., & DeCaprio, J. A. (2003). Targeted disruption of p185/Cul7 gene results in abnormal vascular morphogenesis. *Proceedings of the National Academy of Sciences of the United States of America, 100*(17), 9855–9860. https://doi.org/10.1073/pnas.1733908100
120. Hassink, R. J., Nakajima, H., Nakajima, H. O., Doevendans, P. A., & Field, L. J. (2009). Expression of a transgene encoding mutant p193/CUL7 preserves cardiac function and limits infarct expansion after myocardial infarction. *Heart, 95*(14), 1159–1164. https://doi.org/10.1136/hrt.2008.150128
121. Armstrong, N. D., Srinivasasainagendra, V., Patki, A., Tanner, R. M., Hidalgo, B. A., Tiwari, H. K., Limdi, N. A., Lange, E. M., Lange, L. A., Arnett, D. K., & Irvin, M. R. (2021). Genetic contributors of incident stroke in 10,700 African Americans with hypertension: A meta-analysis from the genetics of hypertension associated treatments and reasons for geographic and racial differences in stroke studies. *Frontiers in Genetics, 12*, 781451. https://doi.org/10.3389/fgene.2021.781451
122. Feng, C., Yang, Y., Yang, S., Tu, X., Wang, Y., Song, Y., Hui, R., & Zhang, W. (2019). Effect of gene–gene and gene–environment interaction on the risk of first-ever stroke and post-stroke death. *Molecular Genetics & Genomic Medicine, 7*(8), e846. https://doi.org/10.1002/mgg3.846
123. Chen, Z., Yu, D., Xu, Z., Li, S., Li, X., Li, J., & Yang, X. (2015). C-reactive protein gene polymorphisms and gene-environment interactions in ischaemic stroke. *Neurological Research, 37*(11), 979–984. https://doi.org/10.1179/1743132815y.0000000053
124. Cheng, Q., Li, Y.-K., Lu, F., Yin, L., Wang, Y.-Z., Wei, W., & Lin, Q. (2017). Interactions between *ACYP2* genetic polymorphisms and environment factors with susceptibility to ischemic stroke in a Han Chinese Population. *Oncotarget, 8*(58), 97913–97919. https://doi.org/10.18632/oncotarget.18430
125. Zheng, X.-Z., Bian, X.-L., Sun, Z.-H., & Wang, H.-D. (2020). Interaction between methylenetetrahydrofolate reductase (MTHFR) gene polymorphisms and environment with susceptibility to ischemic stroke in Chinese population. *Annals of Indian Academy of Neurology, 23*(4), 491–491. https://doi.org/10.4103/aian.aian_192_19
126. Qureshi, I. A., & Mehler, M. F. (2010). Emerging role of epigenetics in stroke. *Archives of Neurology, 67*(11), 1316–1322. https://doi.org/10.1001/archneurol.2010.275
127. Phillips, C. M., Stamatovic, S. M., Keep, R. F., & Andjelkovic, A. V. (2023). Epigenetics and stroke: Role of DNA methylation and effect of aging on blood–brain barrier recovery. *Fluids and Barriers of the CNS, 20*(1), 14. https://doi.org/10.1186/s12987-023-00414-7
128. Kumar, A., Misra, S., Nair, P., & Algahtany, M. (2021). Epigenetics mechanisms in ischemic stroke: A promising avenue? *Journal of Stroke and Cerebrovascular Diseases, 30*(5), 105690. https://doi.org/10.1016/j.jstrokecerebrovasdis.2021.105690

129. Zhou, S., Zhang, Y., Wang, L., Zhang, Z., Cai, B., Liu, K., Zhang, H., Dai, M., Sun, L., Xu, X., Cai, H., Liu, X., Lu, G., & Xu, G. (2016). CDKN2B methylation is associated with carotid artery calcification in ischemic stroke patients. *Journal of Translational Medicine, 14*(1), 1–8. https://doi.org/10.1186/s12967-016-1093-4
130. Cullell, N., Soriano-Tárraga, C., Gallego-Fábrega, C., Cárcel-Márquez, J., Torres-Águila, N. P., Muiño, E., Lledós, M., Llucià-Carol, L., Esteller, M., Castro, M., Montaner, J., Fernández-Sanlés, A., Elosua, R., Delgado, P., Martí-Fábregas, J., Krupinski, J., Roquer, J., Jiménez-Conde, J., & Fernández-Cadenas, I. (2022). DNA methylation and ischemic stroke risk: An epigenome-wide association study. *Thrombosis and Haemostasis, 122*(10), 1767–1778. https://doi.org/10.1055/s-0042-1749328
131. Peng, J., Ghosh, D., Zhang, F., Yang, L., Wu, J., Pang, J., Zhang, L., Yin, S., & Jiang, Y. (2022). Advancement of epigenetics in stroke. *Frontiers in Neuroscience, 16*, 981726. https://doi.org/10.3389/fnins.2022.981726

Experimental Models of Stroke: Seeing the Unseen

3

Hamza Yassaad, Manar Kahoul,
Mohamed Mahmoud Rabie Ali, Lobna Balouta,
and Wael Mohamed

Abstract

Stroke is a major global health problem which is often manifested by mixed clinical picture. This summary shows a remarkable implication of experimental models in studying stroke injury seen in human patients in both ischemic and hemorrhagic stroke, Experimental models provide researchers with a controlled environment where the detrimental processes can be delineated and provided with potential approaches to be tested, in this way, early science will be close to clinical research and there will be a reduction in the costs of drug development. The rodent models, for instance, rats and mice, are the most widely used animals due to their high availability, prospective genetics and experimental methods for promoting stroke such as MCAO. On the flip-side, these species' restricted brain complexity only allows for the use of large animals like non-human primates,

H. Yassaad
Department of Dentistry, Faculty of Medicine, University of Constantine 3, Constantine, Algeria

M. Kahoul
Faculty of Medicine, University of Constantine 3, Constantine, Algeria

M. M. R. Ali
Beni Seuf University, Beni Seuf, Egypt

L. Balouta
Faculty of Exact Sciences and Natural and Life Sciences, University of Mohammed Khaider, Biskra, Algeria

W. Mohamed (✉)
Basic Medical Science Department, Kulliyyah of Medicine, International Islamic University Malaysia (IIUM), Kuantan, Pahang, Malaysia

Clinical Pharmacology Department, Menoufia Medical School, Menoufia University, Menoufia, Egypt

W. Mohamed, M. B. Mat-Nor (eds.), *Nutrition and Stroke*, Nutritional Neurosciences, https://doi.org/10.1007/978-981-97-6818-9_3

which provide stronger anatomical resemblance to humans although there is ethical skepticism and more expensive maintenance. Varied techniques are used in the models to produce different engine types of the stroke. In focal ischemia models, most commonly the MCAO, only specific brain regions that are responsible for its pathology are affected, whereas global ischemia models which mimics complete stroke employs multiple pathways such as 4 VO. Moreover, experiments simulating hemorrhagic stroke, a different but important type of stroke, where the bleeding in the brain is restricted, also contribute to our understanding of the pathophysiology of stroke of this category imaging techniques particularly MRI and CT scans, play a crucial role in experimental stroke research, this chapter highlights their significance in advancing our understanding of stroke mechanisms and improving clinical management. In addition, in this chapter, we discuss the critical issue of translating findings from animal models of stroke to human conditions. Like any experimental research, such translation encounters significant challenges due to interspecies differences, model complexity, and clinical relevance, highlighting the discrepancies between preclinical and clinical outcomes and proposing strategies to bridge this gap considering future directions that can enhance the clinical relevance of preclinical findings.

Keywords

Stroke · Ischemic stroke · Animal modelsh · Middle Cerebral Artery Occlusion (MCAO) · Cerebrovascular disease · Cerebral hemorrhage

3.1 Introduction

In the first place, Stroke remains one of the most contributing factors to deaths around the world. It occurs to about 13.7 million individuals each year, killing approximately 5 million annually. It is the sudden stop of blood flow to the brain that results from the decreased perfusion from the blood vessels to the brain. In the front, blood can be controlled from the internal carotid artery, whereas the two vertebral arteries behind at the back form a circle of willis. There are two types of stroke: ischemic, which is characterized by inadequate blood and reinforced by oxygen in the brain, and hemorrhagic, occasioned by the bleeding or leaking of the blood vessels [1].

According to the United Nations, approximately 85% of strokes are ischemic [2] and are responsible for three millions deaths annually [3]. The most common cause of stroke is the formation of a blood clot, known as a thrombus, in a blood vessel that supplies blood to the brain. This may occur due to atherosclerosis, where fatty deposits form in the blood vessels and restrict blood flow. Another cause of ischemic stroke is an embolism [4], Imagine tiny blockages, like crumbs in a pipe, traveling through the bloodstream and getting lodged in the delicate vessels of the brain. This is what can happen with blood clots or debris from other parts of the body [5]. When these blockages occur, they cut off the vital supply of glucose (sugar) and oxygen to brain cells, starving them of the fuel they need to function. This sets off a chain reaction of events called the ischemic cascade, a complex series of molecular processes triggered by insufficient oxygen and glucose. This cascade, like a domino effect, ultimately leads to the degeneration and death of brain cells through various mechanisms, including oxidative stress (cell damage from free radicals), inflammation, and overstimulation by glutamate (a brain chemical) [6]. As research by Farina et al. [6] highlights, the brain's dependence on a steady supply of glucose and oxygen for its high-energy functions makes it particularly vulnerable to these disruptions in blood flow [6].

The major highway in our brain called the middle cerebral artery. This critical artery, branching off the carotid artery, delivers oxygen and essential nutrients to key areas for movement, sensation, and language, like the frontal lobe and parts of the temporal and parietal lobes. When a blockage occurs in this middle cerebral artery (MCA occlusion), it's like a traffic jam cutting off vital supplies. This can lead to a range of symptoms, often sudden and severe, that worsen quickly over minutes or hours [7]. Patients might experience hemiparesis (weakness on one side of the body) or even monoparesis (weakness in a single limb), numbness on one side, and difficulty with speaking or understanding language (aphasia or dysphasia) [7].

It's important to note that not all blood flow disruptions are full-blown strokes. Sometimes, there's a temporary interruption, like a brief traffic slow down. This is called a transient ischemic attack (TIA), often referred to as a "mini-stroke." TIAs typically last from a few minutes to a few hours, and the good news is that symptoms usually resolve on their own without lasting damage [8]. However, a TIA serves as a serious warning sign. The risk of a full stroke is significantly higher, especially in the days and weeks following a TIA [8]. There's another type of stroke, called global ischemic stroke, that affects a much larger area. Imagine a city-wide blackout instead of a single highway closure. This occurs when blood flow to a large portion or even the entire brain is disrupted, starving brain cells of oxygen and glucose. This can cause permanent damage, and the most common culprit is a heart attack [9]. Finally, let's talk about hemorrhagic strokes, which are caused by bleeding in the brain rather than blockage. These strokes can occur within the brain tissue itself (intracerebral hemorrhage) or in the spaces surrounding the brain (subarachnoid hemorrhage). While less common than ischemic strokes, they can still be very serious [10].

High blood pressure, or hypertension, is the biggest culprit behind intracerebral hemorrhage (ICH). Over time, chronically high pressure damages tiny arteries in the brain, a process called hyperplastic arteriolosclerosis. Imagine these arteries as rubber bands. With constant strain, the muscle in the bands weakens and gets replaced by scar tissue, making them less elastic and more prone to ballooning. These weak spots, called Charcot-Bouchard aneurysms, can burst and cause bleeding in the brain. Another major risk factor for ICH is cerebral amyloid angiopathy (CAA), especially for people over 70. CAA is like having a sticky protein, called amyloid, gum up the walls of arteries in the brain. This buildup can weaken and damage the vessels, increasing the risk of rupture and bleeding. Blood thinners, while crucial for certain medical conditions, can also be a double-edged sword. Taking them increases the risk of bleeding in the brain, a situation doctors fear [11]. The brain's plumbing system can also have congenital defects called cerebral vascular malformations (CVMs). These are basically abnormal tangles of blood vessels. Some, like arteriovenous malformations (AVMs) and cavernous malformations (CMs), can leak or burst, leading to ICH. Finally, brain tumors can also be a source of bleeding. Both cancerous growths that originate in the brain (primary) and those that spread from elsewhere (metastatic) can rupture and cause hemorrhages. While some bleeding is obvious, in other cases, a tumor might be hiding behind the initial signs of bleeding seen on scans [10] (Fig. 3.1).

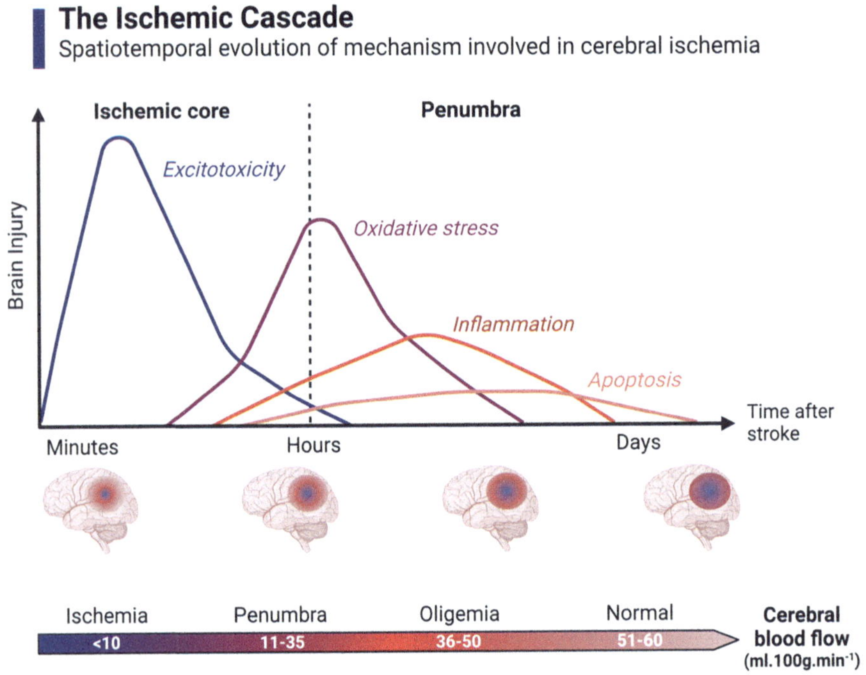

Fig. 3.1 The Ischemic Cascade. The ischemic cascade graph represents the sequence of events that occur in the brain during an ischemic stroke. Excitotoxicity: excessive activation of glutamate receptors leads to the overstimulation of neurons, causing cell damage and death. Oxidative stress: increase in reactive oxygen species (ROS) production can damage cellular membranes, proteins and DNA. Inflammatory response: release of pro-inflammatory cytokines and chemokines, attracting immune cells to the affected area

3.2 Experimental Models of Stroke

The heterogeneity and complexity of stroke disease have pushed scientists to develop several in vivo and in vitro models to understand the pathophysiological mechanisms of the disease and to investigate potential therapies. An important role of experimental stroke models in stroke research is to provide information on the processes involved in the development, progression, and regeneration of ischemic brain lesions. Essentially, include the introduction of a stroke-like injury in animals, such as rodents, rabbits, and non-human primates, under various methods: Ischemic stroke experimentals are created by means of the obstruction of blood flow to the newborn include either operation ligation of the major cranial arteries, or with the injections of embolic material. Haemorrhagic stroke models function by the following means: bleeding in the brain (either through rupture of vessels (vessels break) or by injection or introducing blood or blood components into the brain tissue. Global cerebral ischemia models: a balloon of sorts in the artery blocks the blood flow whether cardiac arrest happens or language pressure only reaches the lowest, for emergency cases, such as cardiac arrest or massive drop in blood pressure [12].

Animal models of study have been used for decades to investigate stroke. Rat and mice models share the important aspect of studying the stroke pathophysiology as well as looking for effective treatments; however, these animals' short lives limit their use. Institutes often use rodent models, as the key consideration for cost and ethics, side by side, with the fact that the condition of each species might be different from the human condition. Also, there are several other factors such as comorbidities and diverse populations that are disregarded in the experiment. Selecting the fit behavioral test for a distinct brain area and treatment, validation of the sensibility for different stroke-induced impairments, and surpassing the ethical considerations and funding limitations are the key ones we need to think through. Model with the closest human experience would be the ideal model for the compensation [13]. Unilateral (MCAO) will serve the representation of stroke symptoms and the compensatory behavior due to this model would mimic the same features. Therefore, it is crucial the tests were sensitive to such differences in sample collection, preparation and analysis algorithms to avoid interfering the results. Future models should endeavour to show better visualizations of the intricate texture of human stroke while addressing other related issues including comorbidities, multiculturalism, and a realistic setting which would bring study results to the forests of human life [14]. Researchers may opt for more practical mammals like rats, mice, and gerbils to serve as their animal models in order to understand the essence of stroke without cost problems, permitting their research. But they rather tend to imitate what a human stroke is like. Profoundly big animals, such as cats, dogs, pigs, and even primates are used within animals used. Conversely, their expenditure and ethical defects limit them.

Using rodents as models for stroke gives an opportunity to study the impact and the possible treatments for the condition, and also to observe behavioral change for the purpose of evaluating brain damage and recovery. Nevertheless, they should exercise caution while planning appropriate tests which might be used for a certain brain region affected or treatment, choose sensitive tests which might detect all impairments stroke caused, as well as figure out ethically acceptable research questions, insufficient funding, and the need for specific training.

It is desired that frameworks identify and mimic the human experience with properly chosen methods, e.g. "unilateral middle cerebral artery occlusion" one and also be prone to the precipitate injuries which stroke may cause. Stroke, which holds the first place among the conditions that cause death and disability throughout the world, grabs patients and healthcare systems in its net. In spite of great successes of stroke research, many difficulties occur while the findings in the animal models become a relevant therapeutic strategy for use in humans. Therein there is a breach that shows the acuteness of a critical examination of inquiry mind of how well experimental stroke models mirror upon humans [15].

3.3 Ischemic Stroke Models

3.3.1 Ischemic Stroke Models In Vivo

The development of animal models has always been aimed at serving stroke research by resembling stroke damage seen in human. Many experimental models are used to study this injury, and the mechanism of cell damage is determined by measuring the effects of different cell types on the 'stress or death' scale on a particular model. The three main categories of in vivo brain models are (1) global ischemia, (2) focal ischemia, and (3) hypoxia/ischemia [16].

3.3.1.1 Global Ischemic Models

A global ischemic model is achieved by occlusion of four arteries (vertebral and carotid arteries) or two arteries (aorta or vertebral artery) with or without causing hypotension during ischemia [17].

Four Vessels Occlusion (4-VO) Model

It is the most widely used ischemia modeling technique in the world, defined by Pulsinelli. This technique involves two surgical stages over a 24-hour period and likely reduces the decrease in cerebral blood flow (CBF) and has a preventive effect. The two-stage surgery is invasive and technically demanding [18].

Stage 1: The first stage of the surgery involves the placement of a ventral neck incision and the occlusion of the carotid arteries. This is achieved by placing a bulldog arterial clamp on the silastic tubing that exits the PE90 tubing inserted into the ventral neck incision.

Stage 2: Two days later, in the second stage of the surgery, the animals are reanesthetized, and the carotid arteries are exposed and clamped again for a specific duration, typically 3–5 min. This step is performed to induce ischemia and simulate the effects of a stroke [19].

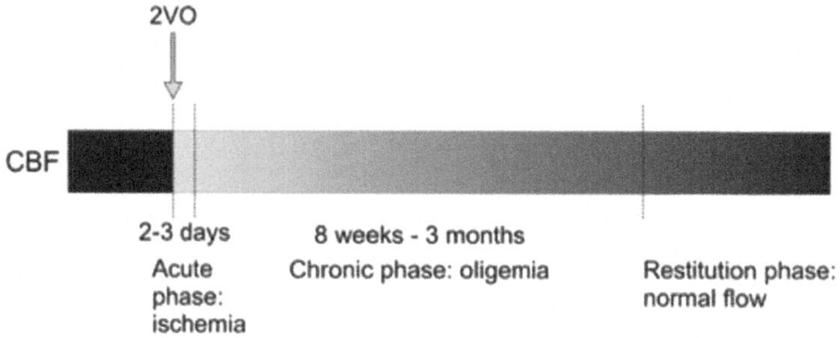

Fig. 3.2 The successive phases of CCH prompted by 2VO model in the rat [22]

Two Vessels Occlusion (2-VO) Model

In 1972 Eklof and Siesjo have invented the two vessels occlusion model and since then several modifications were practiced on it, the realization of this model will take place on the carotid arteries of rats, using two different techniques each technique will provide different results in both cerebral perfusion and brain damage.

Firstly, cerebral blood flow through the carotid arteries will be reduced by legating these two vessels only. The observation noted was really interesting, the ligation has reduced the cerebral blood pressure half of normal however brain metabolism was prevented and no considerable change in tissue dynamics have been discussed. This could be explained by the great ability of the circle of willis in rats to maintain blood flow to the brain in serious complications such as strokes. The varieties in cerebral blood flow after inducing the ischemia have been marked and divided into three phases, acute phase (hypoperfusion lasts at least 2–3 days), chronic hypoperfusion phase (8 weeks to 3 months), response phase. Consequently this model in its chronic phase can be an example of human chronic cerebral hypoperfusion neurodegenerative diseases such as dementia and simulate other pathophysiological changes including persistent oxidative stress [20].

The second technique is used to induce transient ischemia and it shares the same steps with the previous described technique it differs only in the creation of low blood pressure to insure the decrement in cerebral blood flow using specific device and medications like phenolamine or trimethapha, blood pressure must be lowered to at least 50mmhg to guarantee brain blood deprivation, which will have place in cerebral cortex, caudoputamen, hippocampus, midbrain, thalamus and globus pallidus after 15 min of maintaining occlusion and hypotension, fluctuating variances in CBF in the ischemic area were also noticed. 2VO model helps study energy metabolism in ischemia, neurotransmitter metabolism, phospholipids, histopathology when its realized in specific conditions [21] (Figs. 3.2 and 3.3).

3.3.1.2 Focal Ischemic Stroke Models

Focal ischemia is characterized by decreased blood flow to a very specific area, a specific brain region. Considering that the center of the injured area have absolutely no blood flow while there is some flow that reaches the borders [23]. The middle cerebral artery is a frequent site of cerebral ischemia in humans, and many cerebral

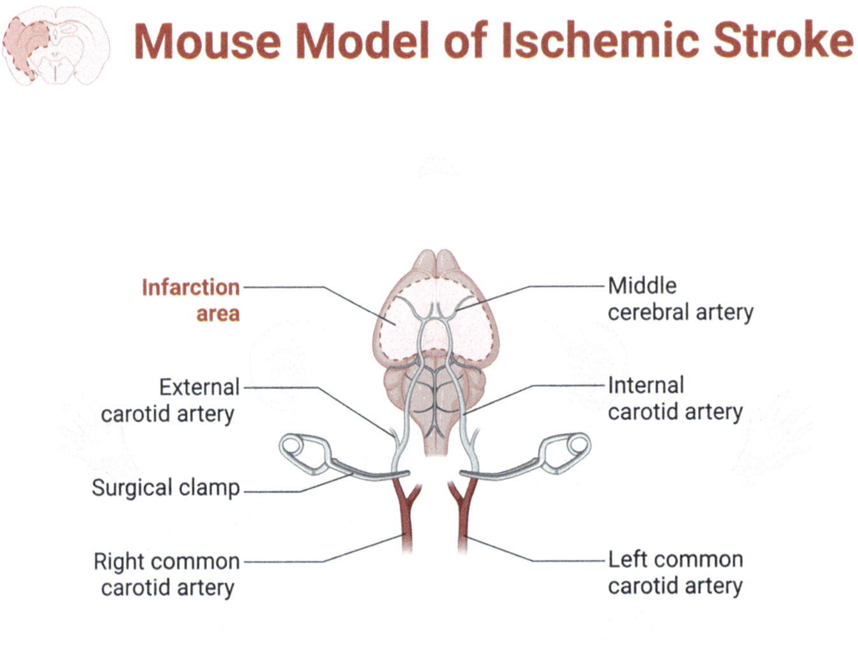

Fig. 3.3 Mouse model of Ischemic Stroke

ischemia models have been developed for MCA and for the most part they are either permanent or temporary (reperfusion) occlusion with MCA occlusion at either the proximal or distal part of the vessel.

Endovascular Filament Middle Cerebral Artery Occlusion

The MCAO intraluminal filament model is considered as the most used surgical model to mimic stroke in humans. This model was first published by Jin-ichi Koizumi in 1986 and modified by Enrique Zea Longa in 1989 [24]. induction of MCAO is accomplished by inserting a thin monofilament into the common carotid artery (CCA) or external carotid artery (ECA). Accessing the CCA requires the vessel to be ligated for incision . Similarly, for filament access, ECA is typically transected in the modified MCAO method . Once the filament enters the vein, it turns into the internal carotid artery (ICA) and to bifurcation of the middle cerebral artery (MCA). Once the filament head reaches the origin of the MCA, it restricts blood flow to the MCA territory and remains in place during ischemia [25]. Although this model allows permanent occlusion of the MCA, a transient MCAO is achieved by removing the monofilament tip to allow recanalization and is therefore clinically

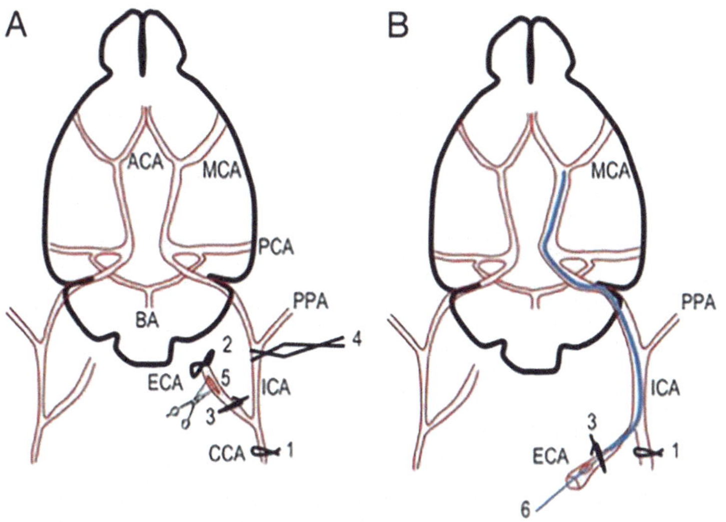

Fig. 3.4 Schematic representation of central cerebral artery occlusion using silicone structured intraluminal monofilament: (**a**) Schematic diagram of rat brain and brainstem showing consecutive sutures and clips in preparation for silicone monofilament placement. (**b**) placement of the monofilament into the MCA, the filament is introduced in the external carotid artery to reach the internal carotid artery and then the circle of willis to occlude the origin of MCA. *ACA* anterior cerebral artery, *BA* basilar artery, *CCA* common carotid artery; (**c**) *Willis* Circle of Willis, *ECA* external carotid artery, *ICA* internal carotid artery, *MCA* middle cerebral artery, *PCA* posterior communicating artery, *PPA* pterygopalatine artery [27]

relevant and has additional benefits [26]. Using the iMCAO model, post stroke effect can be assessed months after initial results. The physical properties of the occluder (tip diameter, length, shape and flexibility) are important for permanent MCAO. This has led to the use of a conventional monofilament coated with silicone to reduce subarachnoid hemorrhage, and because its length extending the entire length of the occluder. The amount of blood remaining in the carotid artery is prevented and the change in infarct size is reduced [27] (Fig. 3.4).

Embolic Models of MCAO

Most human strokes are of thrombo or cardioembolic origin, the reason why animal models of thromboembolism may mimic the human brain better than other brain models. Besides the thromboembolic model allows investigation of the potential neurotoxic effects of Recombinant Tissue Plasminogen Activator as demonstrated by experimental studies. A widely used thromboembolic model based on emboli formation for MCA occlusion. Clots were first obtained outside the body using autologous blood. Pre-prepared nodes of a certain diameter and length are placed inside a thin catheter placed in the internal carotid artery (ICA) and advanced to the origin of the MCA, in a similar way to intraluminal filament model [28]. In addition to size and deposition, the composition of the tissue fibrin-poor (red) or fibrin-rich (white) largely determines the duration of spontaneous reperfusion. The use of autologous blood clots to occlude more distal intra cranial arteries show high variability in infarct size compared to intraluminal surgical MCA occlusion so lesions vary significantly in size and location. A novel mouse model of thromboembolic stroke has recently been described. Direct autologous thrombus formation in the exposed MCA lumen, induced by injection of pure thrombin, resulted in an increased incidence of ischemic injury. The administration of RTPA showed a significant decrease in brain lesions. Such technique can be used to overcome low-frequency ischemic infarctions. However, insertion of arterial catheters to inject thrombus can cause endovascular damage and inflammatory processes. It should be evaluated whether such spectator injuries could be prevented by improving clot delivery techniques [29].

Endothelin-1 Model of MCAO

Endothelin (ET) is a potent vasoconstrictor peptide of 21 amino acids, encoded by three genes, and although it was first discovered as a peptide produced by the endothelium of blood vessels [30]. Originally, endothelin-1 was used to induce ischemia by application directly to the MCA surface after craniotomy; This caused significant vasoconstriction and severely ischemic cerebral cortex and striatum on the injected side of the brain [31]. Later this model was modified by using a stereotactic ET-1 injection proximal to the MCAO to produce focal cerebral ischemia allowing rapid activation of the process and control artery constriction by controlling the delivered agent without the need for manipulating extracranial vessels as well as gradual reperfusion rates that is very similar to human reperfusion [32]. Because of the large number of lesions that can occur due to differences in vascular response to ET-1, Ansari and colleagues attempted to overcome this by using laser Doppler flowmetry to monitor cerebral ischemia during ET-1 infusion [28].

Photothrombosis Models of MCAO

The use of photothrombosis model has proved its efficiency in preclinical research to study neuroprotective mechanisms and in vivo outcomes. Low variability in the infarction size, and its capacity to control the infarct size and location has made it widely used as a research model [33]. The principle of this model is based on intravascular photooxydation to create a sort of stress inside the lumen of the arteries leading to a chain reaction of blood clotting processes and the formation of thrombus inside the vessels to occlude cerebral arteries, this can be achieved by the injection of a photoactive dye (Rose benga, erythrocin B) which is a substance that changes its properties or induce a reaction when exposed to light, the injection will have place on intraperitonealy (into mice) or intravenously (rats) after that, the intact skull will be irradiated which activate the dye causing in endothelial damage and the initiation of the coagulation cascade in both pial and intraparenchymal vessels, the operation will be controlled using MRI and computed tomography CT [34]. Photothrombotic infarct causes long-term sensorimotor deficit and long-term survival in chronic stroke studies. Infarct size may not respond to neuroprotective treatment. However, the absence of neuroprotection in the acute phase allows distinction between neuroprotective and neuroregenerative effects, and as a result, neuroregenerative drugs can be better evaluated. Together, photochemically induced infarctions are particularly suitable for analyzing the effectiveness of neuroregenerative therapies [35] (Table 3.1).

Table 3.1 Summarizes the advantages and disadvantages of ischemic stroke models

Ischemic stroke models	Advantages	Disadvantages
Intraluminal filament middle cerebral artery occlusion model	– Limiting ischemic time is a very important aspect of keeping a patient fit – Removal of the embolus for reperfusion without the need of transcranial operation is the reason why this tool is convenient to use – The height of the operator and the wide-open window in the requirements for operators are the two factors that make this technology so easy to use and learn – Moral of the story the dangerous car makes the animals not apply for the job and no one would be injured at all – Direct surgical procedure with no second period of operation and no cerebral hemorrhage – The capability to imitate different physiological reactions in cerebral ischemia patients after embolization and thrombolysis is one of the main functions – NHP research has been also a detention area on human stroke [36]	– Since observed benefits of this method are not fully reproducible compared to the direct occlusion of cerebral vessels – The factors that influence the results of the fusion, these include weight, strain, age, sex of the animal, and the size of the filament – The surgical skill and the unmatched precision as a result of endothelial damage and filament location are the main factors that have an impact on the functional outcome – The inconsistency of functional outcome which results in a significant difference in lesion size and location – The huge diversity in technical details between labs, like the different durations of occlusion is the second reason why this task becomes difficult [37]
Embolic models	The "Mimics" most closely resembles pathogenesis of human stroke Low reproducibility of infarcts The case study on the thrombolytic usage is suitable for research about thrombolytic Agents [14].	Spontaneous recanalization High variability of lesion size [14]

(continued)

Table 3.1 (continued)

Ischemic stroke models	Advantages	Disadvantages
Endothelin-1 model	– Infarcts of different sizes in almost any brain areas are possible – Subcortical stroke The restoration and plasticity mechanisms in chronic stroke are the ways in which the brain can recover from the stroke and reorganize itself in order to overcome the damage caused by the stroke (Transient ischemia only) – Short invasiveness Ischemia not controllable is easy to be used – Induction of ischemic lesion in cortical on demeanors promotes the anatomical change of the neurons in the brain areas or subcortical regions – Induction of astrocytosis and axonal sprouting, are some of the major effects of the breakdown of blood-brain barrier what could be the fact that may become the major cause of the wrong interpretation of the Low mortality [15]	– Minimal edema – Endothelin-1 and endothelin-1 receptors that are also found on neurons are the subject of the sentence Astrocytes—might be the reason for the interference with the post-stroke recovery mechanisms – Ischemia time is not a controllable factor – Induction of the ischemic lesion in the cortical is a possible reason for the occurrence of limbic-cortical circuit dysfunction during limbic-cortical conversion in the context of ischaemic lesion in the cortical or subcortical regions – Observation of astrocytosis and axonal sprouting has been the key to the success of this theory The possible of the second verification of the results therefore, the interpretation of the result may be more complicated [15].
Photothrombosis models	Very reproducible; low mortality Enables well-defined localization of an ischemic lesion Highly reproducible, Low invasiveness [12]	Requires injection of photochemical dye; early vasogenic edema The main cause of early vasogenic edema is the exogenous or indirect complement system which is activated through exoantigens Uncharacteristic for human stroke Improper for the neuroprotective paragraph investigating Agents [12].

3.3.2 Ischemic Stroke Models In Vitro

Translatability factors in focusing on the connection between the impact of research discoveries, and the realization, or implementation, of these findings. Research about stroke within the benefit of experiments of cell based can be often a part of this process.

- Identifying the causes of stroke: Scientists can do this experiment to understand whether the lack of oxygen or over exaggerated excitement of brain cells lead to cell death or brain damage among people that have had a stroke [38].
- Testing potential therapeutic strategies: Cell models represent a good tool to screen and test new drugs or treatment modalities that can be used to prevent cell damage, to promote tissue repair or to achieve neurological improvement after a stroke.
- Developing methods for stroke prevention: Through elucidating the cellular and molecular processes incorporated in stroke development, researchers determine which preventive strategies can be useful to diminish the risks of stroke [39].

Investigating ischemic strokes in vitro, researchers built models such brain slice, putting particular emphasis on a hippocampal slice, and primary neuronal/glial cultures that they might have been obtaining from various brain regions of embryonic or neonatal rats and mice. The methods which are the most used are based on this approach, and they are praised for the ability to imitate ischemic damage. To recreate ischemia-like conditions.

In vitros settings observe significant differences from in vivo stroke models throughout many of their properties. Usually, a longer duration of the anoxic or hypoxic stressor in vitro is required to produce the neuronal death. With the milder depletion of glucose and delayed release of [glutamate] as compared to the in vivo stroke, blood vessel and blood flow in vitro lack vital structural and functional components of the damage process present in vivo, such as the invasion of inflammatory cells. In the same manner, glial cells' composition and reactivity when they are in in vitro culture are different than what they are in the whole brain [16].

Although from different perspectives, it is fascinating to observe these similarities of the reaction of the cells under a substrate stress and also in severe stroke which occurs in the mentioned conditions. This cooperation brings forth the more precise fundamental knowledge about the response of the specific cell types to ischemia as an alternative to the in vivo models which have high complexity of the brain during the stroke. Through the complementary use of vitro and in vivo experiments, which aims to understand the pathology of ischemic stroke, it has become more sophisticated [40].

3.3.3 Types of Experimental Cell Models for Stroke Research

3.3.3.1 Neuronal Cell Cultures

A primary focus is the in vitro neuronal cell culture that is performed in a designated laboratory where the neurons are subjected to possible conditions causing a stroke.

3.3.3.2 Glial Cell Cultures

These models actually preliminary culture of glial cells being the main cells that provide all necessary grounds for neurons growing good. It can sign the roadmap to their possible function in stroke pathophysiology as well as in rehabilitation [41].

3.3.3.3 Co-cultures

These models utilize different cell types such as neurons, glial cells and so on to model a more complicated environment within the brain tissue. These models are able to mimic conditions of the real environment where complex interactions of cellular constituents within the brain take place.

3.3.3.4 Blood-Brain Barrier Models

These in vitro barriers reproduce the physical boundary that separates the blood from the brain, where researchers can study the effect of stroke on this barrier and develop means by which drugs can be delivered to the brain's tissue [41].

3.3.4 Hemorrhagic Stroke Models

It is without a doubt that appropriate animal models that at least partially replicate the characteristics of human stroke would contribute to a better understanding of the diseases as stroke as well as the development and evaluation of therapeutic approaches. Since the 1960s, there have been experimental ICH models that utilize cerebral blood vessel avulsion, balloon inflation, or intracerebral injection of autologous blood or bacterial collagenase. The pathophysiology of ICH in patients has been mimicked using a variety of species, including rodents, rabbits, cats, dogs, pigs, baboons, and other primates. The intracerebral injection of autologous blood or bacterial collagenase into the brain are the two most popular preclinical study models.

3.3.4.1 Autologous Blood Injection Model

A single injection is the simplest way to introduce blood and cause hematoma formation in the brain. To create a hematoma model, blood is stereotactically injected into the striatum from a superficial vessel at various volumes. In patients with ICH, the quick accumulation of intraparenchymal blood is significant. Both in the vicinity of the hematoma and in the surrounding brain, CBF is decreased. The amount of injected blood varies across investigations and is consistent with the size of a typical human hematoma. This change is significantly volume-dependent and is not followed by appreciable changes in cerebral perfusion pressure. For good repeatability of hematoma volumes, a gradual injection of 50 L blood with a Hamilton syringe over a period of 5 min is advised. Some investigators have also performed double or multiple injections, which produces consistent neurological deficits, brain swelling, and cortical hypoperfusion, this technique has been adapted to mice and is used widely [42].

3.3.4.2 Collagenase Injection Models

Certain enzyme are secreted by cells to perform a particular task, such as the collagenase enzyme, which is secreted at the areas of inflammation by mononuclear cells. In order to study the pathophysiology of ICH, Rosenberg et al. developed the collagenase induced ICH model, which typically evolved over hours. Collagen is found in the basal lamina of blood vessels in brain tissue, so hematoma expansion and vasogenic edema following ICH have been thought to result from elevated local concentrations of collagenase released from injured cells. This repeatable model is used in preclinical research to examine the consequences of hematoma and brain edema after spontaneous intracerebral bleeding that occurs over several hours. Both models yield consistent hemorrhagic infarcts and are basic methods for preclinical ICH research with intrastriatal injection of autologous blood (30 μL) or bacterial collagenase (0.075U) leading to reproducible neurofunctional deficits in mice [43].

3.3.4.3 Balloon Inflation Models

Balloon inflation models are used to study the mass effects of a hematoma and its removal on brain injury. It is an unusual experimental model of ICH. A modified procedure has been developed in piglets, in which supratentorial ICH is induced with a balloon introduced into the right striatum through a burr hole [43].

3.3.4.4 Subarachnoid Hemorrhage Models

Subarachnoid hemorrhage it is bleeding in the space that surrounds the brain. it happens when aneurysm on the surface of the brain ruptures leading to increase in pressure inside the skull and the brain. This can cause brain cell damage and many complications as disabilities.

3.3.4.5 Blood Injection Model

An average of 300 μL whole blood is injected into the cisterna magna in the single-hemorrhage model to induce SAH. In many double-hemorrhage models, the second injection with autologous arterial blood is given 48 h after the first. The single- and double-hemorrhage models have been performed in dogs, rabbits, rats, and mice with acceptable mortality rates, and the models are reproducible because a fixed amount of blood is injected into the subarachnoid space [44].

3.3.4.6 Endovascular Puncture Model

The ECA and all of its branches are located, split, cauterized, and dissected during endovascular puncture. A suture is passed through the ICA, the ECA, and the MCA, where the vessel is punctured. After that, the suture is removed via the ICA and into the ECA, enabling perfusion and creating a SAH that as nearly resembles the clinical scenario as feasible. Rats are the primary animal used in endovascular puncture models, while mice have also been employed in a few investigations, although mice have also been used in a few studies [45].

3.3.4.7 Prechiasmatic Model

In the prechiasmatic model, a rat that has been anesthetized is given a craniotomy in order to insert a needle into the prechiasmatic cistern. The needle is positioned 7.5 mm anterior to the bregma on the midline, with a 30° slant. The ICP is maintained at the level of mean arterial blood pressure by injecting the required volume of autologous blood over an appropriate amount of time. Cerebrospinal fluid mixed with saline is given to the control animals the craniotomy is not sealed before injection, its decompressive effect influences the ICP [46].

3.3.4.8 Double Transgenic Mouse (R±/A±) Model

The human renin and human angiotensinogen are overexpressed in the double transgenic mice (R+/A+), which results in higher mean arterial pressure (MAP) and chronic hypertension. Despite this, there is no evidence that these transgenic mice spontaneously develop ICH, and they live normal lives. In a study, Iida et al. used a high-salt diet and L-NAME to cause severe hypertension in R+/A+ mice, which led to ischemic and hemorrhagic strokes in the cerebral cortex. In almost 30% of transgenic mice fed only a high-salt or L-NAME diet, symptoms of ICH appeared before 24 weeks. In this model, oxidative stress was discovered to be crucial to the pathogenesis of ICH [47].

3.3.4.9 Spontaneous ICH in Acute/Chronic Hypertension

In a study on chronic hypertension in rats, Wakisaka et al. used subcutaneous angiotensin II and L-NAME administration. Angiotensin II or norepinephrine subcutaneous injections were used to cause acute hypertension in the mice. After a week, mice were checked for stroke symptoms three times a day and their systolic blood pressure was monitored every day. According to the findings, multiple ICHs were found in the thalamus, brainstem, basal ganglia, and cerebral cortex of 73% of the mice in the angiotensin II group, which also showed indications of stroke. Greater volume ICH was linked to acute angiotensin II administration than to acute norepinephrine injection. According to the study, increased brain oxidative stress and matrix metalloproteinase (MMP) activity contribute to the development of ICH. Angiotensin II and L-NAME's effects on the kidneys [47] (Table 3.2).

3.3.5 Cerebrovascular Pathologies Unraveled Using Zebrafish Models

Cerebrovascular disease research remains vibrant and alive on the research front; in this context, zebrafish become indispensable partners of a multi-faceted nature, providing precious insights into hemorrhagic and ischemic strokes [49].

Table 3.2 Summarizes the key pros and cons associated with each model

Hemorrhagic stroke model	Advantages	Disadvantages
Collagenase injection	Highly reproducible [36]	Amplification of inflammatory response at high doses of bacterial collagenase [36]
Autologous blood injection	– Recapitulates a single bleed, easily reproducible – Simulate the main pathophysiological traits in the human condition such as mass effect, brain edema, neurological deficits, and neuroinflammation – Allows investigation of the neuroinflammatory response, including innate immune cell infiltration, activation of brain resident immune cells, oxidative DNA damage, and pro-inflammatory cytokine release, which can be potential therapeutic targets [48]	– Not spontaneous – Hematoma expansion, which occurs in 1/3 of patients during the first 24 h post-ictus won't be explored using this model – Can be used only to investigate the direct effects of haemotoxicity on the brain [48]
Microballoon inflation	– Very helpful to reduce cerebral blood flow and increase the intracranial pressure – Can generate systemic impacts on cerebral perfusion pressure [36]	– Time constrained due to brain damage that can occur – Secondary brain injury caused by blood elements and blood brain barrier disruption cannot be evaluated [36]
Thrombin injection	– Learning the effect of thrombin [36]	Thrombin is the only blood element that can be investigated [36].

Hemorrhagic Models: For one, the transparency of zebrafish embryos is effectively used to dissect out the intimate nature of intracranial hemorrhages and model pathological conditions through deliberate injury of vessels and genetic manipulation. These models will provide certainty on the complex nature of endothelial dysfunction, the disruption of the blood-brain barrier, and inflammatory responses—the very steps in understanding hemorrhagic stroke progression. Furthermore, models of zebrafish provide grounds for experimentation of possible therapeutic interventions, targeting interventions, such as pharmacological agents aimed at inter-coagulation pathway cross-talk and vascular stability [50].

Ischemic Models: While the former serves to complement hemorrhagic studies, zebrafish models become a solid starting ground in which researchers can understand the post-stroke aftermath in greater depth. Induced by arterial ligation or pharmacological occlusion, these models will elucidate the sequence of events following cerebral vessel obstruction, encompassing neuronal cell death, neuroinflammation, and tissue repair processes. Zebrafish larva's regenerative powers open a route toward the understanding of endogenous neuroprotective mechanisms and screening new therapeutic candidates to augment recovery after ischemic injury. Using high-throughput screening, researchers will identify promising candidates to promote neuronal survival and functional restoration after ischemic injury [51].

3.3.6 The Role of Experimental Models in Treatment

Numerous treatments and research projects have been carried out to find cures for hemorrhagic stroke outcomes. One of those medication is Trans-sodium crocetinate (TSC) Trans-sodium crocetinate (TSC) is a new carotenoid chemical medication that can improve the transportation of small molecules in water. TSC enhances oxygen and glucose transport and raises oxygenation levels in ischemic brain tissue. During an ongoing ischemic period, TSC also reduces the severity of an ischemic stroke. The current study looked at how TSC affected rat models of hemorrhagic and ischemic stroke. A rat model of intracerebral hemorrhage (ICH) especially collagenase-injection model was used to examine the effects of TSC. TSC treatment also improved neurological outcomes as measured by a multi-scale assessment and an automated gait analysis [48]. One of the most significant benefits of this medicine is reduced edema and hemorrhagic volume in a rat model of ICH, TSC's effects on hematoma volume, hemispheric edema, and hemorrhage volume in both vehicle-treated and TSC-treated mice were assessed using the collagenase-injection model. The TSC-treated group's hematoma size was shown to be on the decreases, along with edema and hemorrhage volume [53].

3.3.6.1 Argon and New Hope

There are many research projects carried out on experimental models to discover the effect of argon in treatment hemorrhagic stroke it was found that argon is a noble gas, has been shown to have neuroprotective effects in animal models, including stroke. Argon administration after individual injury improved cell survival and neuronal function. The neuroprotective effects of argon may be of potential clinical use in stroke treatment. Further research is needed to understand its mechanisms in nerve function [53]. The neuroprotective function of argon is not fully understood, but it is known to activate the γ-aminobutyric acid (GABA) receptor by acting at the benzodiazepine binding site. GABA receptor agonists can confer neuroprotection after ischemia via the phosphoinositide 3- kinase pathway or by abating the phosphorylation of the NR2A subunit of the N-methyl D-aspartic acid (NMDA) receptor. Akt serine/threonine kinases, which are mediators of cell survival in brain ischemia, are important mediators of cell survival. Argon can exert a neuroprotective effect via the PI-3K/Akt pathway, which could inhibit the ASK1-JNK cascade by activating the GABA receptor. Glutamate receptors, mainly metabotropic and ionotropic, play a crucial role in neuronal development, addiction, learning, and memory. Over-excessive release of glutamate can lead to excitotoxicity, triggering calcium-dependent signaling cascade reactions that eventually lead to neuronal death (Li et al. P65). It was discovered that 75% vol/vol argon prevented the death of neurons, but neon, krypton, and helium had no impact and increased cell death. According to research, argon (at dosages of 25, 50, or 75% vol) was neuroprotective in a hippocampus organotypic slice treated to OGD, and this protection could be sustained with a delayed treatment [54].

3.3.7 Hemorrhagic Stroke and Regenerative Medicine

Therapeutic advancements in stem cells (SCs) have shown promising potential in treating hematological disorders (HS). Recent studies have shown that hUCMSCs, human umbilical cord Wharton's jelly-derived mesenchymal stem cells, can reduce hematoma volumes in rat models of moderate ICH. However, these cells do not show significant functional improvements in rat models. Further research is needed to understand the mechanisms behind these therapeutic benefits. Despite these advancements, the use of SCs in animal models offers promising clinical applications, as they can help mitigate stroke-induced peripheral inflammation and improve survival rates, motor function, and hematoma volume, Stem cell treatments for ischemic stroke (HS) have shown promising results, with IL-17A promoting angiogenesis and functionality, and IL-17A enhancing bone marrow mesenchymal stem cells (SCs) differentiation into neuron-like cells. Research on stem cell death mechanisms, such as ferroptosis and Tongfu Xingshen capsule (TXC), has also enriched SC applicability for HS. Electroacupuncture (EA) treatment for ICH in rats enhances neuron-like differentiation of MSCs and SC expression of BDNF and NGF. Future studies may use these findings to improve exogenous NSC functionality and viability with TXC and EA. Oxyhemoglobin-induced neuron injury model, in which, the cell types used in this model is Exosomes from endothelial progenitor cells (EPC-Exs) overexpressing miR-137. The dosage is 1 × 109 exosomes incomplete medium. Findings are miR-137 amplifies the ability of EPC-EXs in lessening apoptosis and mitochondrial dysfunction in an in-vitro model of Hs [55].

3.3.8 Imaging Techniques and Stroke Experimental Models

Imaging techniques such are MRI, CT can play an important role in studies of stroke recovery through serial in vivo whole-brain imaging that measures functional and/or anatomical changes in the damaged brain [56].

3.3.8.1 Magnetic Resonance Imaging (MRI)

MRI is a non-invasive contrast free method which allows the evaluation of MCAO success by detecting blood flow into the MCA. MRA is used to guide and confirm the placement of the filament in its appropriate site within the middle cerebral artery indirectly based on the lack of blood flow in this area [57].

3.3.8.2 Computed Tomography

Computed tomography (CT) is a diagnostic anatomical method that provides detailed images of internal structures by detecting the variety of X-rays absorption by different tissues due to the variation of tissue densities. X-ray images are available from different angles and multiple systems and are produced to create a 3D image. Although CT has high spatial and temporal resolution, it lacks the superior soft tissue differentiation achieved by other methods such as MRI. An advantage of using contrast-enhanced CT angiograms (CTA) is that it can provide information about spatial geometry, lumen patency, and vascular connections. Researchers use micro-CT angiography to visualize vascular anatomy and gross morphology, often with a vascular contrast agent [58].

3.3.8.3 Ultrasound

Ultrasound (US) is a conventional imaging technique that uses acoustic waves emitted by a transducer to create images of internal structures based on acoustic reflections. US can also be used to measure biological parameters such as blood velocity using Doppler imaging and vessel wall using motion mode (M mode) imaging. The main advantages of US are the absence of ionizing radiation, high imaging speed, and portability does not involve the necessity for specialized shielding measures as required for MRI. high-frequency systems component of anatomical or brightness mode (B-mode) imaging and US Doppler provides high temporal and spatial resolution and is designed primarily for use in small animal research [58].

3.3.8.4 Positron Emission Tomography (PET)

PET imaging is an excellent and minimally invasive tool that have been utilized for metabolic-biochemical-molecular in vivo studies, Note that neuroimaging with 15O-labeled tracers is considered one of the most effective methods for identifying brain regions where brain oxygen consumption is reduced. However, this type of tracer is not widely available in most hospitals. Conversely, −2-[18F]-fluoro-2-deoxy-D-glucose (18F-FDG) is the commonly used tracer in clinical practice with a short half-life. in this situation, glucose oxidation is the main source of energy in the brain, and FDG is transported to tissues and phosphorylated to FDG-P, but unlike glucose it does not undergo significant metabolism, FDG remains trapped in the brain tissue, and the amount of FDG trapped is directly related to the metabolism of glucose in that specific area. Small animal PET is a technology that makes it possible to explore biological processes in vivo and helps the translation of findings from animal models of diseases—such as stroke—to human patients. Small animal PET with 18F-FDG in particular is sensitive to metabolic changes following stroke injury, however it does not provide good illustration of brain structures because of the spatial resolution in the order of 1.5 mm hence the combination of MRI and PET images to facilitate the observation of functional impairment [59].

3.4 Translational Relevance of Experimental Models

One key area is the development of more sophisticated models. Future models should consider the occurrence of genetic susceptibility, comorbidity, environmental exposures, and related factors to generalize better to the human experience [60]. Utilizing Large Animal Models Closely similar species to human in neuroorganism, such as great apes or even chimpanzees, may be the most translatable researches.

Nevertheless, ethical issues and economic limits restrict the adoption of neonaps. Another crucial aspect is the Focus on Functional Outcomes: It is necessary to target conducive diagnostics with an aim to measure the effectiveness of stroke treatment beyond the basic neurological measures. The underlying aim of stroke treatment is oriented to improving the quality of life for patients [60]. Integration of Human Data is also essential. Firstly input can be broken down into three parts. First, integrating human genetic and clinical data into preclinical research will emphasize planning and selection of the most relevant models for intervention development [60]. It is of crucial importance to perfect the translational applicability of experimental models and this way the stroke treatment related progress will be accelerated. We find in the general area of multimodal, multi-model, and multi-data-type, as well as the focus on clinically relevant outcomes as having the best chance and what seems to be the most likely transition from scientific discovery to application of health care. By this, including the interest of joint actions in the basic research is also the other side of that coin. Scientists, clinicians, and industry partners is crucial for translating promising preclinical findings into effective clinical therapies that ultimately benefit stroke patients [61].

3.5 Future Directions and Emerging Techniques

In this rapidly developing field of stroke research in experimental animals, several promising future directions and emerging techniques are poised to drive significant advancements in our understanding of stroke pathophysiology, treatment strategies, and neuroregeneration. These innovative approaches promise to revolutionize the way we study and address stroke-related issues. Advances in genomics and molecular profiling open up a new frontier in the study of stroke. By tailoring interventions to an individual's genetic and molecular makeup, researchers are developing personalized treatment strategies. Animal models genetically modified to mimic specific human traits will play a critical role in unraveling the effectiveness of these personalized therapies [62]. This approach promises to enhance treatment precision and improve outcomes for stroke patients. Another exciting avenue is neuroinflammation. Following stroke, activation of immune cells and the release of pro-inflammatory mediators in the central nervous system characterizes neuroinflammation. It plays an important role in the progression of stroke-induced brain damage [63]. Emerging techniques aimed at modulating neuroinflammation present exciting avenues for stroke research, with potential implications for the development of novel therapeutic interventions. Animal models are indispensable tools for the unraveling of the complex mechanisms, underlying neuroinflammation and for testing novel strategies aimed at mitigating its detrimental effects [39]. Post-stroke neuroinflammation involves the activation of resident microglia, infiltration of peripheral immune cells, and release of inflammatory cytokines [64]. While acute inflammation is a protective response that aids in the removal of damaged tissue, chronic activation exacerbates neuronal injury and hampers recovery processes. The intricate balance between neuroinflammation and repair mechanisms needs to be understood for the development of effective modulation strategies [65].

Animal models allow for the examination of the effects of targeting specific immune cell populations involved in neuroinflammation [66]. Techniques such as antibody-based depletion of immune cells or genetic manipulation of microglia offer insights into the consequences of dampening or altering immune responses [67]. By dissecting the roles of different immune cell subsets, researchers can uncover potential therapeutic targets to regulate neuroinflammatory processes. Neurotrophic factors and anti-inflammatory therapeutics are among the relatively new approaches to the treatment of CMD. Any kind of animal researches can be focused on the experiments where neurotrophic elements like brain-derived neurotrophic factor (BDNF) or nerve growth factor (NGF) will be used to control microglial activation and eventual nerve cell survival [68].

The application of anti-inflammatory agents, such as minocycline or cytokine inhibitors, can also be tried in animal models to see if they can help identify the ability to decrease free radical-induced toxicity [69]. New cellular therapies have been used in research on stroke. Such new cellular therapies are stem cell-based therapies. The most used of these therapies are MSCs (Tregs). These are valuable because of their property of being immunomodulatory [70]. These cells have the ability to reduce excessive inflammation and improve tissue repair. Animal models provide the platform to study the mechanism through which these cells interact with the immune system and influence the post-stroke inflammatory milieu [71].

In the pursuit of precision in addressing neuroinflammation, targeted immune modulation has emerged as a promising avenue. Techniques such as optogenetics or chemogenetics offer an unprecedented degree of control over immune cell activity, providing researchers with valuable insights into the mechanisms underlying neuroinflammation and its behavioral consequences [72]. Animal models are crucial in establishing the safety and efficacy of emerging neuroinflammation modulation strategies before transitioning them to clinical trials [39]. The ability to finely tune the immune response while avoiding detrimental effects is a challenge that animal-based research seeks to address. Through systematic investigations in animal models, researchers can identify optimal intervention parameters and better understand the potential long-term consequences of manipulating neuroinflammation. The investigation of the intricate way in which brain activity is linked to blood flow in the cerebral area is a requirement for the proper determination of stroke-related brain damage. New imaging technologies, such as fMRI and advanced optical imaging, can capture premise-level neural dynamics in easily available animal models in real time. The dissection of these fine connections creates important avenues to create medications that will do less damage to the brain and also prevent acquired disabilities after a stroke [73].

Following a stroke that has occurred and induced brain damage is considered to be the most complicated challenge of neuroregeneration. Recent approaches like stem cell therapies, organ transplant, and gene editing looks promising in reviving dead tissue and healing injured or lost neural pathways. As they come among these, the pulp stem cells (DPSCs) have bestowed the honor with their capability of neural regeneration [74].

3.6 Dental Pulp Stem Cells (DPSCs) and Neural Regeneration

DPSCs, which originate from the retro tooth pulp, have proved to be one of the best neurogenic cell types. The above-mentioned stem cells that exhibit multipotent properties can be classified into neurogenic lineage which provides hope in the development of neuron regenerative therapies. Animal models of stroke have proved that the transplantation of DENT op progenitor cells (DPSCs) can enhance functional recovery to the improved level and support neuro plasticity. The DPSCs are able to secrete neurotropic factors and cytokines into the surrounding milieu, which enhances the proliferation and recovery of neuronal cells. Moreover, their equally, immunomodulatory characteristics control neuro-inflammation thus assisting in the full recovery after a stroke. DPSC-based animal studies have brought the mechanisms of action of DPSCs into light, in addition to showing that DPSCs protect cells from damage and promote axonal growth as well. The use of DPSCs with the scaffolds or stimulation of growth factors accompanied by delivery systems increases the power of regeneration. These matrixes induce sculptural physicality and lead axonal outgrowth along their contours as neural pathways emerge from the stroke restructuring process. There is of course, much to be done in the application of DPSCs for stroke research and beyond which is still in its infancy but holds great promise for the future. Continuous refinement of transplantation procedures and the choice of differentiation methods besides detailed work on chronic effects of DPSC-induced treatments in experimental animals will therefore form an important part of this transition between the lab and the clinic (Fig. 3.5).

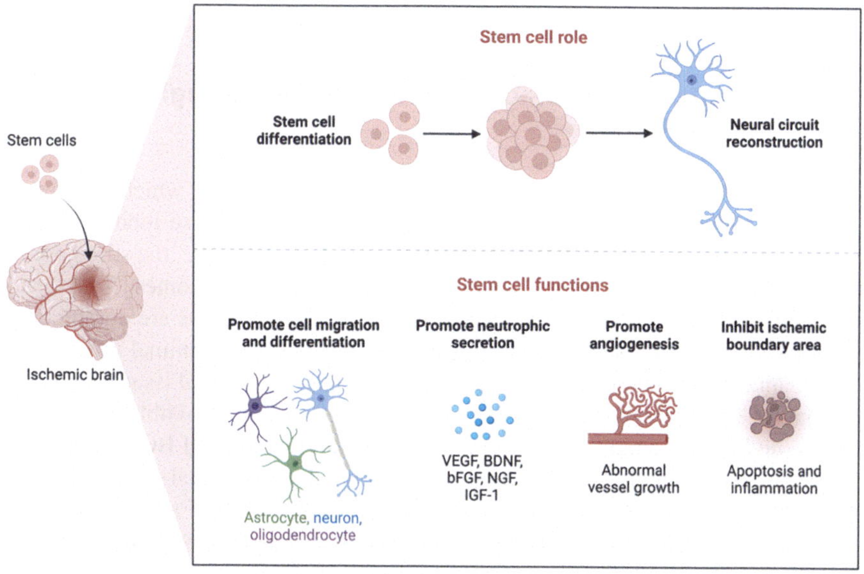

Fig. 3.5 Stem cell and ischemic stroke

3.7 Neuromodulation and Brain-Computer Interfaces

The integration of neuromodulation techniques and brain-computer interfaces (BCIs) is an exciting frontier in the research of stroke. These novel approaches utilize the plasticity of the brain for the recovery and restoration of lost functions post-stroke. Animal models are essential platforms for fine-tuning and validating these sophisticated approaches [75]. Neuromodulation; the process of direct control of neurological component; is used to restore or enhance their function. Traditionally in stroke research techniques, animal models now examined, such as transcranial magnetic stimulation, transcranial direct current stimulation, and deep brain stimulation, are applied. These techniques may help to regulate neuropathic excitability; promote plasticity of dendrites, and control network connectivity in the brain. Animal neuroscience yields valuable knowledge on optimum parameters and design, most appropriate area and time of neuromodulation. They also provide the details on the functioning of the neurobiological matrix that is involved in the post-stroke recovery. After discovering the distorted relationship of inhibiting and excitement of the neural pathways, the researchers are able to design neuromodulation plans in order to elevate the favorable activities by acting on them [76].

Brain-Computer Interfaces (BCIs) are interfaces between the brain and external devices. BCIs have the potential to circumvent damaged neural pathways, thereby recovering lost motor functions in stroke survivors. Animal models are crucial for developing and testing BCI technologies. Researchers can implant electrode arrays in experimental animals in discrete brain regions to record and interpret the neural activity. Thus, algorithms analyze these signals to decode the user's intent and translate it into device commands. Animal studies offer insights into the feasibility, accuracy, and adaptability of BCIs for stroke survivors [77].

3.8 Advancing Neuromodulation and BCIs Through Animal Models

Animal models are a means to be in control of the environment in which the effects of neuromodulation and BCI interventions can be studied. These models enable researchers vary many factors such as the location of the electrode, the stimulation. Parameters and rehabilitation protocols to achieve the best outcomes. Long-term animal studies provide clues whether the neuromodulatory effects are durable and Whether the recovery is sustained in place. On the other hand, animal models are significant tools in the process since they help in safety and ethical issues. Through preclinical studies, researchers will be able to evaluate the possible risks, side effects, and the long-term consequences of neuromodulation and BCI technology on neural tissue. The field will continue changing, and the animal-based research will probably develop the neuromodulation techniques and BCI applications further. These developments would eventually close the gap between basic neuroscience discoveries and their clinical applications for stroke survivors [78].

References

1. Kuriakose, D., & Xiao, Z. (2020). Pathophysiology and treatment of stroke: Present status and future perspectives. *International Journal of Molecular Sciences, 21*(20), 7609. https://doi.org/10.3390/ijms21207609
2. Murphy, S. J., & Werring, D. J. (2020). Stroke: Causes and clinical features. *Medicine (United Kingdom), 48*(9), 561–566. https://doi.org/10.1016/j.mpmed.2020.06.002
3. Feigin, V. L., Brainin, M., Norrving, B., et al. (2022). World Stroke Organization (WSO): Global Stroke Fact Sheet 2022. *International Journal of Stroke, 17*(1), 18–29. https://doi.org/10.1177/17474930211065917
4. Washington, H. H., Glaser, K. R., & Ifejika, N. L. (2021). CE: Acute ischemic stroke. *American Journal of Nursing, 121*(9), 26–33. https://doi.org/10.1097/01.NAJ.0000790184.66496.1d
5. Lee, K. F., Chen, J. H., Teng, C. C., Shen, C. H., Hsieh, M. C., Lu, C. C., Lee, K. C., Lee, L. Y., Chen, W. P., Chen, C. C., Huang, W. S., & Kuo, H. C. (2014). Protective effects of Hericium erinaceus mycelium and its isolated erinacine A against ischemia-injury-induced neuronal cell death via the inhibition of iNOS/p38 MAPK and nitrotyrosine. *International Journal of Molecular Sciences, 15*(9), 15073–15089. https://doi.org/10.3390/ijms150915073
6. Farina, M., Vieira, L. E., Buttari, B., Profumo, E., & Saso, L. (2021). The Nrf2 pathway in ischemic stroke: A review. *Molecules, 26*(16), 5001. https://doi.org/10.3390/molecules26165001
7. Nogles, T. E., & Galuska, M. A. (2024). Middle cerebral artery stroke. In: *StatPearls* [Internet]. StatPearls Publishing. [Updated 3 Apr 2023]. Available from: https://www.ncbi.nlm.nih.gov/books/NBK556132/
8. Shin, T. H., Lee, D. Y., Basith, S., Manavalan, B., Paik, M. J., Rybinnik, I., Mouradian, M. M., Ahn, J. H., & Lee, G. (2020). Metabolome changes in cerebral ischemia. *Cells, 9*(7), 1630. https://doi.org/10.3390/cells9071630
9. Johansson, S., Povlsen, G., & Edvinsson, L. (2012). Expressional changes in cerebrovascular receptors after experimental transient forebrain ischemia. *PLoS One, 7*(7), e41852. https://doi.org/10.1371/journal.pone.0041852
10. Smith, S. D., & Eskey, C. J. (2011). Hemorrhagic stroke. *Radiologic Clinics, 49*(1), 27–45.
11. Testai, F. D., & Aiyagari, V. (2008). Acute hemorrhagic stroke pathophysiology and medical interventions: Blood pressure control, management of anticoagulant-associated brain hemorrhage and general management principles. *Neurologic Clinics, 26*(4), 963–985.
12. Yao, H., Ago, T., Kitazono, T., & Nabika, T. (2017). NADPH oxidase-related pathophysiology in experimental models of stroke. *International Journal of Molecular Sciences, 18*(10), 2123.
13. Schaar, K. L., Brenneman, M. M., & Savitz, S. I. (2010). Functional assessments in the rodent stroke model. *Experimental & Translational Stroke Medicine, 2*, 1–11.
14. Fluri, F., Schuhmann, M. K., & Kleinschnitz, C. (2015). Animal models of ischemic stroke and their application in clinical research. *Drug Design, Development and Therapy, 9*, 3445–3454.
15. Sommer, C. J. (2017). Ischemic stroke: Experimental models and reality. *Acta Neuropathologica, 133*(2), 245–261.
16. Woodruff, T. M., Thundyil, J., Tang, S. C., Sobey, C. G., Taylor, S. M., & Arumugam, T. V. (2011). Pathophysiology, treatment, and animal and cellular models of human ischemic stroke. *Molecular Neurodegeneration, 6*(1), 11. https://doi.org/10.1186/1750-1326-6-11
17. Kaya, A. H., Erdogan, H., & Tasdemiroglu, E. (2017). Searching evidences of stroke in animal models: A review of discrepancies. *Turkish Neurosurgery, 27*(2), 167–173. https://doi.org/10.5137/1019-5149.JTN.15373-15.2
18. Atlasi, M. A., Naderian, H., Noureddini, M., Fakharian, E., & Azami, A. (2013). The morphology of rat hippocampus CA1 neurons following modified two and four vessel global ischemia models. *Archives of Trauma Research, 2*(3), 124–128. https://doi.org/10.5812/atr.10240
19. O'Neill, M. J., & Clemens, J. A. (2000). Rodent models of global cerebral ischemia. *Current Protocols in Neuroscience, 12*(1). https://doi.org/10.1002/0471142301.ns0905s12
20. Li, Y., & Zhang, J. (2021). Animal models of stroke. *Animal Models and Experimental Medicine, 4*, 204–219. https://doi.org/10.1002/ame2.12179

21. Orsu, P., & Srihari, Y. (2022). Experimental animal models of cerebral ischemic reperfusion injury. In *Preclinical animal modeling in medicine*. IntechOpen. https://doi.org/10.5772/intechopen.97592
22. Farkas, E., Luiten, P. G. M., & Bari, F. (2007). Permanent, bilateral common carotid artery occlusion in the rat: A model for chronic cerebral hypoperfusion-related neurodegenerative diseases. *Brain Research Reviews, 54*(1), 162–180. https://doi.org/10.1016/j.brainresrev.2007.01.003
23. Traystman, R. J. (2003). Animal models of focal and global cerebral ischemia. *ILAR Journal, 44*(2), 85–95. https://doi.org/10.1093/ilar.44.2.85
24. Li, Y., Tan, L., Yang, C., He, L., Liu, L., Deng, B., Liu, S., & Guo, J. (2023). Distinctions between the Koizumi and Zea Longa methods for middle cerebral artery occlusion (MCAO) model: A systematic review and meta-analysis of rodent data. *Scientific Reports, 13*(1), 10247. https://doi.org/10.1038/s41598-023-37187-w
25. Trotman-Lucas, M., & Gibson, C. L. (2021). A review of experimental models of focal cerebral ischemia focusing on the middle cerebral artery occlusion model. *F1000Research, 10*, 242. https://doi.org/10.12688/F1000RESEARCH.51752.2
26. Lemmerman, L. R., Harris, H. N., Balch, M. H. H., Rincon-Benavides, M. A., Higuita-Castro, N., Arnold, W. D., & Gallego-Perez, D. (2022). Transient middle cerebral artery occlusion with an intraluminal suture enables reproducible induction of ischemic stroke in mice. *Bio-Protocol, 12*(3). https://doi.org/10.21769/BioProtoc.4305
27. Rousselet, E., Kriz, J., & Seidah, N. G. (2012). Mouse model of intraluminal MCAO: Cerebral infarct evaluation by cresyl violet staining. *Journal of Visualized Experiments: JoVE, 69*, 4038. https://doi.org/10.3791/4038
28. McCabe, C., Arroja, M. M., Reid, E., & Macrae, I. M. (2018). Animal models of ischaemic stroke and characterisation of the ischaemic penumbra. *Neuropharmacology, 134*(Pt B), 169–177. https://doi.org/10.1016/j.neuropharm.2017.09.022
29. Bacigaluppi, M. (2010). Animal models of ischemic stroke. Part two: Modeling cerebral ischemia. *The Open Neurology Journal, 4*(1), 34–38. https://doi.org/10.2174/1874205x01004010034
30. Schiffrin, E. L. (2001). Role of endothelin-1 in hypertension and vascular disease. *American Journal of Hypertension, 14*(6 II), 83S–89S. https://doi.org/10.1016/s0895-7061(01)02074-x
31. Mecca, A. P., Regenhardt, R. W., O'Connor, T. E., Joseph, J. P., Raizada, M. K., Katovich, M. J., & Sumners, C. (2011). Cerebroprotection by angiotensin-(1-7) in endothelin-1-induced ischaemic stroke. *Experimental Physiology, 96*(10), 1084–1096. https://doi.org/10.1113/expphysiol.2011.058578
32. Ansari, S., Azari, H., Caldwell, K. J., Regenhardt, R. W., Hedna, V. S., Waters, M. F., Hoh, B. L., & Mecca, A. P. (2013). Endothelin-1 induced middle cerebral artery occlusion model for ischemic stroke with laser Doppler flowmetry guidance in rat. *Journal of Visualized Experiments: JoVE, 72*, 50014. https://doi.org/10.3791/50014
33. Li, H., Zhang, N., Lin, H. Y., Yu, Y., Cai, Q. Y., Ma, L., & Ding, S. (2014). Histological, cellular and behavioral assessments of stroke outcomes after photothrombosis-induced ischemia in adult mice. *BMC Neuroscience, 15*, 58. https://doi.org/10.1186/1471-2202-15-58
34. Seong, D., Yi, S., Han, S., Lee, J., Park, S., Hwang, Y. H., Kim, J., Kim, H. K., & Jeon, M. (2022). Target ischemic stroke model creation method using photoacoustic microscopy with simultaneous vessel monitoring and dynamic photothrombosis induction. *Photoacoustics, 27*, 100376. https://doi.org/10.1016/j.pacs.2022.100376
35. Schmidt, A., Hoppen, M., Strecker, J. K., Diederich, K., Schäbitz, W. R., Schilling, M., & Minnerup, J. (2012). Photochemically induced ischemic stroke in rats. *Experimental and Translational Stroke Medicine, 4*(1), 13. https://doi.org/10.1186/2040-7378-4-13
36. Lin, X., Wang, H., Chen, J., Zhao, P., Wen, M., Bingwa, L. A., Jin, K., Zhuge, Q., & Yang, S. (2022). Nonhuman primate models of ischemic stroke and neurological evaluation after stroke. *Journal of Neuroscience Methods, 376*, 109611. https://doi.org/10.1016/j.jneumeth.2022.109611

37. Sutherland, B. A., Neuhaus, A. A., Couch, Y., Balami, J. S., Deluca, G. C., Hadley, G., Harris, S. L., Grey, A. N., & Buchan, A. M. (2016). The transient intraluminal filament middle cerebral artery occlusion model as a model of endovascular thrombectomy in stroke. *Journal of Cerebral Blood Flow and Metabolism, 36*(2), 363–369. https://doi.org/10.1177/0271678X15606722
38. Horky, L., et al. (2019). Animal models of stroke. *Neurotherapeutics, 16*(2), 342–353. https://www.ncbi.nlm.nih.gov/pmc/articles/PMC3155396/
39. Domínguez-Oliva, A., Hernández-Ávalos, I., Martínez-Burnes, J., Olmos-Hernández, A., Verduzco-Mendoza, A., & Mota-Rojas, D. (2023). The importance of animal models in biomedical research: Current insights and applications. *Animals, 13*, 1223. https://doi.org/10.3390/ani13071223
40. Mergenthaler, P., & Meisel, A. (2012). Do stroke models model stroke? *Disease Models & Mechanisms, 5*(6), 718–725.
41. Lee, M. C., Jin, C. Y., Kim, H. S., Kim, J. H., Kim, M. K., Kim, H. I., et al. (2011). Stem cell dynamics in an experimental model of stroke. *Chonnam Medical Journal, 47*(2), 90.
42. Yan, T., Chopp, M., & Chen, J. (2015). Experimental animal models and inflammatory cellular changes in cerebral ischemic and hemorrhagic stroke. *Neuroscience Bulletin, 31*, 717–734.
43. Krafft, P. R., Rolland, W. B., Duris, K., et al. (2012). Modeling intracerebral hemorrhage in mice: Injection of autologous blood or bacterial collagenase. *Journal of Visualized Experiments, 67*, e4289. https://doi.org/10.3791/4289. Published 2012 Sep 22.
44. Yan, T., Chopp, M., Chen, J.. op.cit.p.726.
45. Ibid. p726.
46. Serbian, D., Durukan, A., & Tatlisumak, T. (2008). Rodent models of hemorrhagic stroke. *Current Pharmaceutical Design, 14*(4), 352–358.
47. Alharbi, B. M., Tso, M. K., & Macdonald, R. L. (2016). Animal models of spontaneous intracerebral hemorrhage. *Neurological Research, 38*(5), 448–455.
48. Withers, S. E., Parry-Jones, A. R., Allan, S. M., & Kasher, P. R. (2020). A multi-model pipeline for translational intracerebral haemorrhage research. *Translational Stroke Research, 11*(6), 1229–1242. https://doi.org/10.1007/s12975-020-00830-z
49. Walcott, B. P., & Peterson, R. T. (2014). Zebrafish models of cerebrovascular disease. *Journal of Cerebral Blood Flow and Metabolism, 34*(4), 571–577. https://doi.org/10.1038/jcbfm.2014.27
50. Chaoul, V., Dib, E. Y., Bedran, J., Khoury, C., Shmoury, O., Harb, F., & Soueid, J. (2023). Assessing drug administration techniques in zebrafish models of neurological disease. *International Journal of Molecular Sciences, 24*(19), 14898. https://doi.org/10.3390/ijms241914898
51. Kim, G. H. J., Mo, H., Liu, H., Wu, Z., Chen, S., Zheng, J., Zhao, X., Nucum, D., Shortland, J., Peng, L., Elepano, M., Tang, B., Olson, S., Paras, N., Li, H., Renslo, A. R., Arkin, M. R., Huang, B., Lu, B., et al. (2021). A zebrafish screen reveals Renin-angiotensin system inhibitors as neuroprotective via mitochondrial restoration in dopamine neurons. *eLife, 10*, e69795. https://doi.org/10.7554/eLife.69795
52. Wang, Y., Yoshimura, R., Manabe, H., Schretter, C., Clarke, R., Cai, Y., et al. (2014). Trans-sodium crocetinate improves outcomes in rodent models of occlusive and hemorrhagic stroke. *Brain Research, 1583*, 245–254.
53. Li, X., Zhang, Z. W., Wang, Z., Li, J. Q., & Chen, G. (2018). The role of argon in stroke. *Medical Gas Research, 8*(2), 64.
54. Ibid. p64.
55. Monsour, M., & Borlongan, C. V. (2023). Emerging regenerative medicine for hemorrhagic stroke: An update on stem cell therapies. *Brain Hemorrhages, 4*(1), 22–26.
56. Dijkhuizen, R. M., van der Marel, K., Otte, W. M., Hoff, E. I., van der Zijden, J. P., van der Toorn, A., & van Meer, M. P. A. (2012). Functional MRI and diffusion tensor imaging of brain reorganization after experimental stroke. *Translational Stroke Research, 3*(1), 36–43. https://doi.org/10.1007/s12975-011-0143-8

57. Gubskiy, I. L., Namestnikova, D. D., Cherkashova, E. A., Chekhonin, V. P., Baklaushev, V. P., Gubsky, L. v., & Yarygin, K. N. (2018). MRI guiding of the middle cerebral artery occlusion in rats aimed to improve stroke modeling. *Translational Stroke Research, 9*(4), 417–425. https://doi.org/10.1007/s12975-017-0590-y
58. Lin, J. B., Phillips, E. H., Riggins, T. E., Sangha, G. S., Chakraborty, S., Lee, J. Y., Lycke, R. J., Hernandez, C. L., Soepriatna, A. H., Thorne, B. R. H., Yrineo, A. A., & Goergen, C. J. (2015). Imaging of small animal peripheral artery disease models: Recent advancements and translational potential. *International Journal of Molecular Sciences, 16*(5), 11131–11177. https://doi.org/10.3390/ijms160511131
59. Sobrado, M., Delgado, M., Fernández-Valle, E., García-García, L., Torres, M., Sánchez-Prieto, J., Vivancos, J., Manzanares, R., Moro, M. A., Pozo, M. A., & Lizasoain, I. (2011). Longitudinal studies of ischemic penumbra by using 18F-FDG PET and MRI techniques in permanent and transient focal cerebral ischemia in rats. *NeuroImage, 57*(1), 45–54. https://doi.org/10.1016/j.neuroimage.2011.04.045
60. Doeppner, T. R., Hermann, D. M., & Doeppner, M. (2017). Rodent models of ischemic stroke: From phenotyping to therapy development. *Journal of Cerebral Blood Flow and Metabolism, 37*(Suppl 1), S24–S47. https://www.ncbi.nlm.nih.gov/books/NBK499997/
61. Carmichael, S. T. (2016). Translational stroke research: A bumpy road to bedside. *Neurotherapeutics, 13*(2), 117–126. https://www.ncbi.nlm.nih.gov/pmc/articles/PMC10649331/
62. Yang, W., Chen, X., Li, S., & Li, X. J. (2021). Genetically modified large animal models for investigating neurodegenerative diseases. *Cell & Bioscience, 11*(1), 218. https://doi.org/10.1186/s13578-021-00729-8
63. Candelario-Jalil, E., Dijkhuizen, R. M., & Magnus, T. (2022). Neuroinflammation, stroke, blood-brain barrier dysfunction, and imaging modalities. *Stroke, 53*(5), 1473–1486. https://doi.org/10.1161/STROKEAHA.122.036946
64. Wang, Y., Leak, R. K., & Cao, G. (2022). Microglia-mediated neuroinflammation and neuroplasticity after stroke. *Frontiers in Cellular Neuroscience, 16*, 980722. https://doi.org/10.3389/fncel.2022.980722
65. Shao, F., Wang, X., Wu, H., Wu, Q., & Zhang, J. (2022). Microglia and neuroinflammation: Crucial pathological mechanisms in traumatic brain injury-induced neurodegeneration. *Frontiers in Aging Neuroscience, 14*, 825086. https://doi.org/10.3389/fnagi.2022.825086
66. Kong, A. H., Wu, A. J., Ho, O. K., Leung, M. M., Huang, A. S., Yu, Y., Zhang, G., Lyu, A., Li, M., & Cheung, K. H. (2023). Exploring the potential of aptamers in targeting neuroinflammation and neurodegenerative disorders: Opportunities and challenges. *International Journal of Molecular Sciences, 24*(14), 11780. https://doi.org/10.3390/ijms241411780
67. Muzio, L., Viotti, A., & Martino, G. (2021). Microglia in neuroinflammation and neurodegeneration: From understanding to therapy. *Frontiers in Neuroscience, 15*, 742065. https://doi.org/10.3389/fnins.2021.742065
68. Alfonsetti, M., d'Angelo, M., & Castelli, V. (2023). Neurotrophic factor-based pharmacological approaches in neurological disorders. *Neural Regeneration Research, 18*(6), 1220–1228. https://doi.org/10.4103/1673-5374.358619
69. Mallah, K., Couch, C., Borucki, D. M., Toutonji, A., Alshareef, M., & Tomlinson, S. (2020). Anti-inflammatory and neuroprotective agents in clinical trials for CNS disease and injury: Where do we go from here? *Frontiers in Immunology, 11*, 2021. https://doi.org/10.3389/fimmu.2020.02021
70. Song, N., Scholtemeijer, M., & Shah, K. (2020). Mesenchymal stem cell immunomodulation: Mechanisms and therapeutic potential. *Trends in Pharmacological Sciences, 41*(9), 653–664. https://doi.org/10.1016/j.tips.2020.06.009
71. Galvan, A., Caiola, M. J., & Albaugh, D. L. (2018). Advances in optogenetic and chemogenetic methods to study brain circuits in non-human primates. *Journal of neural transmission (Vienna, Austria: 1996), 125*(3), 547–563. https://doi.org/10.1007/s00702-017-1697-8

72. Phillips, N. L. H., & Roth, T. L. (2019). Animal models and their contribution to our understanding of the relationship between environments, epigenetic modifications, and behavior. *Genes, 10*(1), 47. https://doi.org/10.3390/genes10010047
73. Liao, L. D., Tsytsarev, V., Delgado-Martínez, I., Li, M. L., Erzurumlu, R., Vipin, A., Orellana, J., Lin, Y. R., Lai, H. Y., Chen, Y. Y., & Thakor, N. V. (2013). Neurovascular coupling: In vivo optical techniques for functional brain imaging. *Biomedical Engineering Online, 12*, 38. https://doi.org/10.1186/1475-925X-12-38
74. Luo, L., He, Y., Wang, X., Key, B., Lee, B. H., Li, H., & Ye, Q. (2018). Potential roles of dental pulp stem cells in neural regeneration and repair. *Stem Cells International, 2018*, 1731289. https://doi.org/10.1155/2018/1731289
75. Yang, S., Li, R., Li, H., Xu, K., Shi, Y., Wang, Q., Yang, T., & Sun, X. (2021). Exploring the use of brain-computer interfaces in stroke neurorehabilitation. *BioMed Research International, 2021*, 9967348. https://doi.org/10.1155/2021/9967348
76. Uzair, M., Abualait, T., Arshad, M., Yoo, W. K., Mir, A., Bunyan, R. F., & Bashir, S. (2022). Transcranial magnetic stimulation in animal models of neurodegeneration. *Neural Regeneration Research, 17*(2), 251–265. https://doi.org/10.4103/1673-5374.317962
77. Mane, R., Chouhan, T., & Guan, C. (2020). BCI for stroke rehabilitation: Motor and beyond. *Journal of Neural Engineering, 17*(4), 041001. https://doi.org/10.1088/1741-2552/aba162
78. Cho, Y. K., Kim, S., Jung, H. H., et al. (2016). Neuromodulation methods for animal locomotion control. *Biomedical Engineering Letters, 6*, 134–147. https://doi.org/10.1007/s13534-016-0234-2

Unveiling Stroke: Biomarkers and Diagnostic Strategies

4

Al-Hassan Soliman Wadan and Wael Mohamed

Abstract

Stroke, a leading cause of sickness and death worldwide, is difficult to diagnose and treat. This book chapter examines biomarkers and diagnostic methods' role in stroke research. This article discusses stroke biomarkers, including proteins, nucleic acids, metabolites, and cellular components, and their ability to detect strokes early, evaluate risk, and predict outcomes. The chapter shows how multimodal approaches may accurately diagnose and classify strokes by combining clinical assessments with biomarker technology. Additionally, it addresses biomarker standardization, timing, and stroke differentiation from similar conditions. This chapter aims to increase stroke pathophysiology understanding and encourage the development of new diagnostic tools and personalized treatment methods to improve patient outcomes. Critically assessing study outcomes and investigating prospects will accomplish this.

Keywords

Stroke · Ischemic stroke · Hemorrhagic stroke · Cerebrovascular disease · Biomarkers · Diagnosis

A.-H. S. Wadan
Faculty of Dentistry, Sinai University, Arish, North Sinai Peninsula, Egypt

Sinai University Research Center, Sinai University, Sinai, Egypt

W. Mohamed (✉)
Basic Medical Science Department, Kulliyyah of Medicine, International Islamic University Malaysia (IIUM), Kuantan, Pahang, Malaysia

Clinical Pharmacology Department, Menoufia Medical School, Menoufia University, Menoufia, Egypt

W. Mohamed, M. B. Mat-Nor (eds.), *Nutrition and Stroke*, Nutritional Neurosciences, https://doi.org/10.1007/978-981-97-6818-9_4

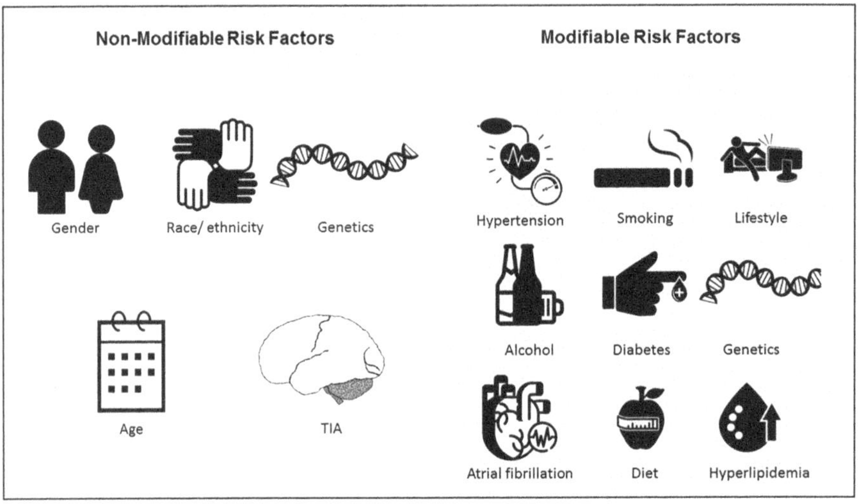

Fig. 4.1 Modifiable and Non-Modifiable Risk Factors for Stroke. This figure illustrates the distinction between modifiable and non-modifiable risk factors associated with stroke. Lifestyle changes or medical interventions can alter or control modifiable risk factors. In contrast, non-modifiable risk factors are inherent characteristics or conditions that cannot be modified. Understanding these risk factors is crucial in developing effective stroke prevention and management strategies

4.1 Introduction

Stroke, a leading cause of morbidity and mortality worldwide, is a prevalent neurological condition that poses significant health challenges [1, 2]. It occurs when the blood supply to a part of the brain is disrupted, impairing oxygen and nutrient delivery to brain cells [1, 2]. There are two main types of strokes: ischemic and hemorrhagic. Ischemic stroke, accounting for approximately 85% of cases [2, 3], occurs due to a blockage or occlusion of a blood vessel supplying the brain [3]. Hemorrhagic stroke, on the other hand, results from bleeding within the brain or on its surface [4]. Both types can cause severe brain damage and functional impairments. Various risk factors contribute to the development of stroke. Non-modifiable factors include age, gender (more common in men), and family history. Modifiable risk factors encompass hypertension, diabetes, smoking, obesity, high cholesterol, atrial fibrillation, sedentary lifestyle, excessive alcohol consumption, and illicit drug use (Fig. 4.1). Controlling these risk factors through lifestyle modifications and appropriate medical interventions can significantly reduce the incidence of stroke [5, 6].

4.1.1 Stroke Etiology and Pathophysiology

Stroke is a diverse condition with various underlying causes [7]. Hypertension is responsible for approximately 50–70% of cases of hemorrhagic stroke [6]. Other causes include cerebral amyloid angiopathy [8], vascular malformations, brain tumors, and coagulation disorders. In ischemic stroke, the causes can be cardioembolism [9], extensive artery atherosclerosis (LAA) [10, 11], small vessel disease, or other known factors such as dissection, mitochondrial disorder [12–14], or genetic mutation [15]. Despite extensive inquiries, an apparent cause of stroke cannot be explained in more than 30% of patients, leading to an ambiguous or cryptogenic originality [16]. Also, certain patients may have multiple potential etiologies, making identifying the possible cause more challenging. The classification scheme for stroke etiology underlines this complexity. Specifically, in lacunar strokes, clinicians depend on indirect instigators like the size and location of the infarction lesion since there are no conclusive methods to image the underlying small vessel lesion. Interestingly, biomarkers can potentially improve the identification of stroke roots [2, 17]. Natriuretic peptides have been studied to determine the etiology of cardioembolic stroke lesions [18] (Fig. 4.2).

The pathophysiology of stroke involves multiple complex mechanisms. For instance, ischemic stroke, atherosclerosis, thrombosis, or embolism leads to the obstruction or blockage of a cerebral artery, causing a shortage in blood supply to an area of the brain tissue [3]. The lack of oxygen and nutrients leads to a cascade of mechanisms such as excitotoxicity, inflammation, oxidative stress, and neural tissue death [19]. Hemorrhagic stroke arises due to the rupture of blood vessels, resulting in bleeding into a brain tissue region, leading to direct tissue damage. Additionally, the leaked blood in a specific brain tissue area led to increased intracranial pressure. The clinical manifestations of stroke are variant depending on the affected area of the brain and the extent of damage that has occurred. Common

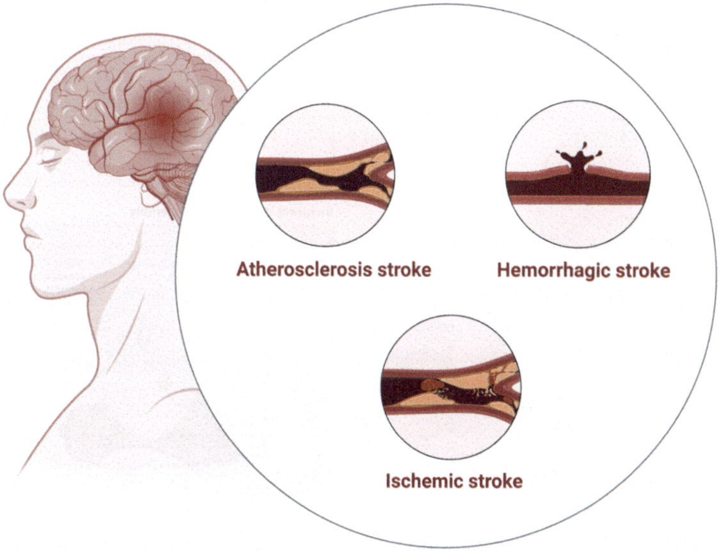

Fig. 4.2 Types of strokes

symptoms include sudden onset of weakness or numbness on one side of the body, difficulty speaking or understanding speech (aphasia), visual disturbances, coordination problems, dizziness, severe headache, and altered consciousness. Prompt recognition of these symptoms is vital to ensure timely medical intervention. Stroke has a significant public health impact globally. It is a leading cause of disability, with survivors often experiencing long-term physical, cognitive, and emotional impairments. The economic burden is substantial due to healthcare costs, rehabilitation, lost productivity, and caregiver burden. Moreover, stroke can impact individuals of all ages, including young adults and children, highlighting the need for early prevention and effective management strategies.

4.2 Ischemic Stroke

In ischemic stroke, the most common underlying mechanism is the occlusion of a cerebral artery, leading to reduced blood flow and subsequent ischemia in the affected area. This can occur through different mechanisms, including thrombosis, embolism, or systemic hypoperfusion. Thrombotic strokes are caused by forming a blood clot (thrombus) within a cerebral artery, often associated with atherosclerosis (Fig. 4.3). The thrombus can wholly or partially occlude the blood vessel,

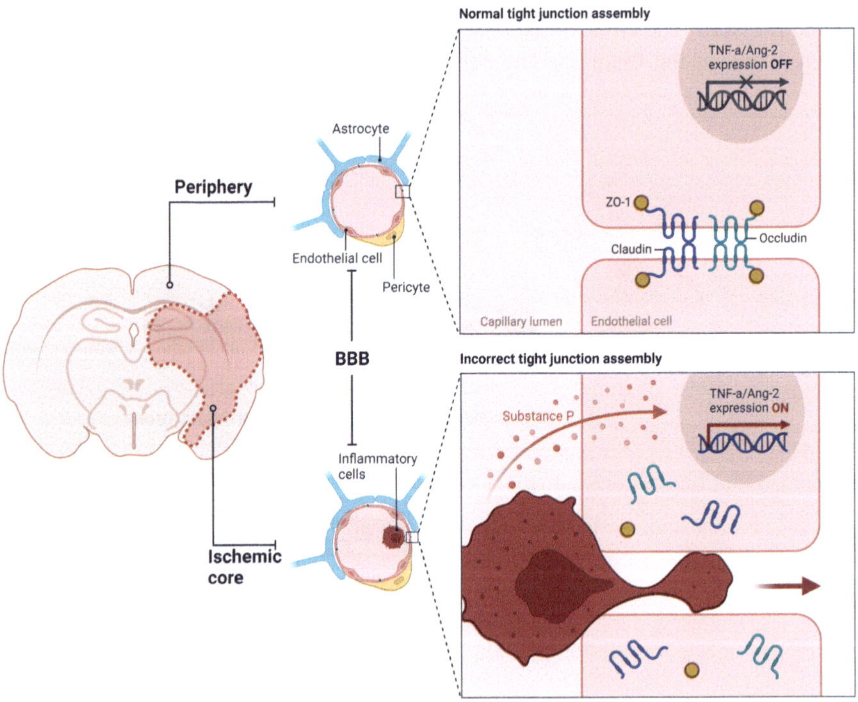

Fig. 4.3 Ischemic stroke pathophysiology

decreasing blood flow to the downstream brain tissue. Embolic strokes occur when a blood clot or other debris (embolus) originating from a different location, such as the heart or carotid arteries, travels through the bloodstream and lodges in a smaller cerebral artery—the blockage caused by the embolus results in the interruption of blood flow to the downstream brain tissue. Systemic hypoperfusion, such as in cases of severe hypotension or cardiac arrest, can also lead to ischemic stroke. In these situations, the reduced blood supply to the brain affects multiple areas simultaneously. Once blood flow to a specific brain region is disrupted, a cascade of pathophysiological events is set in motion. Initially, there is a reduction in the supply of oxygen and glucose to the affected tissue, leading to energy failure and impaired cellular metabolism. Without energy, the brain cells cannot maintain average ion gradients and neurotransmitter homeostasis, resulting in depolarization and release of excitatory neurotransmitters, particularly glutamate. The excessive release of glutamate triggers excitotoxicity, a process where prolonged activation of glutamate receptors leads to an influx of calcium ions into the cells. The high calcium levels within the neurons trigger different enzymes and pathways causing cellular damage, including generating ROS particles, activating the enzymes of proteases, and causing dysfunctional mitochondria.

The buildup of ROS particles and the shortage of energy supply led to oxidative stresses, further aggravating neuronal damage consequences. Oxidative stresses endorse lipid peroxidation, DNA damage, and protein degradation, contributing to the overall pathophysiological torrent. Inflammatory events play a noteworthy role in the advance of ischemic stroke. The ischemic insult induces the pro-inflammatory cytokines release, chemokines, and immune cells adduction. These pro-inflammatory mediators aggravate tissue damage, disrupt the BBB, and trigger the recruitment of leukocytes to the site of insult. The inflammatory response also encompasses activating glial cells, particularly microglial and astrocytic cells [12]. The activated microglia secrete inflammatory signals and contribute to further neuronal injury. Together, astrocytes form the glial scar, which acts as a physical barrier to bind the spread of injury but can also obstruct tissue repair. Besides, the disruption of BBB physiology permits peripheral immune cells penetration into the brain tissue, which further progresses to the inflammatory response and increased tissue damage.

4.3 Hemorrhagic Stroke

Hemorrhagic stroke occurs due to bleeding within the brain (intracerebral hemorrhage) or around it (subarachnoid hemorrhage). Intracerebral hemorrhage often stems from a ruptured blood vessel, usually due to hypertension or vascular abnormalities causing direct damage to brain tissue that compresses nearby structures. The occurrence of blood within the brain tissue and surrounding it triggers a cascade of harmful events. The blood proves lethal to neurons and glial cells, releasing substances that provoke inflammation and cell death. Moreover, blood accumulation and building up within the brain advances intracranial pressure, exacerbating the impairment of blood circulation and neuronal survival in neighboring areas.

Subarachnoid hemorrhage typically arises from the rupture of an intracranial aneurysm, resulting in bleeding into the space between the arachnoid and pia mater. This sudden release of blood sets off a chain of three physiological events, including first, the activation of the coagulation cascade, second, the release of vasoactive substances, and lastly, the recruitment of inflammatory cells. The extravasation and accumulation of blood and subsequent release of vasoactive substances induce arterial constriction known as vasospasm, which progresses the impairment of blood flow to the brain and contributes to secondary brain damage.

Hypertension, a crucial modifiable risk factor for cerebrovascular disease, contributes to about 54% of strokes globally. Despite this significant impact, there is a dearth of systematic research into the shared genetic susceptibility between hypertension and stroke at a genome-wide level. To address this gap, a study by Ji et al. gathered SNPs associated with essential hypertension and stroke from the NHGRI-EBI GWAS catalog. Subsequently, genotype imputation was performed using data from the 1000 Genomes Project. The study then compared the SNPs and mapped genes of the two diseases, followed by functional clustering and analysis of enriched GO terms and KEGG pathways [20]. The study revealed a single shared SNP (rs3184504) and 11 common genes between hypertension and stroke. After genotypic imputation, the count increased to 129 shared SNPs and 16 shared genes. These genes showed significant enrichment across 10 GO terms, mainly associated with lipoprotein and triglyceride metabolism. Additionally, the KEGG analysis highlighted glycerolipid metabolism as a significantly enriched pathway in both diseases. These findings strongly indicate that the gene network responsible for lipid metabolism and blood circulation plays a significant role in the shared genetic basis of hypertension and stroke [20, 21].

4.4 Classification Systems Used in Stroke Diagnosis (Table 4.1)

Classification systems are crucial for stroke diagnosis, offering a standardized framework for categorizing stroke subtypes and guiding management. The Trial of ORG 10172 in Acute Stroke Treatment (TOAST) and the Atherosclerosis, Small Vessel Disease, Cardiac Pathology, Other Causes (ASCO) system are commonly used as guiding systems for stroke classification. Interestingly, TOAST categorizes ischemic strokes by etiology, while ASCO provides a comprehensive classification of stroke subtypes. The TOAST system is now widely employed in clinical settings even initially invented for research purposes. It classifies ischemic strokes into five key subcategories: (1) Large artery atherosclerosis, (2) Cardioembolism, (3) Small vessel occlusion (SVO), (4) Other determined etiology, and (5) Undetermined etiology. This categorization system aids clinicians in investigating the root cause of ischemic stroke, and hence, facilitating treatment decisions. For instance, distinguishing between extensive artery atherosclerosis and cardio-embolism influences the choice of antithrombotic agents or anticoagulants.

Table 4.1 Classification systems used in the diagnosis of Stroke

Classification system	TOAST	ASCO
Ischemic stroke subtypes	Large artery atherosclerosis	Atherosclerosis
	Cardioembolism	Small vessel disease
	Small vessel occlusion	Cardiac pathology
	Other determined etiology	Other causes
	Undetermined etiology	–
Hemorrhagic stroke subtypes	N/A (TOAST focuses on ischemic stroke subtypes)	Hemorrhage (intracerebral, subarachnoid)
	–	Vascular malformation (aneurysm, AVM)
	–	Vasculitis
	–	Arterial dissection
	–	Hypercoagulable states

Note: TOAST mainly covers ischemic stroke, while ASCO includes a wider range of etiologies, including hemorrhagic stroke and less common causes.

In contrast, the ASCO classification system offers a more reliable approach to stroke categorization and has a broader spectrum of etiologies. It outlines four primary categories: (1) Atherosclerosis, (2) Small vessel disease, (3) Cardiac pathology, and (4) Other causes. Each category is further subdivided into specific etiologies, allowing for a more reliable characterization of stroke subcategories. This system facilitates the identification of less common stroke causes like vasculitis, arterial dissection, or hypercoagulable conditions. The implementation of these classification systems standardizes stroke diagnosis and enhances communication among healthcare professionals. They establish a strict and specific language for researchers, clinicians, and stroke specialists, ensuring accurate and consistent reporting of stroke subtypes.

4.5 Stroke Secondary to Covid-19

In a recent study, researchers found that COVID-19, caused by SARS-CoV-2, a novel coronavirus, has been linked to severe lung and brain tissue damage [22]. In the respiratory system, especially the lungs, COVID-19 can lead to severe acute respiratory distress syndrome, characterized by generalized alveolar injuries, severe interstitial inflammation, and harsh fibrosis [23]. The viral infection mainly targets a certain receptor expressed in lung epithelial cells known as the angiotensin-converting enzyme 2 (ACE2) receptor, leading to the uncontrolled release of cytokines, immune dysregulation, and activation of several coagulation pathways [24]. The infections progress till they contribute to the development of a proinflammatory microenvironment and compete with alveolar tissue damage.

Moreover, COVID-19 infections have been linked to the development of other neurological manifestations such as anosmia, loss of smell, ageusia, encephalopathy, and stroke. It may directly invade the neural tissues through the olfactory pathways or attract systemic inflammatory responses that lead to neuroinflammation, disruption of the BBB, and brain tissue damage (Figs. 4.4 and 4.5) [12]. Understanding the ways, the lung and brain tissue are damaged by COVID-19 infections is crucial for developing targeted cures and improving patient prognostic outcomes [25, 26].

Stroke secondary to COVID-19 has become an upsetting complication among people with COVID-19 infection [27]. Recent research has highlighted the association between COVID-19 and a higher degree of stroke, especially in severe cases. The reasons behind this link are diverse and involve factors like higher blood clotting, impaired blood vessel functions, generalized inflammation, and the virus directly affecting endothelial cells. Emerging in individuals with COVID-19, Numerous case studies and observations have detailed different types of strokes, such as ischemic stroke, hemorrhagic stroke, and cerebral venous sinus thrombosis

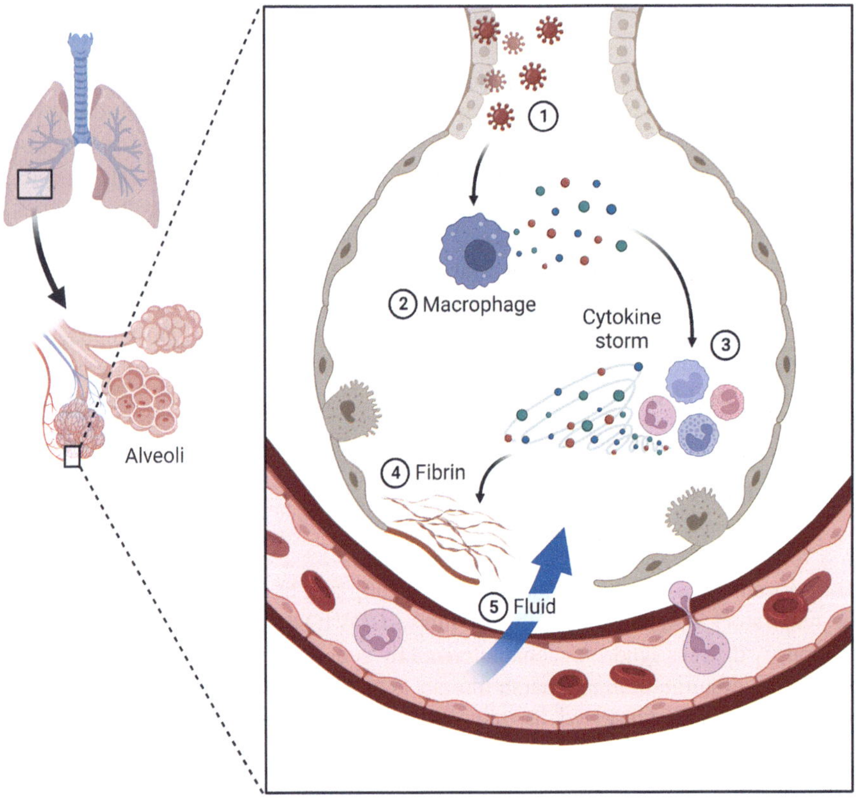

Fig. 4.4 COVID-19 impacts the lungs alveoli. The virus targets specific lung cells called pneumocytes. After the virus infects pneumocytes (1), immune cells like macrophages detect the virus (2) and release cytokines, signaling molecules that attract more immune cells, resulting in an inflammatory response termed a "cytokine storm" leading to severe pneumocytes damage (3). Additionally, the virus triggers fibrin formation, worsening tissue damage (4). As blood vessels weaken, fluid leaks into the lungs, leading to respiratory failure (5)

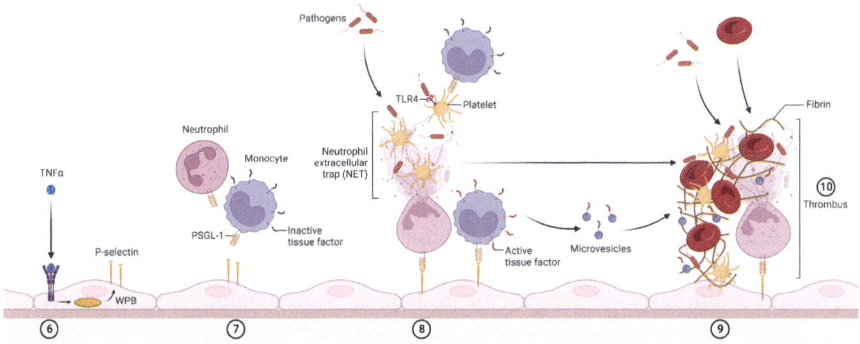

Fig. 4.5 The impact of COVID-19 on the brain tissue leads to ischemic stroke. Although it takes some time, it involves a complex cascade. When the COVID-19 virus infects the cells, it activates monocytes which leads to the release of proinflammatory cytokines. Proinflammatory cytokines initiate a storm of cytokines that recruit neutrophils and activate tissue factor (TF), which binds to coagulation factor VII (FVII), thus upregulating blood clot formation. Additionally, the viral infection triggers endotheliopathy, releasing von Willebrand factor (vWF) and activating platelets, intensifying clotting. Persistent neutrophil recruitment fosters the creation of neutrophil extracellular traps (NETs), further aiding clot formation. Moreover, NETs and platelet aggregation hinder natural clot dissolution mechanisms like tissue factor pathway inhibitors (TFPI). Consequently, these interrelated processes contribute to the prothrombotic state seen in COVID-19 infection

[9, 23, 28]. The nearer the identification and treatment of stroke symptoms in COVID-19 patients, the more vital results for improving patient outcomes we get. This requires a collaborative effort involving imaging, the prevention of blood clots, and the observation of specific interventions like mechanical clot removal or the use of blood thinners [22].

4.6 Cryptogenic Stroke

Cryptogenic stroke occurs when the definitive etiology or origin remains unidentified despite thorough examination [16]. In cryptogenic strokes, thorough investigations identify the potential causes such as cardiac abnormalities, blood clotting disorders, or blood vessel narrowing. Despite extensive examinations and procedures, no definite cause is pinpointed. It is crucial to eliminate and pull out known stroke causes before labeling it cryptogenic. These strokes present treatment and prevention challenges due to the elusive manner of cause. Prolonged monitoring and further evaluation are advised to identify risk factors or recurrent events. Advances in medical imaging and diagnostics, like extended cardiac monitoring and genetic testing, might reveal previously cryptogenic stroke causes [29].

Subclinical atrial fibrillation (sAF) may be a missing link in the development of cryptogenic stroke. An ASSERT study examined the association between sAF and clinical outcomes [30]. The study enrolled 2451 patients with hypertension and sinus node (SN) or atrioventricular-node (AVN) disease who had recently received a pacemaker or cardioverter-defibrillator. Subclinical atrial tachyarrhythmias occurred in 10.1% of patients during the first 3 months. Patients with subclinical arrhythmic episodes had higher rates of subsequent clinically notable AF and ischemic stroke or systemic embolism compared to those without such episodes. The risk of stroke or embolism was highest in patients with other stroke risk factors. Continuous atrial overdrive pacing did not reduce the incidence of AF.

While poststroke depression is a complex disorder with numerous fundamental origins that require additional investigations, explicit modifications in the brain pathology and neurophysiology probably contribute to certain sorts of poststroke depressive conditions [31]. Despite the inadequate understanding of the pathophysiology of poststroke depression, this chapter offers a different hypothesis concerning the upregulation of pro-inflammatory cytokines owing to brain ischemia in different cerebral regions related to mood disorders. After strokes, the amplified secretion of pro-inflammatory cytokines, such as IL-1β, TNF-α, or IL-18, may induce an inflammatory pathway [32], especially in the limbic areas, while triggering indoleamine 2,3-dioxygenase (IDO) [33, 34]. As a result, Serotonin is depleted in paralimbic regions like the ventral lateral frontal cortex, polar temporal cortex, and basal ganglia [35]. Serotonin depletion in these areas can cause physiological dysfunction and the progression of poststroke depression [36].

4.7 Animal Models of Stroke

Researchers often use animal models in research, including induced, spontaneous, negative, and orphan types of models [37]. In the induced model, researchers deliberately create the animal disease condition to study its effects [38]. On the contrary, the spontaneous model involves choosing an animal that naturally develops a disease like the one under study [39]. Negative animal models explore the resistance mechanisms associated with a particular disease condition [40]. Orphan models are used to gain insight into the pathology of newly identified diseases in humans [41]. Multiple animal models have been created to investigate stroke pathophysiology, offering advantages over studying stroke in humans or *in vitro*. The unpredictability, diverse clinical manifestations, and various localizations characterize the nature of stroke in humans. Particularly for the study of brain tissues. Brain atlases, such as those available at websites like http://www.med.harvard.edu/AANLIB/ and https://portal.brain-map.org, comprehensively characterize brain structures specific to the species of interest. Stereotaxic techniques introduce biomaterials or cells into coordinates within the target tissue. This allows for the study of micro lesions with high precision. Furthermore, the localization of these targeted interventions can be verified using lesion cartography based on magnetic resonance imaging (MRI), ensuring accurate positioning and assessment of the biomaterial's effects.

Investigating pathophysiology often requires direct access to brain tissue, feasible in animal models but not in humans. Moreover, existing imaging techniques cannot characterize the initial minutes of a stroke pathology [42]. Moreover, certain

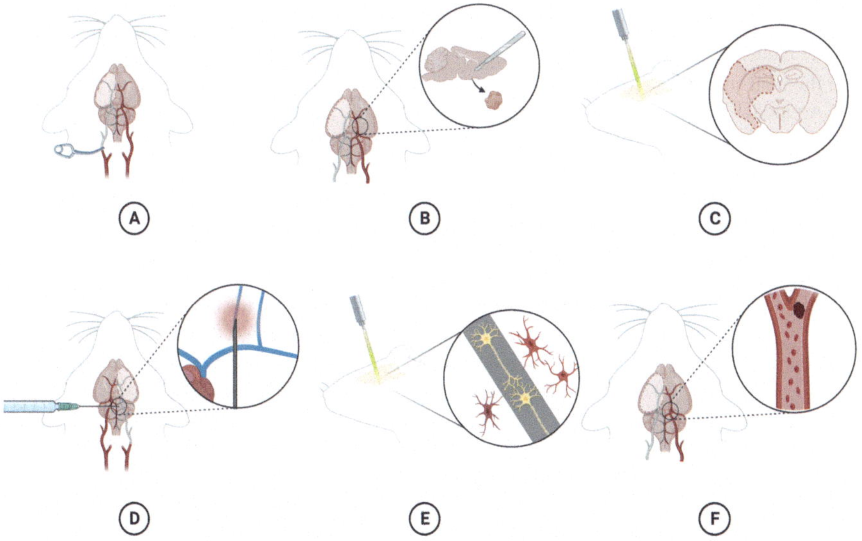

Fig. 4.6 Illustration of available Stroke Models: (**a**) Intraluminal suture MCAo model, (**b**) Craniectomy, (**c**) Photo-thrombosis model, (**d**) Endothelin-1 model, (**e**) Neurorehabilitation, and (**f**) Embolic stroke model

aspects of stroke, like vasculature and perfusion, cannot be adequately studied using *in vitro* models [42] (Fig. 4.6, Table 4.2).

4.8 Stroke Biomarkers

Biomarkers are objective indicators to assess normal or pathological processes [52], evaluate responses to medical interventions, and predict outcomes [53]. They can refer to molecules present in body fluids blood [54, 55], cerebrospinal fluids [56, 57], urine [58] but also to physical measurements on tissues (e.g., imaging [59, 60], electrophysiology [26, 61]. Molecular biomarkers include proteins, metabolites, lipids, and ribonucleic acids (RNA) [61, 62]. They can be used alone or in combination panels [63], scores, or indices to improve their diagnostic accuracy or capacity to estimate disease risk or clinical outcome. Several blood biomarkers are used to aid clinical decisions. For example, high-sensitive cardiac troponin T guides the diagnosis of myocardial infarction, d-dimers are informative for the diagnosis of pulmonary embolism [64], plasma creatinine is used to assess and monitor kidney function [65], antibodies targeting acetylcholine receptors help to diagnose myasthenia gravis [66, 67], B-type natriuretic peptide (BNP) is used to assess heart failure [68], and C-reactive protein levels reflect the response to antibiotic therapy in bacterial infection [69] (Fig. 4.7).

No blood biomarker has been used to diagnose stroke [54, 70, 71]. This is partly because the characteristics required are challenging, including high sensitivity and

Table 4.2 Advantages and disadvantages of the available stroke animal models

Stroke model	Advantages	Disadvantages
Intraluminal suture MCAo model	Mimics human stroke, exhibits penumbra, highly reproducible [43]	Hyper–/hypothermia, increased hemorrhage, not suitable for thrombolysis studies ([43]; M. et al., 2011)
Craniotomy model	High long-term survival rates, visual confirmation of successful MCAo [39]	Highly invasive, requires surgical skills [38]
Photo-thrombosis model	Well-defined lesion localization, highly reproducible	Causes early vasogenic edema, not suitable for investigating neuroprotective agents
Endothelin-1 model	Less invasive induction of ischemic lesions in cortical regions [44]	The duration of ischemia is not controllable and induces astrocytosis and axonal sprouting [45]
Embolic stroke model	Mimics human stroke pathogenesis [46]	Low reproducibility of infarcts spontaneous recanalization [47]
Neurorehabilitation	Rapid establishment of independence in daily activities improves cognitive, language, and motor skills [48]	Lack of in-depth studies on neurorehabilitation efficacy [49]
Biomaterial testing	Reduction in lesion volume promotes neural reorganization and reduces secondary damage [50]	Long-term experiments are challenging due to material degradation [51]

specificity in a heterogeneous disorder and the need for a rapid turnaround. Several reviews have summarized biomarkers studied to date in stroke [72, 73]. The current chapter intends to be a partial description of stroke biomarkers. It is focused on blood biomarkers that show promise for translation into clinical practice and describes newly reported markers that could add to routine stroke care. Avenues for discovering new biomarkers and future research are discussed [17, 71, 74, 75].

Early and accurate diagnosis is pivotal for ensuring effective treatment outcomes across various medical conditions, notably stroke. Timely identification and intervention not only enhance patient outcomes but also mitigate morbidity, mortality, and healthcare costs. In stroke cases, prompt diagnosis facilitates swift initiation of suitable treatments like thrombolysis, mechanical thrombectomy, or medical management, optimizing tissue salvage and functional recovery potential.

4.8.1 GFAP as a Diagnostic Biomarker in Stroke

Glial fibrillary acidic protein (GFAP) is primarily expressed in astrocytes, a glial cell in the central nervous system. In recent years, GFAP has gained significant attention as a potential biomarker in stroke diagnosis and prognosis. This chapter aims to provide a comprehensive overview of GFAP and its role in stroke. GFAP

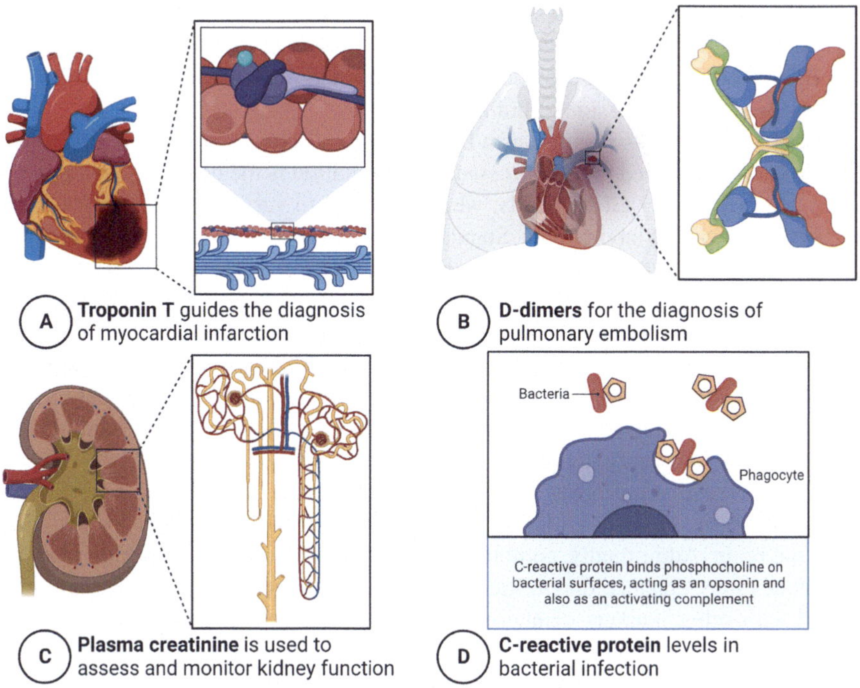

Fig. 4.7 Several blood biomarkers are used to aid clinical decisions. Biomarkers guide the diagnosis and assessment of specific medical conditions. (**a**) Troponin T aids in the diagnosis of myocardial infarction; (**b**) D-dimers assist in the diagnosis of pulmonary embolism; (**c**) Plasma creatinine facilitates the assessment of kidney functions; (**d**) C-reactive proteins aid in concluding a bacterial infection

serves as an intermediate filament protein that provides structural support to astrocytes and plays a crucial role in maintaining their cytoskeletal integrity. In response to brain injury or pathological conditions, the expression and release of GFAP are upregulated in astrocytes, leading to increased levels of GFAP in the cerebrospinal fluid (CSF) and blood.

The concentration of GFAP in serum is prompt and detectable within a few hours of hemorrhagic stroke onset. Conversely, in the case of ischemic stroke, GFAP levels remain detectable in serum for an extended duration. Hemorrhagic strokes are characterized by an abrupt disruption of the blood-brain barrier, leading to the early onset of cerebral injury [76].

GFAP has emerged as a promising biomarker for the diagnosis of stroke, particularly in the detection of hemorrhagic strokes. Hemorrhagic strokes are characterized by the rupture of blood vessels within the brain, leading to bleeding and subsequent release of blood components, including GFAP, into the surrounding tissue and body fluids. Moreover, GFAP levels have been found to correlate with the severity and extent of brain injury in stroke patients. In addition to its diagnostic value, GFAP has shown potential as a prognostic biomarker in stroke. Several studies have suggested that elevated GFAP levels are associated with worse clinical outcomes and increased mortality rates in stroke patients. Studies have shown that changes in GFAP levels over time can provide valuable information about the efficacy of interventions and the progression of brain injury.

While GFAP shows promise as a biomarker in stroke diagnosis and prognosis, certain limitations need to be addressed. One limitation is the need for standardized assays for measuring GFAP levels, which can lead to variability in results across different studies and laboratories. Standardization of assays and reference ranges is crucial for reliable and accurate interpretation of GFAP measurements. Another limitation is the potential for GFAP elevation in conditions other than stroke, such as traumatic brain injury, brain tumors, and neurodegenerative diseases. This highlights the need for careful interpretation of GFAP levels in the context of the patient's clinical presentation and other diagnostic information. Future research should focus on validating the diagnostic and prognostic utility of GFAP in large-scale, multicenter studies encompassing diverse stroke populations.

Further investigation is also needed to determine the optimal timing for GFAP measurement in stroke, considering the dynamic nature of its release and clearance from the brain. Additionally, combining GFAP with other biomarkers and imaging techniques may enhance diagnostic accuracy and prognostic value. Integration of GFAP with other markers of neuronal injury, inflammation, and oxidative stress could provide a more comprehensive and holistic understanding of stroke pathophysiology. In conclusion, GFAP is promising as a biomarker in stroke diagnosis and prognosis. Its measurement in CSF or blood may aid in differentiating hemorrhagic strokes from ischemic strokes and provide valuable prognostic information. However, further research and standardization are required to establish GFAP as a reliable and widely used biomarker in clinical practice, ultimately improving stroke management and patient outcomes.

4.9 Challenges and Limitations with Stroke Biomarkers

The use of stroke biomarkers is hindered by several problems and restrictions, which restrict their extensive clinical application and effectiveness. The issues include a wide spectrum, including both technological obstacles and clinical intricacies. To tackle these difficulties, it is necessary to have joint endeavors with diverse teams consisting of physicians, researchers, and industrial partners. Progress in assay technology, methods for discovering biomarkers, and data analytics provide hope for surpassing existing constraints and fully harnessing the potential of stroke biomarkers to enhance patient treatment and outcomes.

1. ***Study design and standardization:*** The main difficulty is in creating well planned research and carrying out extensive trials involving several centers. Standardizing biomarker tests across many labs and guaranteeing their repeatability are crucial but challenging endeavors. Inconsistencies in test procedures,

reagents, and equipment might result in variable outcomes, posing challenges in establishing the validity and reliability of biomarkers.

2. ***Timing of Biomarker Measurement:*** The promptness of measuring biomarkers is crucial in diagnosing and treating strokes. The amounts of biomarkers might vary based on the duration of time that has passed from the beginning of symptoms, which can complicate the interpretation of their importance. Prompt identification of stroke biomarkers is essential for prompt intervention, but any delays in the collection of samples or the execution of assay methods might undermine their diagnostic precision and therapeutic significance.
3. ***Clinical Utility and Validation:*** The process of confirming the practical use of stroke biomarkers is difficult, despite encouraging first results. To justify their incorporation into ordinary clinical practice, biomarkers need to have significant prognostic or diagnostic value in extensive and heterogeneous groups of patients. Long-term studies are necessary to demonstrate the reliability of biomarkers in predicting stroke risk, determining prognosis, and assessing therapy response.
4. ***Differential Diagnosis:*** Differentiating stroke from other neurological diseases might be difficult because of the similarity in clinical symptoms. Identifying biomarkers that can accurately distinguish stroke from illnesses that mirror it is greatly sought after but sometimes difficult to achieve. The presence of cross-reactivity with non-stroke pathologies or non-specific biomarker increase in many illness states might complicate the interpretation and restrict the accuracy of diagnosis [77–79].
5. ***Stroke Heterogenicity:*** The occurrence of stroke involves a wide range of different causes and clinical characteristics. The performance of biomarkers may differ across various stroke subtypes, including as ischemic and hemorrhagic strokes, as well as among diverse patient groups, including those of different ages, with comorbidities, or with genetic predispositions. It is crucial to customize biomarker panels to accommodate this variability to optimize diagnostic precision and prognostic significance [22, 80, 81].

4.10 Conclusions

Biomarkers have a double function in healthcare, as they assist in the process of diagnosing medical conditions and serve as targets for therapeutic treatments. Biomarkers serve as diagnostic tools by providing objective signs of normal biological processes, pathogenic processes, or reactions to treatment interventions. They enable the early identification of diseases, evaluation of potential risks, and tracking of illness progression or the effectiveness of therapy. Biomarkers include a range of substances, including proteins, nucleic acids, metabolites, and cellular components. Each of these compounds provides distinct insights into the pathophysiology of diseases. In addition, biomarkers play a crucial role in directing therapeutic interventions by assisting in treatment choices, identifying patient subgroups who are likely to benefit from certain medicines, and evaluating the response to therapy. Targeted treatments seek to manipulate the activity or expression of

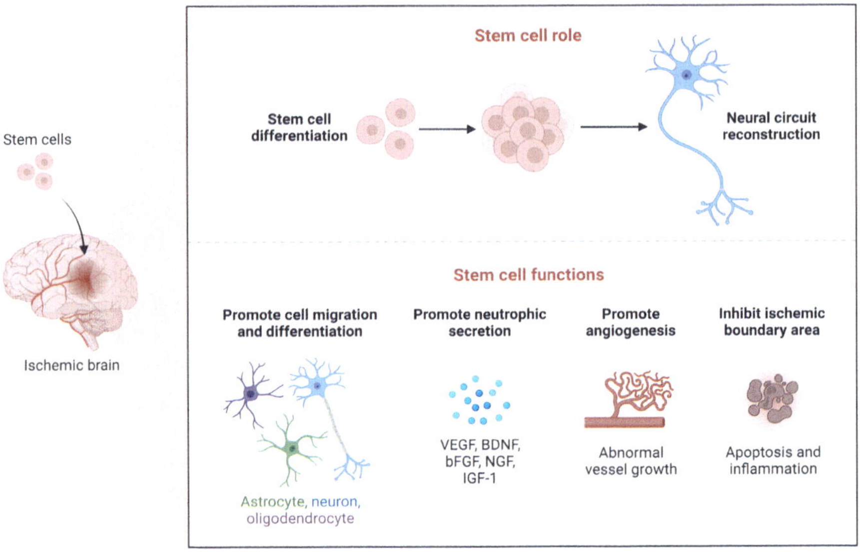

Fig. 4.8 Potential therapeutic targets include stem cell therapy

biomarkers to restore the normal physiological state, relieve symptoms, or halt the course of illness. Biomarker-guided treatments facilitate individualized and precise therapy by clarifying the fundamental biological causes of illness, allowing for tailored methods that meet the specific requirements of each patient. To summarize, biomarkers are essential instruments in contemporary healthcare, providing diagnostic, prognostic, and therapeutic benefits for many medical disorders. Integrating them into clinical practice improves disease management techniques, boosts patient outcomes, and stimulates innovation in drug discovery and precision medicine projects (Fig. 4.8).

References

1. Gasull, T., & Arboix, A. (2022). Molecular mechanisms and pathophysiology of acute stroke: Emphasis on biomarkers in the different stroke subtypes. International Journal of Molecular Sciences, 23(16), 9476. https://doi.org/10.3390/ijms23169476.
2. Guo, Y., Li, P., Guo, Q., Shang, K., Yan, D., Du, S., & Lu, Y. (2014). Pathophysiology and biomarkers in acute ischemic stroke—A review. Tropical Journal of Pharmaceutical Research, 12(6), 1097. https://doi.org/10.4314/tjpr.v12i6.35.
3. Kuriakose, D., & Xiao, Z. (2020). Pathophysiology and treatment of stroke: Present status and future perspectives. International Journal of Molecular Sciences, 21(20), 7609. https://doi.org/10.3390/ijms21207609.
4. Luo, Y., Reis, C., & Chen, S. (2019). NLRP3 inflammasome in the pathophysiology of hemorrhagic stroke: A review. Current Neuropharmacology, 17(7), 582–589. https://doi.org/10.2174/1570159X17666181227170053.

5. Aigner, A., Grittner, U., Rolfs, A., Norrving, B., Siegerink, B., & Busch, M. A. (2017). Contribution of established stroke risk factors to the burden of stroke in young adults. Stroke, 48(7), 1744–1751. https://doi.org/10.1161/STROKEAHA.117.016599.
6. Guzik, A., & Bushnell, C. (2017). Stroke epidemiology and risk factor management: CONTINUUM: Lifelong Learning in Neurology, 23(1), 15–39. https://doi.org/10.1212/CON.0000000000000416.
7. Dong, H., Liu, S., Jing, L., Tian, M., Sun, J., Pang, Y., Xing, L., & Xu, Y. (2020). Hypertension among hemorrhagic stroke patients in Northeast China: A population-based study 2017–2019. *Medical Science Monitor, 26*, e926581. https://doi.org/10.12659/MSM.926581
8. Yamada, M. (2015). Cerebral amyloid angiopathy: Emerging concepts. Journal of Stroke, 17(1), 17. https://doi.org/10.5853/jos.2015.17.1.17.
9. Fonseca, A. C., & Coelho, P. (2021). Update on biomarkers associated to cardioembolic stroke: A narrative review. Life, 11(5), 448. https://doi.org/10.3390/life11050448.
10. Goyal, M., Menon, B. K., Van Zwam, W. H., Dippel, D. W. J., Mitchell, P. J., Demchuk, A. M., Dávalos, A., Majoie, C. B. L. M., Van Der Lugt, A., De Miquel, M. A., Donnan, G. A., Roos, Y. B. W. E. M., Bonafe, A., Jahan, R., Diener, H.-C., Van Den Berg, L. A., Levy, E. I., Berkhemer, O. A., Pereira, V. M., & Jovin, T. G. (2016). Endovascular thrombectomy after large-vessel ischaemic stroke: A meta-analysis of individual patient data from five randomised trials. The Lancet, 387(10029), 1723–1731. https://doi.org/10.1016/S0140-6736(16)00163-X.
11. Ohwaki, K., Yano, E., Nagashima, H., Hirata, M., Nakagomi, T., & Tamura, A. (2004). Blood pressure management in acute intracerebral hemorrhage relationship between elevated blood pressure and hematoma enlargement. Stroke, 35(6), 1364–1367. https://doi.org/10.1161/01.str.0000128795.38283.4b.
12. Mohamed, W., Kumar, J., Alghamdi, B. S., Soliman, A.-H., & Toshihide, Y. (2023). Neurodegeneration and inflammation crosstalk: Therapeutic targets and perspectives. IBRO Neuroscience Reports, 14, 95–110. https://doi.org/10.1016/j.ibneur.2022.12.003.
13. Alshial, E. E., Abdulghaney, M. I., Wadan, A.-H. S., Abdellatif, M. A., Ramadan, N. E., Suleiman, A. M., Waheed, N., Abdellatif, M., & Mohammed, H. S. (2023). Mitochondrial dysfunction and neurological disorders: A narrative review and treatment overview. *Life Sciences, 334*, 122257. https://doi.org/10.1016/j.lfs.2023.122257
14. Ellakwa, D. E.-S., Mushtaq, N., Khan, S., Jabbar, A., Abdelmalek, M. A., Wadan, A.-H. S., Ellakwa, T. E., & Raza, A. (2024). Molecular functions of microRNAs in colorectal cancer: Recent roles in proliferation, angiogenesis, apoptosis, and chemoresistance. *Naunyn-Schmiedeberg's Archives of Pharmacology.* https://doi.org/10.1007/s00210-024-03076-w
15. Boehme, A. K., Esenwa, C., & Elkind, M. S. (2017). Stroke risk factors, genetics, and prevention. Circulation Research, 120(3), 472–495. https://doi.org/10.1161/circresaha.116.308398.
16. Yaghi, S., & Elkind, M. S. V. (2014). Cryptogenic stroke. Neurology: Clinical Practice, 4(5), 386–393. https://doi.org/10.1212/CPJ.0000000000000086.
17. Husseini, N. E., & Laskowitz, D. T. (2010). Clinical application of blood biomarkers in cerebrovascular disease. Expert Review of Neurotherapeutics, 10(2), 189–203. https://doi.org/10.1586/ern.09.151.
18. Regenhardt, R. W., Das, A. S., Lo, E. H., & Caplan, L. R. (2018). Advances in understanding the pathophysiology of lacunar stroke: A review. *JAMA Neurology, 75*(10), 1273.
19. Soliman, A.-H., & Mohamed, W. (2023). Nutrigenomics and trace elements: Hopes and hypes for Parkinson's treatment. In W. Mohamed & R. Sandhir (Eds.), Trace elements in brain health and diseases (pp. 47–69) Springer Nature Singapore. https://doi.org/10.1007/978-981-99-1513-2_3.
20. Ji, L. D., Hu, S. P., Li, J. Y., Yao, B. B., Shen, Q. J., & Xu, J. (2018). Shared genetic etiology of hypertension and stroke: Evidence from bioinformatics analysis of genome-wide association studies. Journal of Human Hypertension, 32(1), 34–39. https://doi.org/10.1038/s41371-017-0012-3.
21. Rankinen, T., Sarzynski, M. A., Ghosh, S., & Bouchard, C. (2015). Are there genetic paths common to obesity, cardiovascular disease outcomes, and cardiovascular risk factors? Circulation Research, 116(5), 909–922. https://doi.org/10.1161/CIRCRESAHA.116.302888.

22. Soliman, A.-H., & Abdellatif, M. (2023). COVID-19 disease treatment: Pivotal challenges in the arena of umbilical cord-mesenchymal stem cells (UC-MSCs). Frontiers in Cell and Developmental Biology, 11, 1146835. https://doi.org/10.3389/fcell.2023.1146835.
23. Swenson, K. E., & Swenson, E. R. (2021). Pathophysiology of acute respiratory distress syndrome and COVID-19 lung injury. Critical Care Clinics, 37(4), 749–776. https://doi.org/10.1016/j.ccc.2021.05.003.
24. Bourgonje, A. R., Abdulle, A. E., Timens, W., Hillebrands, J., Navis, G. J., Gordijn, S. J., Bolling, M. C., Dijkstra, G., Voors, A. A., Osterhaus, A. D., Voort, P. H., Mulder, D. J., & Goor, H. (2020). Angiotensin-converting enzyme 2 (ACE2), SARS-CoV -2 and the pathophysiology of coronavirus disease 2019 (COVID -19). *The Journal of Pathology, 251*(3), 228–248. https://doi.org/10.1002/path.5471
25. Mohamed, W., & Sandhir, R. (Eds.). (2023). Trace elements in brain health and diseases. Springer Nature https://doi.org/10.1007/978-981-99-1513-2.
26. Saby, J. N., Mulcahey, P. J., Zavez, A. E., Peters, S. U., Standridge, S. M., Swanson, L. C., Lieberman, D. N., Olson, H. E., Key, A. P., Percy, A. K., Neul, J. L., Nelson, C. A., Roberts, T. P. L., Benke, T. A., & Marsh, E. D. (2022). Electrophysiological biomarkers of brain function in CDKL5 deficiency disorder. Brain Communications, 4(4), fcac197. https://doi.org/10.1093/braincomms/fcac197.
27. Jone, P.-N., John, A., Oster, M. E., Allen, K., Tremoulet, A. H., Saarel, E. V., Lambert, L. M., Miyamoto, S. D., De Ferranti, S. D., on behalf of the American Heart Association Leadership Committee and Congenital Cardiac Defects Committee of the Council on Lifelong Congenital Heart Disease and Heart Health in the Young, & Council on Hypertension, and Council on Peripheral Vascular Disease. (2022). SARS-CoV-2 infection and associated cardiovascular manifestations and complications in children and young adults: A scientific statement from the American Heart Association. *Circulation, 145*(19), e1037–e1052. https://doi.org/10.1161/CIR.0000000000001064
28. Harpaz, D., Seet, R. C. S., Marks, R. S., & Tok, A. I. Y. (2020). Blood-based biomarkers are associated with different ischemic stroke mechanisms and enable rapid classification between cardioembolic and atherosclerosis etiologies. Diagnostics, 10(10), 804. https://doi.org/10.3390/diagnostics10100804.
29. Acampa, M., Lazzerini, P. E., Lattanzi, S., & Rubiera, M. (2023). Cryptogenic stroke as a working diagnosis: The need for an early and comprehensive diagnostic work-up. BMC Neurology, 23(1), 153. https://doi.org/10.1186/s12883-023-03206-6.
30. Healey, J. S., Connolly, S. J., Gold, M. R., Israel, C. W., Van Gelder, I. C., Capucci, A., Lau, C. P., Fain, E., Yang, S., Bailleul, C., Morillo, C. A., Carlson, M., Themeles, E., Kaufman, E. S., & Hohnloser, S. H. (2012). Subclinical atrial fibrillation and the risk of stroke. New England Journal of Medicine, 366(2), 120–129. https://doi.org/10.1056/NEJMoa1105575.
31. Spalletta, G., Bossù, P., Ciaramella, A., Bria, P., Caltagirone, C., & Robinson, R. G. (2006). The etiology of poststroke depression: A review of the literature and a new hypothesis involving inflammatory cytokines. Molecular Psychiatry, 11(11), 984–991. https://doi.org/10.1038/sj.mp.4001879.
32. Barone, F. C., & Feuerstein, G. Z. (1999). Inflammatory mediators and stroke: New opportunities for novel therapeutics. Journal of Cerebral Blood Flow & Metabolism, 19(8), 819–834. https://doi.org/10.1097/00004647-199908000-00001.
33. Licinio, J., & Wong, M.-L. (1999). The role of inflammatory mediators in the biology of major depression: Central nervous system cytokines modulate the biological substrate of depressive symptoms, regulate stress-responsive systems, and contribute to neurotoxicity and neuroprotection. Molecular Psychiatry, 4(4), 317–327. https://doi.org/10.1038/sj.mp.4000586.
34. Schiepers, O. J. G., Wichers, M. C., & Maes, M. (2005). Cytokines and major depression. Progress in Neuro-Psychopharmacology and Biological Psychiatry, 29(2), 201–217. https://doi.org/10.1016/j.pnpbp.2004.11.003.
35. Loubinoux, I., Kronenberg, G., Endres, M., Schumann-Bard, P., Freret, T., Filipkowski, R. K., Kaczmarek, L., & Popa-Wagner, A. (2012). Post-stroke depression: Mechanisms, translation and therapy. Journal of Cellular and Molecular Medicine, 16(9), 1961–1969. https://doi.org/10.1111/j.1582-4934.2012.01555.x.

36. Emsley, H. (2003). An early and sustained peripheral inflammatory response in acute ischaemic stroke: Relationships with infection and atherosclerosis. Journal of Neuroimmunology, 139(1–2), 93–101. https://doi.org/10.1016/S0165-5728(03)00134-6.
37. Musuka, T. D., Wilton, S. B., Traboulsi, M., & Hill, M. D. (2015). Diagnosis and management of acute ischemic stroke: Speed is critical. *CMAJ: Canadian Medical Association Journal = Journal de l'Association Medicale Canadienne, 187*(12), 887–893. https://doi.org/10.1503/cmaj.140355
38. Fagundes, D. J. T., & Omar, M. (2004). Animal disease model: Choice's criteria and current animals specimens. Acta Cirurgica Brasileira, 19(1), 59–65. https://doi.org/10.1590/s0102-86502004000100010.
39. Fluri, F. S., Michael, K., Kleinschnitz, & Christoph. (2015). Animal models of ischemic stroke and their application in clinical research. Drug Design, Development and Therapy, 9(NA), 3445–3454. https://doi.org/10.2147/dddt.s56071.
40. Broughton, B. R. S. R., David, C., Sobey, & Christopher, G. (2009). Apoptotic mechanisms after cerebral ischemia. Stroke, 40(5), 331–339. https://doi.org/10.1161/strokeaha.108.531632.
41. Bogousslavsky, J., Van Melle, G., & Regli, F. (1988). The Lausanne Stroke Registry: Analysis of 1,000 consecutive patients with first stroke. *Stroke, 19*(9), 1083–1092. https://doi.org/10.1161/01.str.19.9.1083
42. Howells, D. W., Porritt, M. J., Rewell, S. S., O'Collins, V., Sena, E. S., van der Worp, H. B., Traystman, R. J., & Macleod, M. R. (2010). Different strokes for different folks: The rich diversity of animal models of focal cerebral ischemia. Journal of Cerebral Blood Flow and Metabolism: Official Journal of the International Society of Cerebral Blood Flow and Metabolism, 30(8), 1412–1431. https://doi.org/10.1038/jcbfm.2010.66.
43. Belayev, L., Busto, R., Zhao, W., Fernandez, G., & Ginsberg, M. D. (1999). Middle cerebral artery occlusion in the mouse by intraluminal suture coated with poly-l-lysine: Neurological and histological validation. Brain Research, 833(2), 181–190. https://doi.org/10.1016/S0006-8993(99)01528-0.
44. Wang, J., Zhang, P., & Tang, Z. (2020). Animal models of transient ischemic attack: A review. Acta Neurologica Belgica, 120(2), 267–275. https://doi.org/10.1007/s13760-020-01295-5.
45. Knezic, A., Broughton, B. R. S., Widdop, R. E., & McCarthy, C. A. (2022). Optimising the photothrombotic model of stroke in the C57BI/6 and FVB/N strains of mouse. Scientific Reports, 12(1), 7598. https://doi.org/10.1038/s41598-022-11793-6.
46. Labat-gest, V., & Tomasi, S. (2013). Photothrombotic ischemia: A minimally invasive and reproducible photochemical cortical lesion model for mouse stroke studies. Journal of Visualized Experiments, 76, 50370. https://doi.org/10.3791/50370.
47. Abeysinghe, H. C. S., Bokhari, L., Dusting, G. J., & Roulston, C. L. (2014). Brain remodelling following endothelin-1 induced stroke in conscious rats. PLoS One, 9(5), e97007. https://doi.org/10.1371/journal.pone.0097007.
48. Li, F., Geng, X., Huber, C., Stone, C., & Ding, Y. (2020). In search of a dose: The functional and molecular effects of exercise on post-stroke rehabilitation in rats. Frontiers in Cellular Neuroscience, 14, 186. https://doi.org/10.3389/fncel.2020.00186.
49. Ruan, J., & Yao, Y. (2020). Behavioral tests in rodent models of stroke. Brain Hemorrhages, 1(4), 171–184. https://doi.org/10.1016/j.hest.2020.09.001.
50. Katz, D. I., Bernick, C., Dodick, D. W., Mez, J., Mariani, M. L., Adler, C. H., Alosco, M. L., Balcer, L. J., Banks, S. J., Barr, W. B., Brody, D. L., Cantu, R. C., Dams-O'Connor, K., Geda, Y. E., Jordan, B. D., McAllister, T. W., Peskind, E. R., Petersen, R. C., Wethe, J. V., & Stern, R. A. (2021). National Institute of Neurological Disorders and Stroke consensus diagnostic criteria for traumatic encephalopathy syndrome. Neurology, 96(18), 848–863. https://doi.org/10.1212/WNL.0000000000011850.
51. Tharakan, B. (Ed.). (2018). In Traumatic and ischemic injury (Vol. 1717). : Springer. https://doi.org/10.1007/978-1-4939-7526-6.
52. Bsat, S., Halaoui, A., Kobeissy, F., Moussalem, C., El Houshiemy, M. N., Kawtharani, S., & Omeis, I. (2021). Acute ischemic stroke biomarkers: A new era with diagnostic promise? Acute Medicine & Surgery, 8(1), e696. https://doi.org/10.1002/ams2.696.

53. Mrozek, S., Dumurgier, J., Citerio, G., Mebazaa, A., & Geeraerts, T. (2014). Biomarkers and acute brain injuries: Interest and limits. Critical Care, 18(2), 220. https://doi.org/10.1186/cc13841.
54. Jickling, G. C., & Sharp, F. R. (2011). Blood biomarkers of ischemic stroke. Neurotherapeutics, 8(3), 349–360. https://doi.org/10.1007/s13311-011-0050-4.
55. Whiteley, W., Tseng, M.-C., & Sandercock, P. (2008). Blood biomarkers in the diagnosis of ischemic stroke: A systematic review. Stroke, 39(10), 2902–2909. https://doi.org/10.1161/STROKEAHA.107.511261.
56. Hjalmarsson, C., Bjerke, M., Andersson, B., Blennow, K., Zetterberg, H., Åberg, N. D., Olsson, B., Eckerström, C., Bokemark, L., & Wallin, A. (2014). Neuronal and glia-related biomarkers in cerebrospinal fluid of patients with acute ischemic stroke. Journal of Central Nervous System Disease, 6, JCNSD.S13821. https://doi.org/10.4137/JCNSD.S13821.
57. Naik, A., Adeleye, O., Koester, S. W., Winkler, E. A., Hartke, J. N., Karahalios, K., Mihaljevic, S., Rani, A., Raikwar, S., Rulney, J. D., Desai, S. M., Scherschinski, L., Ducruet, A. F., Albuquerque, F. C., Lawton, M. T., Catapano, J. S., Jadhav, A. P., & Jha, R. M. (2023). Cerebrospinal fluid biomarkers for diagnosis and the prognostication of acute ischemic stroke: A systematic review. International Journal of Molecular Sciences, 24(13), 10902. https://doi.org/10.3390/ijms241310902.
58. Kurvits, S., Harro, A., Reigo, A., Ott, A., Laur, S., Särg, D., Tampuu, A., the Estonian Biobank Research Team, Alasoo, K., Vilo, J., Milani, L., Haller, T., & the PRECISE4Q consortium. (2023). Common clinical blood and urine biomarkers for ischemic stroke: An Estonian Electronic Health Records database study. *European Journal of Medical Research, 28*(1), 133. https://doi.org/10.1186/s40001-023-01087-6
59. Harston, G. W. J., Rane, N., Shaya, G., Thandeswaran, S., Cellerini, M., Sheerin, F., & Kennedy, J. (2015). Imaging biomarkers in acute ischemic stroke trials: A systematic review. AJNR: American Journal of Neuroradiology, 36(5), 839–843. https://doi.org/10.3174/ajnr.A4208.
60. Lu, J., Mei, Q., Hou, X., Manaenko, A., Zhou, L., Liebeskind, D. S., Zhang, J. H., Li, Y., & Hu, Q. (2021). Imaging acute stroke: From one-size-fit-all to biomarkers. Frontiers in Neurology, 12, 697779. https://www.frontiersin.org/articles/10.3389/fneur.2021.697779.
61. Maas, M. B., & Furie, K. L. (2009). Molecular biomarkers in stroke diagnosis and prognosis. Biomarkers in Medicine, 3(4), 363–383. https://doi.org/10.2217/bmm.09.30.
62. Li, P., Teng, F., Gao, F., Zhang, M., Wu, J., & Zhang, C. (2015). Identification of circulating microRNAs as potential biomarkers for detecting acute ischemic stroke. Cellular and Molecular Neurobiology, 35(3), 433–447. https://doi.org/10.1007/s10571-014-0139-5.
63. Laskowitz, D. T. (2005). Panel of biomarkers predicts stroke. Annals of the New York Academy of Sciences, 1053(1), 30–30. https://doi.org/10.1196/annals.1344.051.
64. Kearon, C., De Wit, K., Parpia, S., Schulman, S., Afilalo, M., Hirsch, A., Spencer, F. A., Sharma, S., D'Aragon, F., Deshaies, J.-F., Le Gal, G., Lazo-Langner, A., Wu, C., Rudd-Scott, L., Bates, S. M., & Julian, J. A. (2019). Diagnosis of pulmonary embolism with D -dimer adjusted to clinical probability. New England Journal of Medicine, 381(22), 2125–2134. https://doi.org/10.1056/NEJMoa1909159.
65. Nankivell, B. J. (2001). Creatinine clearance and the assessment of renal function., Australian Prescriber, 24(1), 15–17.
66. Lazaridis, K., & Tzartos, S. J. (2020). Autoantibody specificities in myasthenia gravis; implications for improved diagnostics and therapeutics. Frontiers in Immunology, 11, 212. https://doi.org/10.3389/fimmu.2020.00212.
67. Meriggioli, M. N., & Sanders, D. B. (2012). Muscle autoantibodies in myasthenia gravis: Beyond diagnosis? Expert Review of Clinical Immunology, 8(5), 427–438. https://doi.org/10.1586/eci.12.34.
68. Doust, J., Lehman, R., & Glasziou, P. (2006). The role of BNP testing in heart failure. *American Family Physician, 74*(11), 1893–1900.
69. Sproston, N. R., & Ashworth, J. J. (2018). Role of C-reactive protein at sites of inflammation and infection. Frontiers in Immunology, 9, 754. https://doi.org/10.3389/fimmu.2018.00754.

70. Katan, M., & Elkind, M. S. (2018). The potential role of blood biomarkers in patients with ischemic stroke: An expert opinion. Clinical and Translational Neuroscience, 2(1), 2514183X1876805. https://doi.org/10.1177/2514183X18768050.
71. Makris, K., Haliassos, A., Chondrogianni, M., & Tsivgoulis, G. (2018). Blood biomarkers in ischemic stroke: Potential role and challenges in clinical practice and research. Critical Reviews in Clinical Laboratory Sciences, 55(5), 294–328. https://doi.org/10.1080/10408363.2018.1461190.
72. Glushakova, O., Glushakov, A., Miller, E., Valadka, A., & Hayes, R. (2016). Biomarkers for acute diagnosis and management of stroke in neurointensive care units. Brain Circulation, 2(1), 28. https://doi.org/10.4103/2394-8108.178546.
73. Whiteley, W. (2011). Identifying blood biomarkers to improve the diagnosis of stroke. The Journal of the Royal College of Physicians of Edinburgh, 41(2), 152–154. https://doi.org/10.4997/JRCPE.2011.207.
74. Kernagis, D. N., & Laskowitz, D. T. (2012). Evolving role of biomarkers in acute cerebrovascular disease. Annals of Neurology, 71(3), 289–303. https://doi.org/10.1002/ana.22553.
75. Simats, A., García-Berrocoso, T., & Montaner, J. (2016). Neuroinflammatory biomarkers: From stroke diagnosis and prognosis to therapy. Biochimica et Biophysica Acta (BBA) - Molecular Basis of Disease, 1862(3), 411–424. https://doi.org/10.1016/j.bbadis.2015.10.025.
76. Amalia, L. (2021). Glial Fibrillary Acidic Protein (GFAP): Neuroinflammation biomarker in acute ischemic stroke. Journal of Inflammation Research, 14, 7501–7506. https://doi.org/10.2147/JIR.S342097.
77. Wadan, A.-H. S., Abdelsattar Ahmed, M., Hussein Ahmed, A., El-Sayed Ellakwa, D., Hamed Elmoghazy, N., & Gawish, A. (2024). The interplay of mitochondrial dysfunction in oral diseases: Recent updates in pathogenesis and therapeutic implications. *Mitochondrion*, 101942. https://doi.org/10.1016/j.mito.2024.101942
78. Tolba, M. M., Jabbar, A., Afzal, S., Mahmoud, M., Zulfiqar, F., El-Soudany, I., Samir, S., Wadan, A.-H. S., Ellakwa, T. E., & Ellakwa, D. E.-S. (2023). A promising RNA nanotechnology in clinical therapeutics: A future perspective narrative review. *Future Science OA, 9*(8), FSO883. https://doi.org/10.2144/fsoa-2023-0067
79. Wadan, A. S., & Liaquat, M. (n.d.). Synaptic modulation by coffee compounds: Insights into neural plasticity. In *Neuroscience of Coffee Part B* (1st ed.). Elsevier B.V. https://doi.org/10.1016/bs.pbr.2024.06.008
80. Sanaeifar, F., Pourranjbar, S., Pourranjbar, M., & Ramezani, S. (2024). Beneficial effects of physical exercise on cognitive-behavioral impairments and brain-derived neurotrophic factor alteration in the limbic system induced by neurodegeneration. *Experimental Gerontology, 195*(August), 112539. https://doi.org/10.1016/j.exger.2024.112539
81. Sayed, Z. S., Khattap, M. G., Madkour, M. A., Yasen, N. S., Elbary, H. A., Elsayed, R. A., Abdelkawy, D. A., Wadan, A.-H. S., Omar, I., & Nafady, M. H. (2024). Circulating tumor cells clusters and their role in Breast cancer metastasis; a review of literature. *Discover Oncology, 15*(1), 94. https://doi.org/10.1007/s12672-024-00949-7

Stroke Prevention and Treatment: Current State and Future Perspective

5

Kamrul Hassan Suman, Shadia Tasnim, Rony Ibne Masud, and Farwa Munir

Abstract

Stroke is a severe public health concern that results in the loss of life and has an adverse socioeconomic impact worldwide. The incidence of stroke has skyrocketed in recent decades, and experts anticipate this trend to continue, particularly in developing nations. Stroke risk can be reduced by many lifestyle changes, such as quitting smoking, maintaining blood pressure, taking antiplatelet medication, engaging in regular physical activity, and consuming a nutrient-rich diet (including folic acid and vitamin B complex supplements). Antiplatelet medications such as clopidogrel, aggrenox, and aspirin have been shown to be safe and effective in preventing non-cardioembolic strokes by lowering cholesterol and blood pressure, whereas anticoagulants such as warfarin, dabigatran, rivaroxaban, apixaban, and edoxaban are commonly used to prevent cardioembolic strokes. This chapter delves into the recent developments in the field of acute stroke treatment, focusing on the importance of swift care, including thrombolytic therapy, carotid endarterectomy, and endovascular procedures. Prehospital

Kamrul Hassan Suman and Shadia Tasnim contributed equally as co-first authors.

K. H. Suman (✉)
Molecular Biology and Interdisciplinary Life Sciences, New Mexico State University, New Mexico, USA
e-mail: khsuman@nmsu.edu

S. Tasnim
Department of Pathology, Bangladesh Agricultural University, Mymensingh, Bangladesh

R. I. Masud
Department of Microbiology and Hygiene, Bangladesh Agricultural University, Mymensingh, Bangladesh

F. Munir
Faculty of Medicine, University of Sultan Zainal Abidin, Terengganu, Malaysia

W. Mohamed, M. B. Mat-Nor (eds.), *Nutrition and Stroke*, Nutritional Neurosciences, https://doi.org/10.1007/978-981-97-6818-9_5

stroke triage (PST) detects and prioritises patients for rapid medical care, and mobile apps like PRESTRO and Destroke offer promising options for stroke care. Stroke survivors can greatly benefit from occupational therapy and rehabilitation treatment as part of their path to recovery and enhanced ability to perform regular tasks. This book chapter basically gives a full look at how to prevent and treat strokes right now, as well as how things might get better in the future and why it's important to take a unified approach to deal with this worldwide public health issue.

Keywords

Stroke · Risk factors · Prevention · Treatment · Thrombolytic therapy · Tissue plasminogen activator · Endovascular interventions · Rehabilitation therapy

5.1 Introduction

Stroke is the primary cause of substantial functional disability worldwide, leading to long-term neurological impairments including reduced physical abilities, unilateral limb weakness or paralysis, difficulty grasping or manipulating objects, and impaired communication [1]. Hippocrates documented the first stroke between 460 and 370 BC, asserting that "stout" carotid arteries impeded cognition [2]. However, today, people are well known for stroke because of its significant impact on health and socioeconomic impact. According to the World Stroke Organization Report 2022, stroke is the third most common cause of death and disability, and it is predicted to cost more than $721 billion globally (0.66% of the global GDP). The number of strokes, deaths, standard strokes, and disability-adjusted life years (DALYs) dramatically rose from 1990 to 2019 [3]. One in four people will have a stroke in their lifetime, according to the 2022 Global Stroke Factsheet, a rise of 50 percent in risk over 17 years. Stroke increased in frequency by 70%, in prevalence by 102%, in mortality by 43%, and in DALYs by 143% from 1990 to 2019. Stroke accounts for 86% of all deaths and 89% of all DALYs worldwide, with the vast majority of these deaths and DALYs occurring in low and middle-income countries (LMIC) [4]. If significant clinical or public health interventions are not adopted, 23 million new cases of stroke and 7.8 million deaths are projected for 2030 [5].

The three primary categories of strokes encompass ischemic, haemorrhagic, and subarachnoid haemorrhage strokes. In contrast to ischemic stroke, which arises from a disruption in cerebral blood flow resulting in abrupt functional impairment, haemorrhagic stroke is frequently characterised by a split of a blood vessel or an anomalous vascular structure [6]. Ischemic strokes constitute 68% of the total number of strokes globally, and haemorrhagic strokes comprise 32% of the overall stroke cases [7]. A little bit different statistics come from the USA, where 10% are haemorrhagic, 87% of strokes are ischemic, and roughly 3% are subarachnoid haemorrhages [8]. The implementation of risk factor management and secondary stroke prevention strategies has led to a decrease in the incidence of subsequent strokes. Despite advances in cardiovascular disease treatment and government initiatives, little effort is being put into secondary stroke prevention strategies [9]. Regrettably, a significant proportion of patients often fail to remember or adhere to preventive measures the primary goal of stroke prevention is the efficient control of risk factors linked to the onset of the condition. Due to its importance in chronic disease prevention and control, stroke treatment has been addressed in multiple ways within a holistic system [10]. The implementation of primary and secondary stroke preventive strategies has been shown to effectively mitigate the occurrence of strokes [11]. Stroke prevention concentrates on modifying social risk factors, whereas stroke management emphasises symptom relief. Stroke has been extensively studied over the past two decades, but there is no indisputable method of treatment or prevention. The primary and secondary objectives of current stroke research include the identification and development of therapeutic interventions [12]. Rather than being seen as a separate issue, efforts to prevent stroke should be seen as an important part of larger efforts to deal with the risk factors for noncommunicable diseases (NCDs). The success of stroke prevention can only be achieved through collaboration with other measures aimed at preventing NCDs [13]. Hackam and Spence [14] observed that taking many medications, including statins, aspirin, and antihypertensive drugs, in addition to making lifestyle changes and engaging in regular exercise, significantly reduced the risk of future cardiovascular events by 80%. Considering the above facts, this chapter highlights the significance of lifestyle modifications and medication in the prevention of stroke, along with the emerging treatment methods (i.e., thrombolytic therapy, endovascular therapy, occupational therapy, and rehabilitation therapy). In addition, our study stipulated acute stroke management, associated challenges, and prospective recommendations for enhancing stroke care.

5.2 Global Stroke Prevalence and Public Health Impact

As a result of demographic transitions, especially in developing nations, the prevalence of strokes has increased considerably and is expected to continue rising over the next few decades [15]. Every 3 min and 14 s, someone encounters a stroke. According to the most recent CDC report, strokes were responsible for one in six heart disease deaths in 2021 [16]. There are significant correlations between stroke, age, and gender. Stroke prevalence is anticipated to rise drastically worldwide as the number of people over 65 rises by roughly nine million per year [17]. More than 795,000 strokes occur annually in the USA, and one occurs every 40 s. Costs associated with stroke in the United States totalled about $56.5 billion in 2018 and 2019 due to issues like medical care, absence from work, and prescription drugs [18]. The total value of lost welfare (VLW) from strokes worldwide in 2019 was $2059.67 billion, or 1.66% of the global GDP. In terms of global VLW/GDP, ischemic stroke accounted for 0.78% (VLW = $964.51 billion), intracerebral haemorrhage for

0.71% (VLW = $882.81 billion), and subarachnoid haemorrhage for 0.17% (VLW = $212.36 billion) [19].

Stroke and socioeconomic status are closely related, as are geographic and demographic factors. East Asia, Southeast Asia, and Oceania had the most significant total stroke VLW/GDP, followed by Central, Eastern, and Central Asia. The developed countries exhibited a VLW/GDP ratio that was lower than the global norm for both total strokes and all stroke subtypes. These findings indicate that LMICs have higher rates of stroke [20–22]. According to Roushdy et al. [23], the occurrence of stroke in Africa has shifted from being rare to becoming a common noncommunicable disease, with annual incidence rates ranging from 250 to 316 per 100,000 adults. The INTERSTROKE study, which involved 22 countries, found that Africa was responsible for almost 66% and 34% of ischemic strokes and hemorrhagic strokes, respectively. However, high-income nations have a 91% ischemic stroke rate and 9% hemorrhagic stroke rate [24]. Roger et al. [25] found that 17.8% of Americans over 45 who had never experienced a stroke or transient ischaemic attack (TIA) reported having at least one stroke symptom. Blacks with fair to poor health, lesser incomes, and less education had more stroke symptoms than Whites.

5.3 Aetiology and Prospective Risk Factors for Stroke

5.3.1 Ischemic Stroke

The most common kind of stroke is an ischemic stroke, which occurs when a neck or brain blood vessel is blocked [26]. Acute ischemic strokes (AIS), of which there are about 85%, are the primary cause of persistent damage among adults globally and the third largest cause of mortality in developed countries [27]. The complexity of the ischemic stroke results in the rapid loss of blood flow to a particular area of the brain, destroying cerebral tissue [28]. Ischemic strokes can appear in pre-existing syndromes due to the diminished blood flow to particular brain areas consistent with test results. It enables medical professionals to be prepared for the area of the brain's vascular system that may be damaged [29]. Neurons undergo rapid cell death due to metabolic deficiency, excessive activity of glutamate receptors, excessive calcium levels, acidosis, and oxidative damage when deprived of glucose and oxygen [30].

A variety of factors may be the causal factors of ischemic stroke. The most common causes are cardiac embolisms or atherothrombosis of the significant cervical or cerebral arteries [27]. Brain blood channel obstruction is the main cause. Blood clots, or thrombosis, in neck or brain blood vessels may cause the blockage [31]. Ischemic brain injury is a major contributor to global adult mortality and the onset of chronic disability [28], is speculated to occur from energy depletion and cell death [28]. Ischemic stroke can be caused by atrial fibrillation, high blood pressure, diabetes, high cholesterol, other heart problems, obesity, stenosis of the carotid artery, smoking and inhaling cigarette smoke, autoimmune disease, use of postmenopausal hormones, a history of a stroke or TIA. Other causes include atherosclerosis of the aorta, the brain itself, arterial dissection outside the brain, and atherosclerosis or intimal hyperplasia of the carotid arteries outside the brain [32]. Additionally, oral contraceptives, excessive alcohol consumption, major cardiovascular risk factors, hypercholesterolemia, atherosclerosis, minor cardiovascular risk factors, cardiac embolism, haematological disorders, and other unidentified factors can contribute to it [33].

5.3.2 Hemorrhagic Stroke

A hemorrhagic stroke occurs when there is a rupture of a blood vessel, leading to the leakage of blood into the brain. Hemorrhagic stroke also falls under subarachnoid haemorrhage (SAH) and intracerebral haemorrhage (ICH). SAH and ICH are caused by bleeding into the brain parenchyma and subarachnoid space [34]. ICH, which is the prevailing form of hemorrhagic stroke, constitutes around 20% of all stroke cases and exhibits a notably elevated susceptibility to premature death and enduring impairments [35]. Primary intraparenchymal hematoma, intraventricular haemorrhage, and SAH are only a few of the symptoms of nontraumatic intracerebral haemorrhage, a potentially lethal subtype of acute stroke [35]. Hemorrhagic stroke is most frequently caused by hypertension [36]. The most common causes of ICH in individuals under the age of 40 include vascular abnormalities, cerebral venous thrombosis, the use of sympathomimetic medications, eclampsia, hypertension, cryptogenic cases, and cerebral microbleeds (CMBs) associated with hypertension and diabetes mellitus [35]. The primary or secondary causes may include vascular congenital abnormalities, dural arteriovenous fistulas (dAVF), the neurological system, cerebral venous thrombosis (CVT), and other elements [35]. Cerebral amyloid angiopathy (CAA) is a significant risk factor for primary lobar intracerebral haemorrhage in the elderly [37]. Smoking, moderate to severe drinking, chronic alcoholism, and chronic liver illness enhance the risk of ICH due to coagulopathy and thrombocytopenia. The danger of ICH from dual antiplatelet therapy rises in patients with low triglyceride and low LDL cholesterol levels [34]. Rupture of an aneurysm, arteriovenous malformation, dural sinus thrombosis, cerebral artery dissection, pituitary apoplexy, vasculitis, and intracranial bleeding during pregnancy are the most prevalent causes of spontaneous SAH [34]. Figure 5.1 depicts stroke risk factors that contribute to stroke development and may hinder the efficacy of therapy.

5.4 Stroke Prevention Strategies

5.4.1 Lifestyle Modifications

Recent research supports the notion that lifestyle habits influence stroke risk. Unhealthy lifestyle choices increase the risk of myocardial infarction and stroke [38]. Changing an unhealthy lifestyle to a healthy lifestyle with weight loss, a

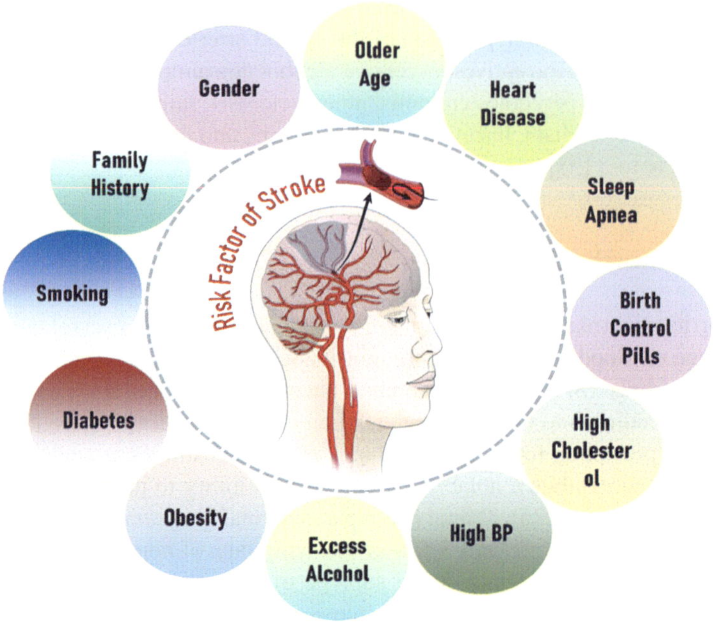

Fig. 5.1 Various risk factors responsible for the development of strokes and occasionally hinder the efficacy of treatment responses

healthy diet, termination of smoking, and regular physical activity improves stroke situation. A diet with nutritional supplement emphasising folic acid, vitamin B complex, and Mediterranean food habit lower the risk of stroke [39]. Human stroke risk can be reduced with suitable dietary items. A healthy lifestyle that includes regular exercise, weight loss, a low-risk diet, abstinence from smoking, and moderate alcohol consumption reduces the risk of stroke by 50% [40]. Patients with a lower risk of stroke and recurrence who also perform daily chores, get sufficient quality sleep, and have a greater level of high-density lipoprotein (HDL) [41].

Antihypertensive medication is not desirable, despite the importance of managing hypertension to reduce stroke risk injuries [42]. Angiotensin-converting enzyme inhibitors with thiazide diuretics give better results than other hypertensive medicines to reduce stroke risk [43]. The administration of sodium-glucose cotransporter-2 inhibitors, pioglitazone, and glucagon-like peptide-1 receptor agonists (such as liraglutide) demonstrated a tendency towards decreasing the occurrence of stroke in diabetic patients whose condition was effectively managed [44–46]. A healthy lifestyle, which encompasses smoking cessation, a low body mass index (BMI), regular exercise, moderate alcohol consumption, and a balanced diet, can substantially reduce the risk of total and ischemic stroke, but hemorrhagic stroke has not been linked to it [47].

5.4.2 Medications and Medical Interventions

A meta-analysis shows that antihypertensive medications used in clinical trials reduce the incidence of stroke recurrence [48]. Pioglitazone showed promise in a recent clinical trial for treating ischemic stroke. Pioglitazone trials for all types of diabetes and insulin resistance demonstrated a reduction in the incidence of stroke and other major vascular illnesses [45]. Taking generic, low-dose lipid- and blood pressure-lowering medications (such as candesartan, amlodipine, and rosuvastatin) along with polypills is found to be the safest method for preventing stroke [49]. Patients with TIA and ischemic stroke exhibit a heightened vulnerability to further occurrences of stroke. Single or dual antiplatelet therapy may be mandatory for subsequent stroke prevention, depending on severity and aetiology. Antiplatelets and anticoagulants are common secondary stroke prevention drugs. Anticoagulation therapy is recommended for secondary prevention in cases of cardioembolic stroke, which includes strokes caused by nonvalvular atrial fibrillation, different cardiopathies, and certain individuals with cervical artery dissection and stroke [50]. Antiplatelets and anticoagulants function by inhibiting platelet activation pathways and the coagulation cascade. Although anticoagulants like warfarin, dabigatran, rivaroxaban, apixaban, and edoxaban are commonly used to prevent cardioembolic strokes, antiplatelets like aspirin, clopidogrel, and aggrenox have been demonstrated to be safe and effective for reducing non-cardioembolic strokes in clinical trials [51]. In the long run, aspirin-based antiplatelet therapy lowers the risk of stroke and the composite endpoint of stroke, MI (myocardial infarction), and vascular mortality [52].

In experimental models of stroke, it has been observed that progenitor cells derived from bone marrow, umbilical cord blood, and adipose tissue have the ability to augment the process of recovery [53]. Endogenous and exogenous cell therapy is used where neural progenitor cells (NPCs) prevent injury to the brain [54]. Patients with severe complications following a stroke may benefit from surgery (carotid endarterectomy) to relieve carotid artery stenosis and reduce their risk of further stroke occurrence. Carotid endarterectomy reduces regenerative stroke, and evidence of moderate quality suggests that it may benefit participants with symptomatic stenosis ranging from 50 to 69% and highly benefit those with stenosis ranging from 70 to 99% [55]. Aspirin is efficacious and should be given prior to, during, and following carotid surgery [56].

5.4.3 Emerging Prevention Approaches

Recent technological advancements have greatly aided stroke prevention. A scientifically approved app, PRESTRO (Prevent Stroke), was employed for the purpose of mitigating stroke risk factors among a cohort of young patients. It provides crucial information about stroke, encourages healthy lifestyle choices, and sets alarms for medication intake and blood pressure. Results showed that patients were almost two times more active than control groups [57]. Patients with a previous stroke history can benefit from using an at-home stroke monitoring app because it increases their comfort level in spotting symptoms. This research led to the development of the mobile app Destroke [58]. Modern neuroimaging has been used to image strokes and monitor patients during and after treatment. Modern neuroimaging techniques, such as advanced magnetic resonance imaging (MRI) and positron emission tomography (PET), can be useful in cases where both primary and secondary preventive measures are advantageous. MR and CT technology is essential to the current generation of clinically recognised perfusion procedures. While CT operations rely on injecting iodine contrast material, MR can be conducted with or without contrast [59].

5.5 Acute Stroke Management/Treatments

5.5.1 Stroke Detection

Acute strokes occur when blood flow to the brain is suddenly cut off, impairing brain function and possibly inflicting brain tissue damage. The stroke should be taken as a medical emergency, which must be treated instantaneously to reduce brain damage and increase recovery prospects. A transient clot is the root cause of a TIA, sometimes known as a "mini-stroke" [60]. The prehospital stroke triage (PST) is a methodical process aimed at detecting and prioritising individuals who have experienced a stroke, with the goal of delivering suitable medical attention and facilitating their transfer to a specialised stroke centre. The approach comprises evaluating the patient's stroke symptoms with the facilities' resources and capabilities to determine the best possible medical centre to treat the patient. Using a PST scale can help direct patients with possible strokes to the most suitable hospitals for their needs [61]. The Japan Urgent Stroke Triage (JUST) score is a predictive tool that can assess various kinds of stroke, such as ICH, SAH, large vessel occlusion (LVO), and cerebral infarction (CI) apart from LVO. According to a study that used the JUST score, implementing it significantly decreased the door-to-puncture time (84 vs. 73 min, $p = 0.03$) but had no impact on the patients' prognosis for acute LVO. Researchers came to the conclusion that PST scales can be used to shorten prehospital planning and management time [62].

5.5.2 Thrombolytic Therapy (Clot-Busting Drugs)

Numerous treatments, including the care of patients within a stroke unit, the administration of aspirin within 48 h of stroke onset, the utilisation of hemi-craniectomy, and the intravenous (IV) administration of thrombolysis within 4.5 h, have been shown to improve outcomes [63]. Thrombolysis has been the most influential of these treatments, prompting a worldwide paradigm shift in stroke management and the development of stroke units to facilitate the safe and successful delivery of thrombolysis [64]. In 1996, the Food and Drug Administration (FDA) granted approval for the intravenous use of recombinant tissue plasminogen activator (rtPA) in the treatment of acute ischemic stroke. Infusions of the previously FDA-approved tissue plasminogen activator have recently taken over as the standard treatment for acute stroke. In a recent large-scale European investigation, an intravenous tissue plasminogen activator proved effective for up to 4.5 h after a stroke [65]. Intra-arterial mechanical or chemical endovascular thrombolysis has demonstrated efficacy in enhancing the patient's condition. The primary focus of initial treatment is on the physical rehabilitation of the damaged area after the acute phase, as highlighted by Veerbeek et al. [66].

Extracranial-intracranial (EC-IC) bypass is a cornerstone treatment for ischemic cerebrovascular disease. Common carotid artery blockage is treated with a Bonnet bypass, and middle cerebral artery and anterior cerebral artery ischemia are treated using EC-IC bypasses. Femoral vein and radial artery intergrafts are used in these procedures [67]. Intravascular thrombi are made up of fibrin monomers that have been cross-linked using lysine side chains. Thrombolytic drugs attach specifically to these lysine cross-linkages, allowing them to dissolve the clot. Because of its strong binding to lysine side chains, plasminogen is highly concentrated near the site of the clot [64]. Once plasminogen has been transformed into plasmin, fibrin breakdown products are released from the thrombus [68]. While tissue plasminogen activator (tPA) has emerged as the established treatment for thrombolytic therapy in cases of acute ischemic stroke, it is worth considering that alternative medications could potentially exhibit greater efficacy. Because of their longer half-lives and greater fibrin specificity compared to tPA, desmoteplase and tenecteplase are superior in preventing systemic bleeding problems. Desmoteplase is preferable to tPA since it does not increase amyloid formation and does not lead to neurotoxicity (at the cellular level). Tenecteplase is the favoured thrombolytic medication for treating myocardial ischemia since it hastens coronary reperfusion [69]. Table 5.1 summarises the most frequently prescribed medications for the treatment of stroke.

5.5.3 Endovascular Therapy (Clot Retrieval)

Recent clinical trials have suggested benefits from the use of endovascular therapies that remove clots mechanically and chemically, and these therapies are used regularly in medical applications [78]. Patients with acute ischemic stroke, major arterial occlusions, and viable brain tissue should undergo mechanical thrombectomy if intravenous thrombolysis fails. Mechanical thrombectomy catheters are used to dislodge thrombi from specific blood vessels. Distal devices physically grab and drag out cerebral thrombi, while proximal devices aspirate occlusive material from cerebral arteries [79]. The technological complexity of distal thrombectomy surpasses that of proximal thrombectomy. Several clinical studies have employed the Merci device, the first distal thrombectomy device approved by the FDA in 2004. Revascularization during neurointervention is graded with the aid of cerebral angiography, which yields multiple systems. Thrombolysis in myocardial infarction

Table 5.1 List of commonly used medications for the treatment of stroke

Medication	Effect	Reference
Aspirin (acetylsalicylic acid)	Antithrombotic treatment reduces the risk of ischemic stroke	[70]
Anticoagulant (warfarin, apixaban, rivaroxaban)	Assist in mitigating excessive bleeding to prevent ischemic and haemorrhagic stroke	[71]
Antiplatelet agents (clopidogrel, ticagrelor)	Help minimise the risk of TIA, non-cardioembolic stroke, and recurrent ischaemic stroke	[72]
Statins (atorvastatin, rosuvastatin)	Combining statins with more intensive low-density lipoprotein cholesterol (LDL-C) reduction is advantageous for the prevention of stroke recurrence	[73]
Vitamin B supplement	Reduces the risk of primary stroke by mitigating small-vessel cerebral damage	[11]
Tissue plasminogen activator (tPA)	It enhances blood flow and breaks up blood clotting for acute ischemic stroke	[74]
Endovascular treatment	Prevents sudden brain artery blockage	[75]
Blood pressure medication (ACE inhibitors, β-blockers)	Effective in antihypertensive treatment, hypertension control, and stroke prevention	[76]
Carotid endarterectomy	Stroke risk is reduced with surgical intervention	[55]
Oral taurine supplementation	Reduce stroke occurrences by preventing genetic alteration	[77]

(TIMI) and thrombolysis in cerebral infarction (TICI) are the most common. In 2015, the TIMI grading system was launched to evaluate the degree of reperfusion in the treatment of acute myocardial infarction [80]. Stentrievers are specialty medical equipment commonly used to treat AIS. The catheter is introduced into the arterial vasculature and navigated towards the site of the cerebral thrombus. The utilisation of Stentriever has the potential to enhance outcomes in the management of acute endovascular stroke by expediting flow restoration and augmenting recanalization [81].

5.6 Rehabilitation Therapy

Rehabilitation treatments play a crucial role in promoting functional recovery and fostering independence among individuals who have experienced an acute stroke. Over two-thirds of stroke survivors receive rehabilitation after hospitalisation. Despite advances in stroke centres and early identification and treatment, few acute stroke patients receive thrombolytic therapy, and many continue to suffer functional limitations years after their stroke [82]. The American Stroke Association and the American Heart Association and recommend a five-category rehabilitation programme. The rehabilitation programme begins with system-level components like organisation and levels of care, followed by prevention and medical management of comorbidities, which references other published recommendations like hypertension. The next two levels are transitions in care and community rehabilitation, which focus on the ICF's (International Classification of Functioning, Disability, and Health) participation level, and sensorimotor impairments and activities (treatments and interventions), both of which focus on the ICF's activity level [83]. During the acute post-stroke phase, between 60% and 80% of survivors have motor deficits in either the upper or lower extremities. The rehabilitation approach offers stroke survivors the optimal opportunity to achieve complete recuperation. In general, a moderate stroke can be expected to recover within six months; however, a more severe stroke may take years to recover fully [84].

Restoration and rehabilitation following stroke provides evidence-based advice on post-stroke deficits, activity restrictions, and participation restraints. Rehabilitation helps people restore lost abilities, learn new skills, and regain independence. Rehabilitation is required for a significant number of stroke patients due to the presence of hemiparesis, upper-limb dysfunction, pain, decreased balance, swallowing difficulties, visual impairments, neglect, and limits in movement, everyday tasks, and cognition [85]. Traditional rehabilitative techniques, such as proprioception, neuromuscular facilitation, brunstrom, motor relearning, bobath, and function-based principles, are the most commonly employed in medical centres. Robotics-aided systems, electrical stimulation, constraint-induced movement therapy, electromyographic biofeedback, virtual reality, partial body weight-supported treadmill training, intermittent compression, and thermal stimulation are just a few of the cutting-edge methods being developed and incorporated into traditional rehabilitation programmes [86]. Based on the patient's functional condition during various stages of recovery following a stroke, the idea of combining beneficial rehabilitation techniques into "a training package" is offered. Stroke patients may have considerable functional gains from integrated sensorimotor rehabilitation regimens with proper temporal arrangements (Chen and Shaw 2014). One of these approaches widely used in Western countries is bobath therapy. Bobath therapy manipulates the key joints of the neck, shoulder, hand, hip, knee, and ankle to normalise the patient's movement pattern and postural tone. Recently, Bobath therapy was reinterpreted as a problem-solving strategy for assessing and treating central nervous system injury-related functional, motor, and postural deficits [87]. Therapists utilise a technique called "selective muscle activation" to assist patients in purposeful movement and attaining correct posture. Concepts of motor control and learning suggest that engaging in movements such as getting in and out of bed, sitting, transferring, standing up, and walking can be beneficial as functional and strengthening exercises [66].

Occupational therapy for stroke survivors aims to enhance their capacity to carry out daily tasks. Occupational therapists use a wide range of approaches, including activity-based therapy, technological aids, adaptive methods, and adaptations to the natural environment. The collaborative process between the adult stroke survivor and the occupational therapist involves mutual agreement on the specific nature and definition of each aim. The adult's self-identified needs and preferences, combined with the latest assessments, scientific knowledge, and therapeutic skills, determine the outcome [88]. In a meta-analysis conducted by Legg et al. [88], it was discovered that the implementation of activities of daily living (ADL)-focused occupational therapy for individuals who have experienced a stroke resulted in enhanced ADL scores and a decreased likelihood of unfavourable outcomes. Despite not reducing the chances of death, dependence, decline, or institutionalisation, they found that occupational therapy increased autonomy in long-term tasks. In this era of technology and AI, robotics are indispensable in the medical field. As with traditional medical practices, robotics have supplanted them. Patients with motor impairments, such as those who have suffered a stroke or spinal cord injury, can benefit from robotic rehabilitation therapy because of the rigorous training it provides. End-effector and exoskeleton forms of robotic apparatus are utilised for motor rehabilitation. Current research supports the use of robot-assisted therapy in conjunction with conventional rehabilitation techniques to enhance motor function in stroke patients [89].

5.7 Conclusion and Future Recommendations

The occurrence of strokes cannot be completely prevented. Nevertheless, effective preventative measures and advancements in treatment interventions have the potential to significantly mitigate this worldwide burden. Extensive clinical research and study have provided valuable insights into effective strategies for prevention and therapy. The prevalence of strokes can be achieved through the combined endeavours of communities and individuals, promoting a proactive attitude towards healthcare. Lifestyle modifications, such as adopting a healthy diet, regular physical exercise, controlling blood pressure, and quitting smoking and consuming alcohol excessively, are effective in reducing the risk of stroke. As our discussion demonstrates, it is important to incorporate secondary stroke preventive strategies and varied treatment protocols, including state-of-the-art biotechnologies and precision medicine. These strategies significantly minimise the danger of stroke and expedite the recovery of patients. Future stroke cures should prioritise customised risk assessment using genetic profiling, high-resolution imaging, and AI-driven predictive algorithms. This could assist healthcare practitioners in promptly identifying patients who are at a heightened risk of stroke and facilitating the implementation of tailored medications and better lifestyle modifications. The following measures have the potential to foster positive outcomes in future stroke management and may be considered:

- Regenerative therapy: Regenerative medicine may restore neural function after a stroke. Stem cell therapy and tissue engineering can restore brain tissue, speeding functional recovery and improving stroke survivors' well-being [90].
- Nanotechnology: Nanoparticles have the potential to selectively deliver therapeutic chemicals to specific locations across the blood-brain barrier [91], thereby enhancing the efficacy of stroke treatment and mitigating adverse effects.
- Neuroprotective drugs: Researchers are currently investigating neuroprotective drugs to reduce cerebral injury during and after stroke. This may involve the

advancement of pharmaceuticals or therapies that augment the brain's capacity for tissue regeneration, reduction of inflammation, and self-healing.

- Behavioural therapies: Behavioural science and psychology advances may help design healthier lifestyle solutions. Gamification, individualised coaching, and virtual reality-based therapies can help people establish and sustain good behaviours, thereby reducing stroke risk.
- Catheter-based therapy: The development of a catheter-based therapy that is both realistic and cost-efficient would facilitate rapid access to the affected regions of the brain, allowing for the extraction of blood clots and the repair of injured tissue. This would result in diminished risks and accelerated recovery periods.
- Big data and predictive modeling: Massive volumes of information on individual genetic material, environments, and lifestyle choices have been analysed using big data and predictive modelling, leading to new insights. Thus, physicians and scientists can improve more precise stroke risk prediction models to improve drugs and treatment protocols.
- Global care access and collaboration: Successful stroke cure development requires cooperation between governments and stakeholders at both national and international levels, including sufficient financial and technological resources to facilitate innovation. The utilisation of telemedicine, mobile health apps, and affordable diagnostic equipment has the potential to enhance the availability of urgent medical advice and interventions, particularly in remote and economically disadvantaged areas.

Author's Contribution Conceptualization (KHS); Visualization (ST); Statistical Data Analysis (ST, RIM and FM); Writing and Editing (KHS, ST, RIM and FM).

Funding This research did not receive any grants from funding agencies in the public, commercial, or not-for-profit sectors.

Declaration of Competing Interest All authors approved the final manuscript and declare that there are no known competing financial interests or personal relationships that could have appeared to influence the work reported in this chapter.

References

1. Pollak, J., Doyle, K. P., Mamer, L., Shamloo, M., & Buckwalter, M. S. (2012). Stratification substantially reduces behavioral variability in the hypoxic-ischemic stroke model. *Brain and Behavior, 2*(5), 698–706. https://doi.org/10.1002/brb3.77
2. Nilsen, M. L. (2010). A historical account of stroke and the evolution of nursing care for stroke patients. *Journal of Neuroscience Nursing, 42*(1), 19–27. https://doi.org/10.1097/JNN.0b013e3181c1fdad
3. Feigin, V. L., Stark, B. A., Johnson, C. O., Roth, G. A., Bisignano, C., Abady, G. G., Abbasifard, M., Abbasi-Kangevari, M., Abd-Allah, F., Abedi, V., Abualhasan, A., Abu-Rmeileh, N. M. E., Abushouk, A. I., Adebayo, O. M., Agarwal, G., Agasthi, P., Ahinkorah, B. O., Ahmad, S., Ahmadi, S., et al. (2021). Global, regional, and national burden of stroke

and its risk factors, 1990–2019: A systematic analysis for the Global Burden of Disease Study 2019. *The Lancet Neurology, 20*(10), 1–26. https://doi.org/10.1016/S1474-4422(21)00252-0
4. Feigin, V. L., Brainin, M., Norrving, B., Martins, S., Sacco, R. L., Hacke, W., Fisher, M., Pandian, J., & Lindsay, P. (2022). World Stroke Organization (WSO): Global Stroke Fact Sheet 2022. *International Journal of Stroke, 17*(1), 18–29. https://doi.org/10.1177/17474930211065917
5. Strong, K., Mathers, C., & Bonita, R. (2007). Preventing stroke: Saving lives around the world. *Lancet Neurology, 6*(2), 182–187. https://doi.org/10.1016/S1474-4422(07)70031-5
6. Bamford, J., Sandercock, P., Dennis, M., Warlow, C., & Burn, J. (1991). Classification and natural history of clinically identifiable subtypes of cerebral infarction. *The Lancet, 337*(8756), 1521–1526. https://doi.org/10.1016/0140-6736(91)93206-O
7. Lozano, R., Naghavi, M., Foreman, K., Lim, S., Shibuya, K., Aboyans, V., Abraham, J., Adair, T., Aggarwal, R., Ahn, S. Y., AlMazroa, M. A., Alvarado, M., Anderson, H. R., Anderson, L. M., Andrews, K. G., Atkinson, C., Baddour, L. M., Barker-Collo, S., Bartels, D. H., et al. (2012). Global and regional mortality from 235 causes of death for 20 age groups in 1990 and 2010: A systematic analysis for the Global Burden of Disease Study 2010. *The Lancet, 380*(9859), 2095–2128. https://doi.org/10.1016/S0140-6736(12)61728-0
8. Benjamin, E. J., Blaha, M. J., Chiuve, S. E., Cushman, M., Das, S. R., Deo, R., De Ferranti, S. D., Floyd, J., Fornage, M., Gillespie, C., Isasi, C. R., Jim'nez, M. C., Jordan, L. C., Judd, S. E., Lackland, D., Lichtman, J. H., Lisabeth, L., Liu, S., Longenecker, C. T., et al. (2017). Heart Disease and Stroke Statistics—2017 update: A report from the American Heart Association. *Circulation, 135*(10), e146–e603. https://doi.org/10.1161/CIR.0000000000000485
9. Chaudhry, H. J., & McDermott, B. (2008). Recognizing and improving patient nonadherence to statin therapy. *Current Atherosclerosis Reports, 10*(1), 19–24. https://doi.org/10.1007/s11883-008-0004-4
10. Chao, B. H., Yan, F., Hua, Y., Liu, J. M., Yang, Y., Ji, X. M., Peng, B., Zhao, G. G., Wang, Y. J., Kang, D. Z., Wang, Y. L., Zeng, J. S., Chu, L., Li, T. X., Xu, Y. M., Liu, M., He, L., Xu, Y., Wu, J., et al. (2021). Stroke prevention and control system in China: CSPPC-Stroke Program. *International Journal of Stroke, 16*(3), 265–272. https://doi.org/10.1177/1747493020913557
11. Sabih, A., Tadi, P., & Kumar, A. (2023). Stroke prevention. In *StatPearls [Internet]*. PMID: 29262078
12. Kuriakose, D., & Xiao, Z. (2020). Pathophysiology and treatment of stroke: Present status and future perspectives. *International Journal of Molecular Sciences, 21*(20), 1–24. https://doi.org/10.3390/ijms21207609
13. Norrving, B., Davis, S. M., Feigin, V. L., Mensah, G. A., Sacco, R. L., & Varghese, C. (2015). Stroke prevention worldwide—What could make it work. *Neuroepidemiology, 45*(3), 215–220. https://doi.org/10.1159/000441104
14. Hackam, D. G., & Spence, J. D. (2007). Combining multiple approaches for the secondary prevention of vascular events after stroke: A quantitative modeling study. *Stroke, 38*(6), 1881–1885. https://doi.org/10.1161/STROKEAHA.106.475525
15. Adogu, P. O. U., Ubajaka, C. F., Emelumadu, O. F., & Alutu, C. O. C. (2015). Epidemiologic transition of diseases and health-related events in developing countries: A review. *American Journal of Medicine and Medical Sciences, 5*(4), 150–157. https://doi.org/10.5923/j.ajmms.20150504.02
16. National Center for Health Statistics. (2023). *Multiple cause of death 2018–2021 on CDC WONDER Database*. Multiple Cause of Death 1999–2019 on CDC Wonder Online Database. https://wonder.cdc.gov/mcd.html
17. Wright, I., & Adams, P. (2012). The World Health Report 2004. Changing history. In *Drapers* (Issues 21-APRIL-2012). https://apps.who.int/iris/handle/10665/42891
18. Tsao, C. W., Aday, A. W., Almarzooq, Z. I., Anderson, C. A. M., Arora, P., Avery, C. L., Baker-Smith, C. M., Beaton, A. Z., Boehme, A. K., Buxton, A. E., Commodore-Mensah, Y., Elkind, M. S. V., Evenson, K. R., Eze-Nliam, C., Fugar, S., Generoso, G., Heard, D. G., Hiremath, S., Ho, J. E., et al. (2023). Heart disease and stroke statistics—2023 update: A

report from the American Heart Association. *Circulation, 147*(8), E93–E621. https://doi.org/10.1161/CIR.0000000000001123
19. Gerstl, J. V. E., Blitz, S. E., Qu, Q. R., Yearley, A. G., Lassarén, P., Lindberg, R., Gupta, S., Kappel, A. D., Vicenty-Padilla, J. C., Gaude, E., Atchaneeyasakul, K. C., Desai, S. M., Yavagal, D. R., Peruzzotti-Jametti, L., Patel, N. J., Aziz-Sultan, M. A., Du, R., Smith, T. R., & Bernstock, J. D. (2023). Global, regional, and national economic consequences of stroke. *Stroke, 54*(9), 2380–2389. https://doi.org/10.1161/STROKEAHA.123.043131
20. Bernhardt, J., Urimubenshi, G., Gandhi, D. B. C., & Eng, J. J. (2020). Stroke rehabilitation in low-income and middle-income countries: A call to action. *The Lancet, 396*(10260), 1452–1462. https://doi.org/10.1016/S0140-6736(20)31313-1
21. Chimatiro, G. L., & Rhoda, A. J. (2019). Scoping review of acute stroke care management and rehabilitation in low and middle-income countries. *BMC Health Services Research, 19*(1), 789. https://doi.org/10.1186/s12913-019-4654-4
22. Owolabi, M. O., Thrift, A. G., Mahal, A., Ishida, M., Martins, S., Johnson, W. D., Pandian, J., Abd-Allah, F., Yaria, J., Phan, H. T., Roth, G., Gall, S. L., Beare, R., Phan, T. G., Mikulik, R., Akinyemi, R. O., Norrving, B., Brainin, M., Feigin, V. L., et al. (2022). Primary stroke prevention worldwide: translating evidence into action. *The Lancet Public Health, 7*(1), e74–e85. https://doi.org/10.1016/S2468-2667(21)00230-9
23. Roushdy, T., Aref, H., Kesraoui, S., Temgoua, M., Nono, K. P., Gebrewold, M. A., Peter, W., Gopaul, U., Belahsen, M. F., Ben-Adji, D., Melifonwu, R., Pugazhendhi, S., Woodcock, N., Mohamed, M. H., Rossouw, A., Matuja, S., Ruanda, M. K., Mhiri, C., Saylor, D., et al. (2022). Stroke services in Africa: What is there and what is needed. *International Journal of Stroke, 17*(9), 972–982. https://doi.org/10.1177/17474930211066416
24. Sanuade, O. A., Dodoo, F. N. A., Koram, K., & De-Graft Aikins, A. (2019). Prevalence and correlates of stroke among older adults in Ghana: Evidence from the Study on Global AGEing and adult health (SAGE). *PLoS One, 14*(3), e0212623. https://doi.org/10.1371/journal.pone.0212623
25. Roger, V. L., Go, A. S., Lloyd-Jones, D. M., Adams, R. J., Berry, J. D., Brown, T. M., Carnethon, M. R., Dai, S., De Simone, G., Ford, E. S., Fox, C. S., Fullerton, H. J., Gillespie, C., Greenlund, K. J., Hailpern, S. M., Heit, J. A., Ho, P. M., Howard, V. J., Kissela, B. M., et al. (2010). Heart disease and stroke statistics—2011 update. *Circulation, 123*(4). https://doi.org/10.1161/cir.0b013e3182009701
26. Wallace, J. B. (2016). Researchers document troubling rise in strokes in young adults, starting at age 25. *The Washington Post, 11*.
27. Lakhan, S. E., Kirchgessner, A., & Hofer, M. (2009). Inflammatory mechanisms in ischemic stroke: Therapeutic approaches. *Journal of Translational Medicine, 7*(1), 1–11. https://doi.org/10.1186/1479-5876-7-97
28. Fan, P. L., Wang, S. S., Chu, S. F., & Chen, N. H. (2023). Time-dependent dual effect of microglia in ischemic stroke. *Neurochemistry International, 169*, 105584. https://doi.org/10.1016/J.NEUINT.2023.105584
29. Hui, C., Tadi, P., & Patti, L. (2021). Ischemic stroke—StatPearls—NCBI Bookshelf. In *StatPearls* (pp. 1–14)
30. Candelario-Jalil, E. (2009). Injury and repair mechanisms in ischemic stroke: Considerations for the development of novel neurotherapeutics. *Current Opinion in Investigational Drugs, 10*(7), 644–654.
31. Randolph, S. A. (2016). Ischemic stroke. *Workplace Health & Safety, 64*(9), 444–444. https://doi.org/10.1177/2165079916665400
32. Silva, G. S., Koroshetz, W. J., González, R. G., & Schwamm, L. H. (2011). Causes of ischemic stroke. In *Acute ischemic stroke* (pp. 25–42). Springer. https://doi.org/10.1007/978-3-642-12751-9_2
33. Varona, J. F., Guerra, J. M., Bermejo, F., Molina, J. A., & Gomez de la Cámara, A. (2007). Causes of ischemic stroke in young adults, and evolution of the etiological diagnosis over the long term. *European Neurology, 57*(4), 212–218. https://doi.org/10.1159/000099161

34. Bokka, S. K., & Das, S. (2020). Hemorrhagic stroke. In *Acute strokes: Signs, symptoms and management* (pp. 131–146). Nova Science. https://doi.org/10.29309/tpmj/2014.21.06.2626
35. Montaño, A., Hanley, D. F., & Hemphill, J. C. (2021). Hemorrhagic stroke. *Handbook of Clinical Neurology, 176*, 229–248. https://doi.org/10.1016/B978-0-444-64034-5.00019-5
36. Kitagawa, K. (2022). Blood pressure management for secondary stroke prevention. *Hypertension Research: Official Journal of the Japanese Society of Hypertension, 45*(6), 936–943. https://doi.org/10.1038/s41440-022-00908-1
37. Castello, J. P., Pasi, M., Kubiszewski, P., Abramson, J. R., Charidimou, A., Kourkoulis, C., DiPucchio, Z., Schwab, K., Anderson, C. D., Gurol, M. E., Greenberg, S. M., Rosand, J., Viswanathan, A., & Biffi, A. (2022). Cerebral small vessel disease and depression among intracerebral hemorrhage survivors. *Stroke, 53*(2), 523–531. https://doi.org/10.1161/STROKEAHA.121.035488
38. Renner, C., Favilla, C. G., & Cucchiara, B. L. (2020). Transient Ischemic Attack. *Decision-Making in Adult Neurology, 347*(21), 1687–1692. https://doi.org/10.1016/B978-0-323-63583-7.00040-0
39. Diener, H.-C., & Hankey, G. J. (2020). Primary and secondary prevention of ischemic stroke and cerebral hemorrhage. *Journal of the American College of Cardiology, 75*(15), 1804–1818. https://doi.org/10.1016/j.jacc.2019.12.072
40. Niewada, M., & Michel, P. (2016). Lifestyle modification for stroke prevention: Facts and fiction. *Current Opinion in Neurology, 29*(1), 9–13. https://doi.org/10.1097/WCO.0000000000000285
41. Huang, Z. X., Lin, X. L., Lu, H. K., Liang, X. Y., Fan, L. J., & Liu, X. T. (2019). Lifestyles correlate with stroke recurrence in Chinese inpatients with first-ever acute ischemic stroke. *Journal of Neurology, 266*(5), 1194–1202. https://doi.org/10.1007/s00415-019-09249-5
42. Ravenni, R., Jabre, J., Casiglia, E., & Mazza, A. (2011). Primary stroke prevention and hypertension treatment: Which is the first-line strategy? *Neurology International, 3*(2), 12. https://doi.org/10.4081/ni.2011.e12
43. Alva, M. L., Gray, A., Mihaylova, B., Leal, J., & Holman, R. R. (2015). The impact of diabetes-related complications on healthcare costs: New results from the UKPDS (UKPDS 84). *Diabetic Medicine, 32*(4), 459–466. https://doi.org/10.1111/dme.12647
44. Bethel, A., Patel, R. A., Merrill, P., Lokhnygina, Y., Buse, J. B., Mentz, R. J., Pagidipati, N. J., Chan, J. C., Gustavson, S. M., Iqbal, N., Maggioni, A. P., Öhman, P., Poulter, N. R., Ramachandran, A., Zinman, B., Hernandez, A. F., & Holman, R. R. (2018). Cardiovascular outcomes with glucagon-like peptide-1 receptor agonists in patients with type 2 diabetes: A meta-analysis. *The Lancet, 6*(2), 105–113. https://doi.org/10.1016/S2213-8587(17)30412-6
45. Lee, M., Saver, J. L., Liao, H. W., Lin, C. H., & Ovbiagele, B. (2017). Pioglitazone for secondary stroke prevention: A systematic review and meta-analysis. *Stroke, 48*(2), 388–393. https://doi.org/10.1161/STROKEAHA.116.013977
46. Zinman, B., Inzucchi, S. E., Lachin, J. M., Wanner, C., Fitchett, D., Kohler, S., Mattheus, M., Woerle, H. J., Broedl, U. C., Johansen, O. E., Albers, G. W., & Diener, H. C. (2017). Empagliflozin and cerebrovascular events in patients with type 2 diabetes mellitus at high cardiovascular risk. *Stroke, 48*(5), 1218–1225. https://doi.org/10.1161/STROKEAHA.116.015756
47. Kurth, T., Moore, S. C., Gaziano, J. M., Kase, C. S., Stampfer, M. J., Berger, K., & Buring, J. E. (2006). Healthy lifestyle and the risk of stroke in women. *Archives of Internal Medicine, 166*(13), 1403. https://doi.org/10.1001/archinte.166.13.1403
48. Katsanos, A. H., Filippatou, A., Manios, E., Deftereos, S., Parissis, J., Frogoudaki, A., Vrettou, A.-R., Ikonomidis, I., Pikilidou, M., Kargiotis, O., Voumvourakis, K., Alexandrov, A. W., Alexandrov, A. V., & Tsivgoulis, G. (2017). Blood pressure reduction and secondary stroke prevention: A systematic review and metaregression analysis of randomized clinical trials. *Hypertension (Dallas, Tex. : 1979), 69*(1), 171–179. https://doi.org/10.1161/HYPERTENSIONAHA.116.08485
49. Brainin, M., Feigin, V., Bath, P. M., Collantes, E., Martins, S., Pandian, J., Sacco, R., & Teuschl, Y. (2019). Multi-level community interventions for primary stroke prevention: A

conceptual approach by the World Stroke Organization. *International Journal of Stroke, 14*(8), 818–825. https://doi.org/10.1177/1747493019873706
50. Tinone, G., Hoshino, M., Lucato, L., & Comerlatti, L. R. (2022). Anticoagulation and stroke. *Arquivos de Neuro-Psiquiatria, 80*(5 Suppl 1), 72–79. https://doi.org/10.1590/0004-282X-ANP-2022-S132
51. Kapil, N., Datta, Y. H., Alakbarova, N., Bershad, E., Selim, M., Liebeskind, D. S., Bachour, O., Rao, G. H. R., & Divani, A. A. (2017). Antiplatelet and anticoagulant therapies for prevention of ischemic stroke. *Clinical and Applied Thrombosis/Hemostasis: Official Journal of the International Academy of Clinical and Applied Thrombosis/Hemostasis, 23*(4), 301–318. https://doi.org/10.1177/1076029616660762
52. Prasad, K., Siemieniuk, R., Hao, Q., Guyatt, G., O'Donnell, M., Lytvyn, L., Heen, A. F., Agoritsas, T., Vandvik, P. O., Gorthi, S. P., Fisch, L., Jusufovic, M., Muller, J., Booth, B., Horton, E., Fraiz, A., Siemieniuk, J., Fobuzi, A. C., Katragunta, N., & Rochwerg, B. (2018). Dual antiplatelet therapy with aspirin and clopidogrel for acute high risk transient ischaemic attack and minor ischaemic stroke: A clinical practice guideline. *BMJ, 363*, k5130. https://doi.org/10.1136/bmj.k5130
53. Shen, L. H., Li, Y., Chen, J., Zacharek, A., Gao, Q., Kapke, A., Lu, M., Raginski, K., Vanguri, P., Smith, A., & Chopp, M. (2007). Therapeutic benefit of bone marrow stromal cells administered 1 month after stroke. *Journal of Cerebral Blood Flow and Metabolism: Official Journal of the International Society of Cerebral Blood Flow and Metabolism, 27*(1), 6–13. https://doi.org/10.1038/sj.jcbfm.9600311
54. George, P. M., & Steinberg, G. K. (2015). Novel stroke therapeutics: Unraveling stroke pathophysiology and its impact on clinical treatments. *Neuron, 87*(2), 297–309. https://doi.org/10.1016/j.neuron.2015.05.041
55. Rerkasem, A., Orrapin, S., Howard, D. P., & Rerkasem, K. (2020). Carotid endarterectomy for symptomatic carotid stenosis. *Cochrane Database of Systematic Reviews, 9*(9), CD001081. https://doi.org/10.1002/14651858.CD001081.pub4
56. Chaturvedi, S., Bruno, A., Feasby, T., Holloway, R., Benavente, O., Cohen, S. N., Cote, R., Hess, D., Saver, J., Spence, J. D., Stern, B., & Wilterdink, J. (2006). Carotid endarterectomy—An evidence-based review: report of the Therapeutics and Technology Assessment Subcommittee of the American Academy of Neurology. *American Journal of Ophthalmology, 141*(1), 238–239. https://doi.org/10.1016/j.ajo.2005.11.031
57. Fruhwirth, V., Berger, L., Gattringer, T., Fandler-Höfler, S., Kneihsl, M., Schwerdtfeger, A., Weiss, E. M., Enzinger, C., & Pinter, D. (2022). Evaluation of a newly developed smartphone app for risk factor management in young patients with ischemic stroke: A pilot study. *Frontiers in Neurology, 12*, 791545. https://doi.org/10.3389/fneur.2021.791545
58. Noch, E. K., Pham, D., Kitago, T., Wuennemann, M., Wortman-Jutt, S., & Falo, M. C. (2023). Qualitative feasibility study of the mobile app Destroke for clinical stroke monitoring based on the NIH stroke scale. *Heliyon, 9*(8), e18393. https://doi.org/10.1016/j.heliyon.2023.e18393
59. Lövblad, K.-O., Bouchez, L., Altrichter, S., Ratib, O., Machi, P., Vargas, M. I., & Sztajzel, R. (2019). The role of advanced neuroimaging techniques in ischemic stroke prevention. *Clinical and Translational Neuroscience, 3*(2), 2514183X1988144. https://doi.org/10.1177/2514183x19881446
60. Ekkert, A., Šliachtenko, A., Grigaitė, J., Burnytė, B., Utkus, A., & Jatužis, D. (2022). Ischemic stroke genetics: What is new and how to apply it in clinical practice? *Genes, 13*(1), 48. https://doi.org/10.3390/genes13010048
61. Ramsay, A. M., Ledger, J., Tomini, S., Hall, C., Hargroves, D., Hunter, P., Payne, S., Mehta, R., Simister, R., Tayo, F., & Fulop, N. (2022). Prehospital video triage of potential stroke patients in North Central London and East Kent: Rapid mixed-methods service evaluation. *Health and Social Care Delivery Research, 10*(26), 1–114. https://doi.org/10.3310/iqzn1725
62. Araki, H., Uchida, K., Yoshimura, S., Kurisu, K., Shime, N., Sakamoto, S., Aoki, S., Ichinose, N., Kajihara, Y., Tominaga, A., Naka, H., Mizoue, T., Sumida, M., Hirotsune, N., Nomura, E., Matsushige, T., Kanazawa, J., Kato, Y., Kawamoto, Y., Kuroki, K., & Morimoto, T. (2021). Effect of region-wide use of prehospital stroke triage scale on management of patients with

acute stroke. *Journal of Neurointerventional Surgery, 14*, 677–682. https://doi.org/10.1136/neurintsurg-2021-017863
63. Behrouz, R., & Birnbaum, L. (2019). *Complications of acute stroke: A concise guide to prevention, recognition, and management*. Springer.
64. Donnan, G. A., Davis, S. M., Parsons, M., Ma, H., Dewey, H. M., & Howells, D. W. (2011). How to make better use of thrombolytic therapy in acute ischemic stroke. *Nature Reviews Neurology, 7*(7), 400–409. https://doi.org/10.1038/nrneurol.2011.89
65. Hacke, W., Kaste, M., Bluhmki, E., Brozman, M., Dávalos, A., Guidetti, D., Larrue, V., Lees, K. R., Medeghri, Z., Machnig, T., Schneider, D., Von Kummer, R., Wahlgren, N., & Toni, D. (2008). Thrombolysis with alteplase 3 to 4.5 hours after acute ischemic stroke. *New England Journal of Medicine, 359*(13), 1317–1329. https://doi.org/10.1056/nejmoa0804656
66. Veerbeek, J. M., Van Wegen, E. E. H., Van Peppen, R., Van Der Wees, P. J., Hendriks, E., Rietberg, M. B., & Kwakkel, G. (2014). What is the evidence for physical therapy poststroke? A systematic review and meta-analysis. *PLoS One, 9*(2), e87987. https://doi.org/10.1371/journal.pone.0087987
67. Otani, N., & Yoshino, A. (2022). Ischemic stroke revascularization. *Advances and Technical Standards in Neurosurgery, 44*, 79–96. https://doi.org/10.1007/978-3-030-87649-4_4
68. Jiménez-Dinamarca, I., Prado, Y., Tapía, P., Gatica, S., Alt, C., Lin, C. P., Reyes-Martínez, C., Feijóo, C. G., Aravena, C., González-Canacer, A., Correa, S., Varela, D., Cabello-Verrugio, C., & Simón, F. (2023). Disseminated intravascular coagulation phenotype is regulated by the TRPM7 channel during sepsis. *Biological Research, 56*(1), 8. https://doi.org/10.1186/s40659-023-00419-4
69. Pan, Y., & Shi, G. (2021). Silver jubilee of stroke thrombolysis with alteplase: Evolution of the therapeutic window. *Frontiers in Neurology, 12*, 593887. https://doi.org/10.3389/fneur.2021.593887
70. Capodanno, D., Mehran, R., Valgimigli, M., Baber, U., Windecker, S., Vranckx, P., Dangas, G., Rollini, F., Kimura, T., Collet, J. P., Gibson, C. M., Steg, P. G., Lopes, R. D., Gwon, H. C., Storey, R. F., Franchi, F., Bhatt, D. L., Serruys, P. W., & Angiolillo, D. J. (2018). Aspirin-free strategies in cardiovascular disease and cardioembolic stroke prevention. *Nature Reviews Cardiology, 15*(8), 480–496. https://doi.org/10.1038/s41569-018-0049-1
71. Chang, T.-Y., Liao, J.-N., Chao, T.-F., Vicera, J. J., Lin, C.-Y., Tuan, T.-C., Lin, Y.-J., Chang, S.-L., Lo, L.-W., Hu, Y.-F., Chung, F.-P., & Chen, S.-A. (2018). Oral anticoagulant use for stroke prevention in atrial fibrillation patients with difficult scenarios. *IJC Heart & Vasculature, 20*, 56–62. https://doi.org/10.1016/j.ijcha.2018.08.003
72. Shah, J., Liu, S., & Yu, W. (2022). Contemporary antiplatelet therapy for secondary stroke prevention: a narrative review of current literature and guidelines. *Stroke and Vascular Neurology, 7*(5), 406–414. https://doi.org/10.1136/svn-2021-001166
73. Lee, M., Cheng, C.-Y., Wu, Y.-L., Lee, J.-D., Hsu, C.-Y., & Ovbiagele, B. (2022). Association between intensity of low-density lipoprotein cholesterol reduction with statin-based therapies and secondary stroke prevention. *JAMA Neurology, 79*(4), 349. https://doi.org/10.1001/jamaneurol.2021.5578
74. Liu, S., Feng, X., Jin, R., & Li, G. (2018). Tissue plasminogen activator-based nanothrombolysis for ischemic stroke. *Expert Opinion on Drug Delivery, 15*(2), 173–184. https://doi.org/10.1080/17425247.2018.1384464.
75. Imahori, T., Koyama, J., Tanaka, K., Okamura, Y., Arai, A., Iwahashi, H., Mori, T., Yokote, A., Matsushima, K., Matsui, D., Kobayashi, M., Hosoda, K., & Kohmura, E. (2020). Impact of introducing endovascular treatment on acute ischemic stroke outcomes: A shift from an era of medical management to thrombectomy in Japan. *Heliyon, 6*(5), e03945. https://doi.org/10.1016/j.heliyon.2020.e03945
76. Turner, J. M., & Kodali, R. (2020). Should angiotensin-converting enzyme inhibitors ever be used for the management of hypertension? *Current Cardiology Reports, 22*(9), 1–8. https://doi.org/10.1007/s11886-020-01352-8
77. Ohsawa, Y., Hagiwara, H., Nishimatsu, S. I., Hirakawa, A., Kamimura, N., Ohtsubo, H., Fukai, Y., Murakami, T., Koga, Y., Goto, Y. I., Ohta, S., & Sunada, Y. (2019). Taurine sup-

plementation for prevention of stroke-like episodes in MELAS: A multicentre, open-label, 52-week phase III trial. *Journal of Neurology, Neurosurgery and Psychiatry, 90*(5), 529–536. https://doi.org/10.1136/jnnp-2018-317964

78. Berkhemer, O. A., Fransen, P. S., Beumer, D., Van Den Berg, L. A., Lingsma, H. F., Yoo, A. J., Schonewille, W. J., Vos, J. A., Nederkoorn, P. J., Wermer, M. J., Van Walderveen, M. A., Staals, J., Hofmeijer, J., Van Oostayen, J. A., Nijeholt, G. J. L. À., Boiten, J., Brouwer, P. A., Emmer, B. J., De Bruijn, S. F., & Dippel, D. W. (2014). A randomized trial of intraarterial treatment for acute ischemic stroke. *New England Journal of Medicine, 372*(1), 11–20. https://doi.org/10.1056/nejmoa1411587
79. Derex, L., & Cho, T. H. (2017). Mechanical thrombectomy in acute ischemic stroke. *Revue Neurologique, 173*(3), 106–113. https://doi.org/10.1016/j.neurol.2016.06.008
80. Smith, E. E., & Schwamm, L. H. (2015). Endovascular clot retrieval therapy. *Stroke, 46*(6), 1462–1467. https://doi.org/10.1161/strokeaha.115.008385
81. Ribó, M., Molina, C. A., Jankowitz, B. T., Tomasello, A., Zaidi, S., Jumaa, M. A., Coscojuela, P., Oakley, J., Álvarez-Sabín, J., & Jovin, T. G. (2013). Stentrievers versus other endovascular treatment methods for acute stroke: Comparison of procedural results and their relationship to outcomes. *Journal of NeuroInterventional Surgery, 6*(4), 265–269. https://doi.org/10.1136/neurintsurg-2013-010748
82. Murie-Fernández, M., & Marzo, M. M. (2020). Predictors of neurological and functional recovery in patients with moderate to severe ischemic stroke: The EPICA Study. *Stroke Research and Treatment, 2020*, 1419720. https://doi.org/10.1155/2020/1419720
83. Weinstein, C. J., Stein, J., Arena, R., Bates, B., Cherney, L. R., Cramer, S. C., DeRuyter, F., Eng, J. J., Fisher, B. E., Harvey, R. L., Lang, C. E., MacKay-Lyons, M., Ottenbacher, K. J., Pugh, S., Reeves, M. J., Richards, L., Stiers, W., & Zorowitz, R. D. (2016). Guidelines for adult stroke rehabilitation and recovery. *Stroke, 47*(6). https://doi.org/10.1161/str.0000000000000098
84. Hartford, W., Lear, S. A., & Nimmon, L. (2019). Stroke survivors' experiences of team support along their recovery continuum. *BMC Health Services Research, 19*(1), 723. https://doi.org/10.1186/s12913-019-4533-z
85. Teasell, R., Salbach, N. M., Foley, N., Mountain, A., Cameron, J. I., De Jong, A., Acerra, N. E., Bastasi, D., Carter, S. L., Fung, J., Halabi, M. L., Iruthayarajah, J., Harris, J. E., Kim, E., Noland, A., Pooyania, S., Rochette, A., Stack, B. D., Symcox, E., et al. (2020). Canadian stroke best practice recommendations: Rehabilitation, recovery, and community participation following stroke. Part One: Rehabilitation and recovery following stroke; 6th Edition Update 2019. *International Journal of Stroke, 15*(7), 763–788. https://doi.org/10.1177/1747493019897843
86. Ghaffar, S. H., Burman, M., & Braimah, N. (2020). Pathways to circular construction: An integrated management of construction and demolition waste for resource recovery. *Journal of Cleaner Production, 244*, 118710. https://doi.org/10.1016/j.jclepro.2019.118710
87. Kollen, B. J., Lennon, S., Lyons, B. E., Wheatley-Smith, L., Scheper, M., Buurke, J. H., Halfens, J. H., Geurts, A. C. H., & Kwakkel, G. (2009). The effectiveness of the Bobath concept in stroke rehabilitation. *Stroke, 40*(4), e89–e97. https://doi.org/10.1161/strokeaha.108.533828
88. Legg, L., Lewis, S. R., Schofield-Robinson, O. J., Drummond, A., & Langhorne, P. (2017). Occupational therapy for adults with problems in activities of daily living after stroke. *The Cochrane Library, 2017*(7), CD003585. https://doi.org/10.1002/14651858.cd003585.pub3
89. Chang, W. H., & Kim, Y. (2013). Robot-assisted therapy in stroke rehabilitation. *Journal of Stroke, 15*(3), 174. https://doi.org/10.5853/jos.2013.15.3.174
90. Chrostek, M. R., Fellows, E. G., Crane, A. T., Grande, A. W., & Low, W. C. (2019). Efficacy of stem cell-based therapies for stroke. *Brain Research, 1722*, 146362. https://doi.org/10.1016/j.brainres.2019.146362
91. Teleanu, D. M., Chircov, C., Grumezescu, A. M., Volceanov, A., & Teleanu, R. I. (2018). Blood-brain delivery methods using nanotechnology. *Pharmaceutics, 10*(4), 269. https://doi.org/10.3390/pharmaceutics10040269

Paediatric/Adolescent Stroke

6

Muhamad Azamin Anuar

Abstract

Childhood stroke is not uncommon. Data and research focussing on childhood stroke remains scarce and needs further study to improve access to health care and investigations including acute interventions and rehabilitation model.

This chapter covers neonatal and paediatric strokes and outline aetiologies, investigations and interventions and rehabilitation management. Wide ranges of pathogenesis brought in this chapter to explain how stroke in neonates and paediatric cohorts develop and pathways in managing them.

Keywords

Childhood stroke · Ischaemic stroke · Haemorrhagic stroke · Rehabilitation

6.1 Childhood and Adolescent Stroke

6.1.1 Introduction

There are different ways to characterising paediatric stroke. One prevalent method of classification is based on age. Stroke incidents ranging from 28 weeks of pregnancy to 28 days post-delivery are generally categorised as perinatal or neonatal

M. A. Anuar (✉)
Department of Paediatrics, Kulliyyah of Medicine, International Islamic University Malaysia, Kuantan, Pahang, Malaysia
e-mail: azamin@iium.edu.my

W. Mohamed, M. B. Mat-Nor (eds.), *Nutrition and Stroke*, Nutritional Neurosciences, https://doi.org/10.1007/978-981-97-6818-9_6

strokes. Childhood stroke is classified as a cerebrovascular incident occurring between 28 days and 18 years of age. Stroke's management and intervention varies depending on the resources available in the countries and further advancement is required to be tailored proportionately among developing to developed countries. This chapter outlines fundamental areas in perinatal/neonatal and childhood strokes.

6.2 Neonatal Stroke: Unravelling the Challenges and Promising Paths to Recovery

6.2.1 Introduction

Neonatal stroke is a relatively rare neurological condition in infants from birth to 28 days of life. It is a serious condition affecting pre-term and term newborns. Despite its scarcity, neonatal stroke can have profound and long-lasting effects on a child's development, compelling prompt recognition and intervention. Unlike stroke in adults and older children, neonatal strokes often present unique diagnostic challenges, as symptoms may be subtle or ascribed to other medical conditions that are more common in newborns. Complete understanding and ability to discriminate characteristics of neonatal stroke, including its risk factors, clinical manifestations, and potential outcomes, is crucial for healthcare professionals to provide timely and precision-focused treatment.

Neonatal stroke stands as a significant contributor to hemiparetic cerebral palsy. It refers to a cerebrovascular incident caused by disrupted blood flow, commonly observed in infants born near or at term within the initial 28 days of life. Confirmation typically requires neuroimaging or pathological studies [1]. The estimated incidence ranging from 25 to 47 in 100,000 live births, which is considered to be relatively uncommon occurrence [1–3]. Its rarity and non-specific presentations make it less well-studied compared to stroke in adults, the consequences for affected infants can be devastating and enduring. Neonatal stroke may present in two primary forms: ischaemic strokes, triggered by a disruption in blood flow to the brain, or haemorrhagic strokes, stemming from bleeding within the brain. The prevalence may be underestimated due to challenges in diagnosis, given the difficulty in recognising symptoms and the fact that some cases may be asymptomatic may be left undiagnosed.

6.2.2 Aetiology and Pathogenesis

The aetiology of neonatal strokes vary and can be attributed to multifactorial or combination or prenatal, perinatal and postnatal factors. Table 6.1 presents a modified depiction of these risks [4]. Due to the broad spectrum of potential causes, it is crucial to adopt a comprehensive strategy for diagnosis and management. This approach ensures the identification of specific underlying factors and the customisation of suitable interventions for each unique case.

Table 6.1 Various risk factors for neonatal stroke

Phase	Prenatal	Perinatal	Postnatal
	Maternal infertility	PROM	Hypoglycaemia
	Pre-eclampsia	Maternal fever / infection	Hypoxic events
	Smoking	Asphyxia	Infection
	IUGR	Resuscitation efforts	Congenital heart disease
	Infection	5-min APGAR <7	Arteriopathies
	Placental infarcts	Shoulder dystocia	Metabolic disease
		Chorioamnionitis	Hypercoagulopathies

While numerous risk factors have been associated to neonatal stroke, the precise pathophysiological mechanisms underlying most cases remain poorly understood. An international conference of experts reached a consensus that neonatal stroke is predominantly characterised as perinatal cerebral injury of ischaemic origin. They defined it as a collection of diverse conditions marked by localised interrupted cerebral blood flow, typically stemming from cerebral arterial or venous thrombosis or embolism, neonatal strokes occur between 20 weeks of gestation and the 28th postnatal day. Diagnosis is typically confirmed through neuroimaging or neuropathological studies [3]. Perinatal arterial ischaemic stroke (PAIS) and cerebral sinovenous thrombosis (CSVT) are two most prevalent neonatal stroke's subtypes (as illustrated in Table 6.2).

PAIS can affect both term and preterm newborns. There is a suggestion that the incidence of PAIS is higher in preterm infants, possibly due to the routine screening practice of cranial ultrasound in preterm infants [5]. The general hypothesis is that thrombi cause arterial occlusion through a patent foramen ovale, allowing them to travel from the placenta or venous circulation.

Additionally, the presence of prothrombotic factors has linked to poor neurological outcome. Studies have demonstrated that prothrombotic factors were observed in over half of the population studied, more frequently in cases than in age- and sex-matched controls [6].

Table 6.1 indicates that infants with PAIS more commonly experience complicated deliveries, instrumental or emergency caesarean sections, low Apgar scores, and hypoglycaemia [7, 8]. Congenital heart disease highly known to pose a significant risk factor for PAIS. The abnormal anatomy associated with congenital heart disease makes it easier for cardiac thrombi to form. Furthermore, other risk factors include infections such as meningitis, sepsis, central blood vessels catheterisation and Hypercoagulopathies, all of which heighten the risk of thrombus formation.

In contrast, CSVT exhibits a lower incidence compared to PAIS, ranging from 0.6 to 12 cases per 100,000 live births [9, 10]. The clinical presentation of CSVT is non-specific, leading to a low threshold for performing neuroimaging and a lack of clinician awareness. Most of infants with CSVT present after 48 h of life, often later in the first week. Symptoms may include seizures, encephalopathy, lethargy, irritability, jitteriness, poor feeding, respiratory distress and apnoea. Seizures account for two-thirds of cases and can manifest either early or late in the first week of life [11].

Table 6.2 Clinical characteristics for PAIS, CSVT, PVHI and HS

	PAIS	CSVT	PVHI	HS
Phase	Term and preterm infants	Term and preterm infants	Predominantly preterm infants	Late preterm or term infants
Incidence	1 in 4000 live births	0.6–12 cases in 100,000 live births	Approximately 25% in extreme preterm infants	Common in <32 weeks gestation. Estimated 0.17 in 1000 live births in term infants
Common presentation	Hemiclonic seizure is most common. Encephalopathy, hypotonia, respiratory and feeding difficulties are also common	Non-specific, often with seizures and encephalopathy. Apnoea is more common in preterm	Non-specific. Commonly found during routine cranial ultrasound	Location often dictates the clinical course which include seizures, fever, poor feeding, apnoea, and respiratory distress
Pathogenesis	Birth trauma and delivery-related anoxia are primary pathogeneses as well as inflammation, infection and placental insufficiency	Abnormal venous blood flow due to risk factor resulting in thrombosis	Impaired venous drainage of the medullary veins in the periventricular white matter.	Universally unclear but postulation of altered haemodynamic especially in congenital heart disease has been suggested
Risk factors /aetiology	Complicated and high risk deliveries, maternal infections and congenital heart disease	Maternal and perinatal risk factors	Very preterm, factor V Leiden, patent ductus arteriosus, thrombocytopenia	Congenital heart disease, fetal distress and assisted delivery
Preferred imaging	MRI with MR angiography	Magnetic resonance venography (MRV) or CT cerebral venography (CTV)	Cranial ultrasound with subsequent MRI imaging with Diffusion Tensor Imaging (DTI)	CT is sufficient for acute phase. MRI is preferred to evaluate extension of injury and prognostication

CSVT is diagnosed when impaired or absent venous blood flow occurs in one of the cerebral sinuses. This condition leads to increased venous pressure, which result in elevated capillary hydrostatic pressure, vasogenic oedema, and eventually haemorrhagic infarction. The specific regions affected depend on the location of the sinuses involved.

Several risk factors are associated with CSVT. Pre-eclampsia is reported as the most frequent maternal risk factor associated with CSVT, involving a hypercoagulopathy state [11, 12]. Other maternal risk factors include chorioamnionitis and gestational diabetes. Gestational or maternal diabetes leads to increased haematocrit level, often resulting in a hypercoagulant state in cerebral sinuses.

Intrapartum factors contributing to CSVT include complicated deliveries such as shoulder dystocia, instrumental-assisted delivery, meconium aspiration, and even intubation at birth [9–13]. CSVT has also been noted in cases of neonatal sepsis and meningitis, congenital heart defects, and infants requiring extracorporeal membrane oxygenation (ECMO) due to internal jugular catheterisation [12, 13]. Infants with prothrombotic factors such as factor V Leiden and MTHFR mutations have been identified as being at a higher risk for CSVT [9, 12].

The last subtypes of neonatal stroke spectrum are periventricular haemorrhagic infarction (PVHI) and haemorrhagic stroke (HS).

PVHI, also referred to as periventricular venous infarction, arises from impaired drainage of the medullary veins in the white matter. The incidence of PVHI is estimated to be between 4% and 5% among infants born under 32 weeks' gestation [14, 15]. Known risk factors for PVHI are akin to those for intraventricular haemorrhage and include inadequate antenatal steroid administration, low gestational age and birth weight, Apgar score below 7 at 5 min, intubation, and significant resuscitation at birth.

Meanwhile HS, even though it is less common in late preterm or term infants, defines as intraventricular, intraparenchymal or subarachnoid haemorrhage. The precise overall incidence of HS remains unknown; however, it has been estimated to occur at a rate of 0.17 in 1000 live births [16, 17]. The clinical presentation often dictated by the haemorrhagic location but overall clinical symptoms are non-specific like other neonatal stroke disorders. Determining the risk factors for HS can be challenging, but some studies have reported an association between assisted deliveries, fetal distress, and post-maturity [18–20].

6.2.3 Pathway for Diagnosis

The majority of infants with neonatal stroke present with seizures within the first 24 h of life, and these seizures can often be the sole initial symptom [21, 22]. While in preterm neonates, 83% presented as apnoea, however, apnoea is one of the common complication as a result of prematurity [23, 24]. Hemiclonic seizures in newborns are highly suspicious due to stroke until proven otherwise. Hence, urgent investigations are mandatory to elicit the cause.

In infants who present with seizures, it is crucial to perform laboratory studies to rule out other pathologies, especially infection, electrolyte imbalances, hypoglycaemia, coagulopathy, and metabolic syndromes. For PAIS, an extensive coagulation profile should be evaluated, including prothrombotic factors [25].

6.2.3.1 Electroencephalogram

An electroencephalogram (EEG), whether it is amplitude-integrated EEG (aEEG) or full neonatal montages EEG, should be conducted in any infant presenting with neonatal seizures. EEG can assist in detecting both clinical and electrographic seizures, as well as in localising the origin of the seizures. In cases of PAIS and CSVT, EEG may reveal persistently abnormal background activity on the affected side due to cerebral oedema following hypoperfusion events.

6.2.3.2 Cranial Ultrasound

Cranial ultrasound is a first choice of imaging method to look for any intracranial bleed, structural abnormality and cerebral hypoxic injury in neonates. It is a quick tool and can be repeated multiple times without additional radiation risk. A series of cranial ultrasound scans is recommended in a high risk group of infants or symptomatic infants. A study that analysed serial cranial ultrasound scans of 47 infants with PAIS found that the sensitivity of detecting the condition improved significantly over time. Initially, within the first 3 days of symptoms onset, the sensitivity was 68%. However, as the infants were monitored, the sensitivity increased to 87% between days 4 and 10 [26]. This demonstrates that serial scanning over time can enhance the accuracy of PAIS detection.

Posterior cerebral strokes are more challenging to detect due to difficulty in getting a good sonographic window via posterior fontanelle. This will need another imaging modality to have a better outlook and structural coverage.

6.2.3.3 Magnetic Resonance Imaging

Magnetic resonance imaging (MRI) is considered the gold standard and the most sensitive imaging modality for detecting of PAIS. CSVT, HS and PVHI can be detected using MRI as well and computed tomography (CT) scan. However, CT scan is less preferred due to its radiation exposure. It is still useful in the event of acute hydrocephalus or raised intracranial pressure that likely to require neurosurgical intervention.

The main sequences utilised for imaging are typically T1-weighted (T1), T2-weighted (T2) and diffusion-weighted imaging (DWI). To comprehend the fundamental principles of MRI sequences, one needs specialised training to fully grasp the physics and their practical applications. MRI revolves around the magnetisation properties of atomic nuclei, which become aligned when exposed to external radiofrequency energy. Subsequently, as these nuclei return to their resting or original alignment during relaxation processes, the resulting signal intensities are captured and converted into grayscale images for analysis. Different tissue relaxation times produce different sequences. T1 (longitudinal relaxation time) and T2 (transverse relaxation time) are very good sequences and commonly facilitate evaluation of the injury on the cortex and white due to changes in the signal intensity over time. On the other hand, DWI is specifically engineered to identify the random movements of water molecules. It holds significant importance in diagnosing PAIS because of its heightened sensitivity in detecting ischaemic lesions during the acute phase, particularly in the initial days as signs on T1 and T2 sequences might not be as apparent. This due to water movement becomes restricted intracellularly during acute stroke resulting in an extremely bright signal on DWI sequence.

MRI, with its capability to examine deeper structures like the basal ganglia, internal capsules, and thalamus in details, allows for the assessment of motor outcome prediction, particularly in the development of spastic cerebral palsy [27, 28]. Furthermore, MRI can offer further insights following PAIS through magnetic resonance angiography (MRA), enabling the evaluation of anatomical details regarding cerebral arteries patency. Special three-dimensional reconstructions can be employed to study the anatomical variations or occlusion of cerebral blood vessels.

6.3 Paediatrics and Adolescent Stroke: An Emerging Condition in Children

6.3.1 Introduction

Stroke in children and adolescents is a rare occurrence. Most epidemiological studies lack sufficient statistical power to thoroughly assess sociodemographic and socioeconomic differences. Stroke in children may be under-recognised due to inconsistencies in diagnosis and variations in access to care.

The reported combined incidence of both ischaemic and haemorrhagic stroke in children, although widely documented in literature, ranges from an estimated 1.3 to 13 children per 100,000 under 18 years old [29–32]. Males generally have a higher prevalence and incidence of strokes compared to females, while black children face a greater risk than Caucasian and Asian children [33]. The mortality rate for strokes in children is approximately 10–25% [33].

In developing countries, haemorrhagic stroke occurs approximately twice as frequently as ischaemic stroke, and it is associated with a higher mortality rate. Historically, the prevalence of ischaemic stroke has been up to five times higher in developing countries compared to developed ones. Approximately 25% of children who experience a stroke will have a recurrent stroke [33].

Stroke in children and adolescents constitutes a heterogeneous entity, characterised by varying causes and presentations. This heterogeneity poses challenges in achieving accurate and timely diagnosis, which consequently leads to uncertainty regarding the optimal approach for acute management. In response to these challenges, the Royal College of Paediatrics and Child Health developed guidelines specifically for paediatric stroke in 2017. These guidelines aim to address the complexities of paediatric stroke, provide clear recommendations for diagnosis and treatment, and establish a standardised framework for managing stroke in children and adolescents [34]. Their ultimate goal in to improve outcomes and ensure consistent care across healthcare settings.

While stroke is relatively uncommon in children, it remains a significant cause of neurological disability in paediatrics. There is currently no universally accepted acute treatment for paediatric stroke, especially concerning thrombectomy. However, since children may face many years living with disability resulting from stroke, timely interventions aimed at improving outcomes can have lasting benefits.

Numerous paediatric centres have established clinical pathways for managing acute stroke [35, 36]. Diagnosis is frequently delayed, with estimates suggesting that it can take up to 24 h from the onset of symptoms to reach a diagnosis. This delay underscores the challenges in promptly identifying and addressing paediatric stroke cases, highlighting the need for improved awareness and streamlined diagnostic processes in clinical settings [37–39].

Among children who survive stroke, approximately 50–80% will experience permanent neurological deficits, with hemiparesis or hemiplegia being particularly common. The long-term complications of paediatric stroke impose a sizeable direct cost, as well as mental and social burdens on families and society as a whole.

6.3.1.1 Clinical Presentation

The signs and symptoms of classic acute stroke in children may resemble those observed in adults, yet they can vary significantly based on the child's age and may lack specificity. This variability underscores the challenge in identifying strokes in paediatric patients, as symptoms can manifest differently and be less straightforward compared to adults. It highlights the importance of considering a wide range of clinical presentations when evaluating and diagnosing stroke in children. The gross structural anatomy of the childhood brain resembles that of an adult, numerous physiological differences exist, leading to significant heterogeneity in stroke presentation.

Unlike neonatal strokes, where seizures are the most common presenting symptoms, older children more commonly present with focal deficit [33]. The predominant symptoms include hemiparesis and hemi-facial weakness in 67–90% of cases, speech or language disturbance in 20–50%, visual disturbance in 10–15%, and ataxia in 8–10%. Additionally, children may present with non-specific symptoms such as headache in 20–50% of cases and altered mental status in 17–38%. Seizures at stroke onset are more frequent in children compared to adults, occurring 15–25% of cases, especially in those under 6 years of age [40–43].

The clinical presentation of stroke varies depending on the child's age, the clinical setting, and the type of stroke involved. Toddlers and infants typically exhibit non-specific symptoms such as irritability, somnolence, lethargy, feeding difficulties, apnoeic spells and hypotonia [44]. It is noteworthy that approximately 30–50% of children presenting with acute neurological symptoms are diagnosed with condition unrelated to vascular issues [44].

As mentioned earlier, delays in diagnosing paediatric stroke are common. Braun et al. conducted a study on diagnostic challenges in paediatric ischaemic stroke and found that correct diagnoses were missed in about 40% of cases, with delays in diagnosis ranging from 15 h to 3 months from the initial presentation [45]. Therefore, it is crucial to be aware of the diverse clinical presentations that may suggest stroke in children. This awareness helps ensure that stroke mimics do not hinder prompt and accurate diagnosis, thereby facilitating timely intervention and management.

6.3.1.2 Risk Factors

There is no single risk factor that predominates paediatric stroke. Diagram 6.1 illuminates cardiac causes, infections, coagulopathies and arteriopathies are significant risk factors. However, other risk factors such as haematological, renal, autoimmune, metabolic diseases, child abuse and head traumas create a heterogeneous patient population.

In adults, ischaemic stroke typically arises from factors like ruptured atherosclerotic plaque or through thromboembolism. However, in children, the leading risk factor for ischaemic events, as highlighted by Mallick et al., is acute systemic disorders such as sepsis, dehydration, hypoxia, or acidosis, which are responsible for 31% of the cases reviewed [46]. Other significant contributors include arteriopathy, chronic systemic disease such as sickle cell disease, and cardiac causes, collectively accounting for between 23–29% of cases studied [46].

Cerebral arteriopathy refers to a diverse range of conditions affecting the arteries in the brain. These include disorders like dissection, vasculitis, Moyamoya disease, varicella vasculitis, complications resulting from radiation therapy, and syndromic conditions such as Sturge-Weber syndrome, arteriovenous malformations and neurofibromatosis type 1. Each of these conditions involves abnormalities or damage to the cerebral arteries, which can lead to various neurological symptoms and complications. Understanding these diverse manifestations is crucial for accurate diagnosis and effective management of cerebral arteriopathy in clinical practice.

Congenital heart disease represents a significant risk factor for stroke, with studies indicating a 19-fold increase in the risk of ischaemic or haemorrhagic stroke in children [47]. This heightened stroke risk is attributed to various factors associated with arrhythmia, anticoagulation therapy, cardiac catheterisation procedures, ventricular assist devices and corrective surgery [46]. These interventions are believed to potentially trigger embolic events during endovascular therapy in this particular subgroup of patients.

Indeed, haematological causes such as factor Leiden V, protein C and S deficiency, and elevated lipoprotein A can contribute to ischaemic stroke in children. Vascular anomalies are responsible for up to 90% of childhood haemorrhagic stroke, along with coagulopathy. These factors underline the importance of thorough haematological evaluation in paediatric stroke cases.

6.3.1.3 Cerebellar Stroke

Cerebellar stroke presents a formidable challenge in neurological care due to its potential for devastating consequences and the anatomical constraints within the posterior cranial fossa. This region, bounded by the parietal, sphenoidal, occipital, and temporal bones and extending from the tentorial incisura to the foramen magnum, provides little space for compensatory mechanisms. As a result, any space-occupying lesion, such as a large cerebellar infarct causing progressive oedema or a rapidly expanding haemorrhage, can lead to downward herniation through the foramen magnum or upward herniation through the tentorial incisura. These conditions can result in life-threatening complications.

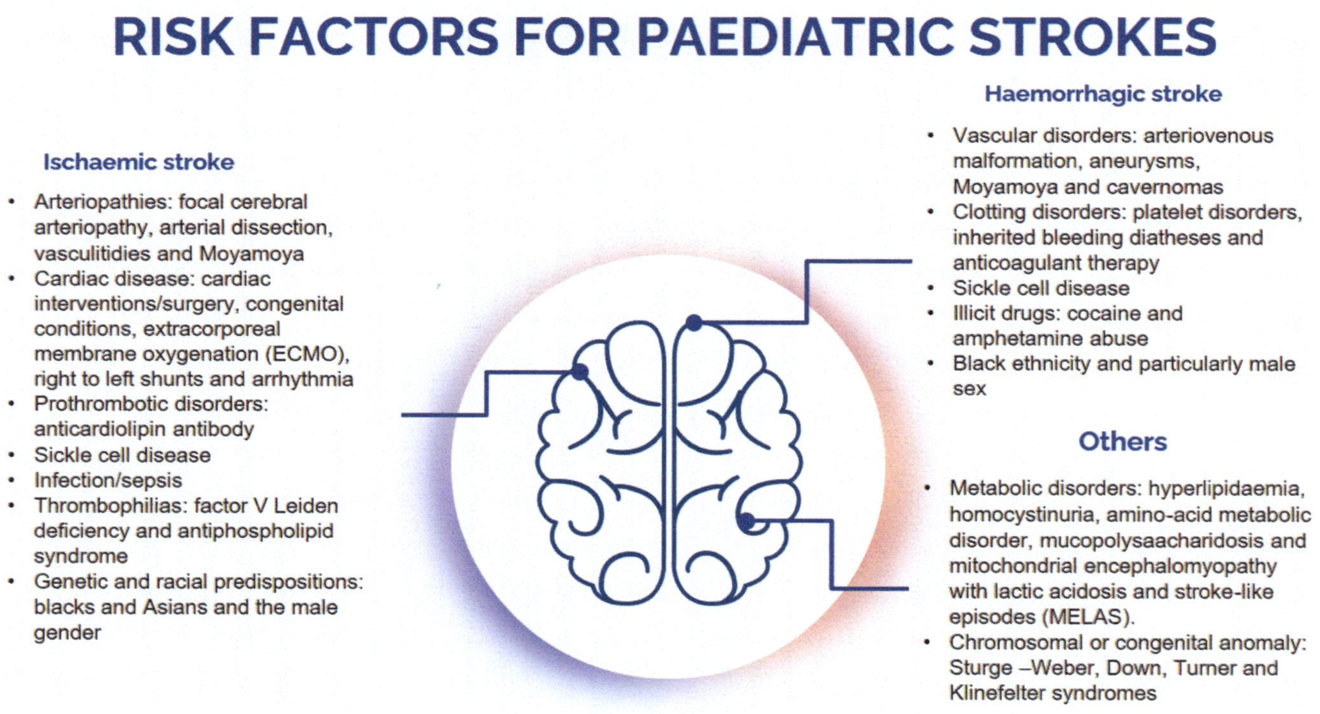

Diagram 6.1 Risk factors for stroke in children (RCPCH guidelines [34] and Mallick et al. 2014 [46])

The cerebellum, essential for coordinating movement and balance, primarily receives its blood supply from the three major cerebellar arteries. These arteries originate from either the vertebral or basilar artery and include the posterior inferior, superior and anterior inferior cerebellar arteries. While childhood stroke accounts for a significant proportion of cases involving posterior circulation (estimated 30–40%), this incidence exceeds that seen in adults [48]. Nevertheless, cerebellar strokes in young individuals often defy easy identification of a specific causes, even after extensive evaluation. Notably, ischaemic strokes affecting the anterior and posterior circulation in adults frequently occur at rest or during routine activities, whereas in young patients, involvement of the posterior circulation may manifest as lateral medullary ischaemia, particularly post- or during strenuous exercise such as swimming.

Symptomatology of cerebellar stroke mirrors that of cerebellar haemorrhages, with gait ataxia and vomiting serving as cardinal findings. Additionally, patients may exhibit veering, leaning, toppling, nystagmus, dysmetria and incoordination. A study conducted by the Swiss Neuropediatric Stroke Registry found that 40% of patients presented with limb weakness, while 30% exhibited gait disturbance and/or speech problems. This highlights the diverse range of clinical manifestations associated with paediatric stroke [49].

During a cerebellar stroke episode, patients commonly experience headache, vomiting, and brainstem or bulbar dysfunction due to cerebellar oedema, potentially leading to acute obstructive hydrocephalus or raised intracranial pressure. Urgent cranial CT scans may not always reveal hypodensity lesions or obliteration of posterior fossa cisterns, underscoring the importance of clinical acumen and a high index of suspicion. In such cases, neurosurgical intervention, such as ventricular drainage, may be essential to relieve life-threatening compression of the brainstem and fourth ventricle.

Unfortunately, promptly diagnosing cerebellar stroke remains a persistent challenge, with only one-third of cases diagnosed within 6 h and over half of cases experiencing delays exceeding 12 h. The elusive nature of cerebellar stroke stems from its propensity to present with subtle or non-specific focal neurologic signs, compounded by the inherent difficulty of conducting a comprehensive clinical [49]. Moreover, the lack of pathognomonic symptoms further complicates the diagnostic process.

In the paediatric cohort, posterior circulation strokes share similar aetiologies with other childhood strokes, including arteriopathies such as dissection, focal transient arteriopathy, and Moyamoya disease [50]. These underlying vascular pathologies emphasise the connotation of a comprehensive assessment and formulate management approach to optimise outcomes in paediatric patients with posterior circulation strokes. Regardless of diagnostic challenges and potential complications, timely intervention and multidisciplinary care are critical for lessening morbidity and mortality in this vulnerable populace.

6.3.1.4 Investigations

Assessment for childhood stroke begins with a foundational step: gathering a detailed medical history. This includes focusing on any underlying conditions, recent trauma, or infections such as varicella within the past year, as well as minor upper respiratory infections. A thorough family history, including a family tree with a special emphasis on first-degree relatives, is perilous to assess the risks of early stroke, coronary heart disease, or venous thrombosis, particularly in the lower limbs.

Moreover, gauging the pregnancy and birth history is essential. This includes determining if the child's mother experienced pre-eclampsia and noting if the child's birth weight falls below third percentile on growth charts, which is often seen in children who have had arterial ischaemic strokes. This comprehensive approach helps to identify potential risks and contributing factors to the child's current neurological condition, aiding in effective diagnosis and management strategies.

In-depth physical examination may give additional information, with specific attention to neurocutaneous stigmata. Skin lesions such as linear sebaceous naevus, port-wine stain or scleroderma "en-coup-de-sabre" may be associated with cerebrovascular disease and should not be disregarded.

First line investigations may include initial clinical assessment commenced with assessing airway patency and closely monitoring vital signs for any abnormalities. Most importantly, obtaining serum glucose levels is critical, as hypoglycaemia can adversely impact cerebral perfusion and lead to altered consciousness. It is imperative to screen for underlying coagulopathy by obtaining full blood count including blood film, renal profiles and coagulation profiles, especially if invasive interventions or thrombolysis are being considered. Also, infection and inflammatory markers should be tested to ensure sepsis can be ruled out. These measures are indispensable for prompt and effective management of cerebrovascular events.

6.3.1.5 Neuroimaging

In countries with limited resources, access to MRI may be challenging. Priority should be given to scheduling imaging promptly after at an event, ideally allowing for intervention if necessary. While CT scans may not reliably show parenchymal infarction within 24 h of an ischaemic stroke, they are essential in emergency situations to rule out haemorrhage and may detect venous sinus thrombosis or other vascular abnormalities.

Ideally, MRI with MRA and MRV, if available, along with appropriate anaesthesia coverage due to the lengthy procedure, is preferred modality for diagnosing the cause of stroke and identifying stroke mimics. The American Heart Association and American Stroke Association endorse brain MRI over CT due to its superior sensitivity, with CT missing diagnoses in 44–83% of cases [51]. The 2017 RCPCH guideline recommends an MRI within 24 h if the initial CT is negative but stoke is still suspected. Diagram 6.2 illustrates number of disorders associated with paediatric stroke. Additionally, further investigation may involve clinical genetics to determine the need for advanced genetic testing such as whole exome or genome sequencing.

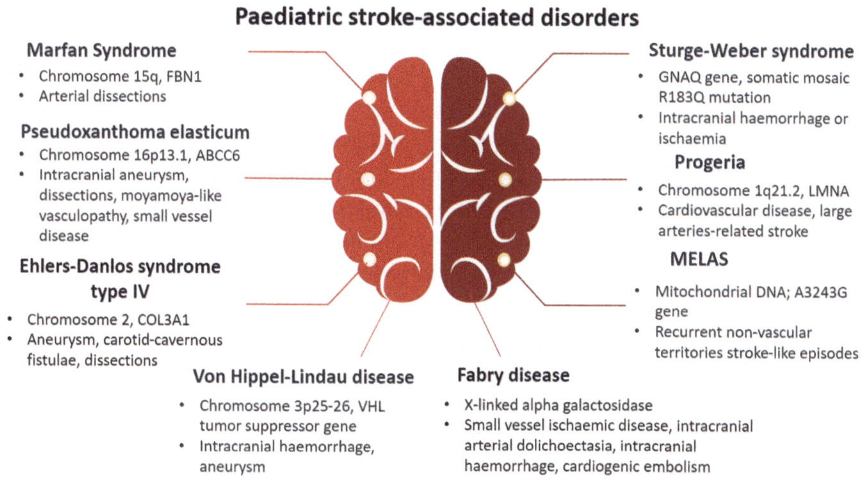

Diagram 6.2 Disorders associated with paediatric stroke [52]

6.3.1.6 Echocardiography and Electrocardiogram

A transthoracic echocardiogram and carotid Doppler are valuable tools for screening cardiac causes for stroke. They can detect poor ventricular function such as cardiomyopathy, as well as identify patent foramen. Significant left-to-right shunts or abnormalities in the aortic valve may explain cerebrovascular events.

While arrhythmias in children are not uncommon. Most do not lead to cerebrovascular events. Supraventricular tachycardia (SVT) is the most frequent childhood arrhythmia, with about half of SVT cases occurring in the first year of life. However, infants may face a higher risk of complications due to lower prevalence and delayed diagnosis. Complications are rare but may include cardiomyopathy, congestive heart failure and even death. Therefore, performing an electrocardiogram should be a routine procedure for any patient presenting with neurological deficits.

6.3.1.7 Delays in Diagnosis

Parents' response times in seeking medical care vary widely, with a median range from 1.7 to 21 h from the onset of symptom, although most seek care within 6 h. The median time for radiological confirmation of diagnosis spans from 15 to 24 h. Children who have strokes while admitted for other illnesses also encounter delays in obtaining radiological confirmation. Major causes of these delays include clinicians' delayed consideration of stroke and challenges in accessing MRI, often necessitating sedation or anaesthesia.

There are various conditions that can mimic stroke in children presenting with acute neurological symptoms in the emergency department. These conditions often trigger the activation of a 'stroke alert' pathway but do not actually involve a stroke. Common stroke mimics include migraine with aura, Bell's palsy, and focal seizures with Todd's paresis. Furthermore, other conditions such as brain tumour, demyelinating disease, cerebellitis, encephalitis, epidural abscess, traumatic brain injury, metabolic disorders, and intoxication can also present with symptoms similar to stroke.

Diagnostic strategies and tools employed in adult to differentiate stroke from stroke mimics have limited effectiveness in children, with a sensitivity of approximately 60% [53, 54]. Although focal deficits are more prevalent in children with actual stroke compared to those with stroke mimics, there is still significant overlap between the two groups. Non-localising symptoms such as headache and altered mental status are equally prevalent in both groups, further complicating the diagnostic process.

6.3.1.8 Management

The 2017 RCPCH Stroke in Childhood guidelines provide clear recommendations for managing acute stroke in children. According to these guidelines, immediate brain imaging is recommended for children and young individuals presenting with [34]:

1. Acute focal neurological deficits
2. Aphasia
3. Reduced level of consciousness (indicated by a Glasgow Coma Scale (GCS) less than 15 or AVPU ('Alert, Voice, Pain, Unresponsive') less than A).

In cases where these symptoms are present, urgent brain imaging should be considered for children displaying [34]:

1. New onset focal seizures,
2. New onset severe headache,
3. Altered mental status, including brief loss of consciousness or changes in behaviour,
4. New onset ataxia, vertigo or dizziness,
5. Sudden onset of neck pain or stiffness,
6. A witnessed acute focal neurological deficit that has resolved.

Hollist et al. provides a simple pathway on managing acute stroke as per Fig. 6.1. At presentation, only half of the non-contrast CT scans showed findings indicative of evolving ischaemic changes [55]. Nevertheless, CT scans remain the preferred modality for acute presentations of focal acute neurology deficits due to their rapidity, ease of preparation, and high sensitivity in identifying acute haemorrhage. This assists clinicians in further refining the differential diagnosis.

During acute management, it is imperative to monitor blood pressure, temperature, oxygen saturation, heart rate and respiratory rate. Regular assessments of GCS and AVPU should be conducted, with any deterioration prompting clinicians to safeguard the airway and initiate cerebral protection protocols. Oral feeding must be withheld until swallow safety has been established. Fluids regime, glucose, and

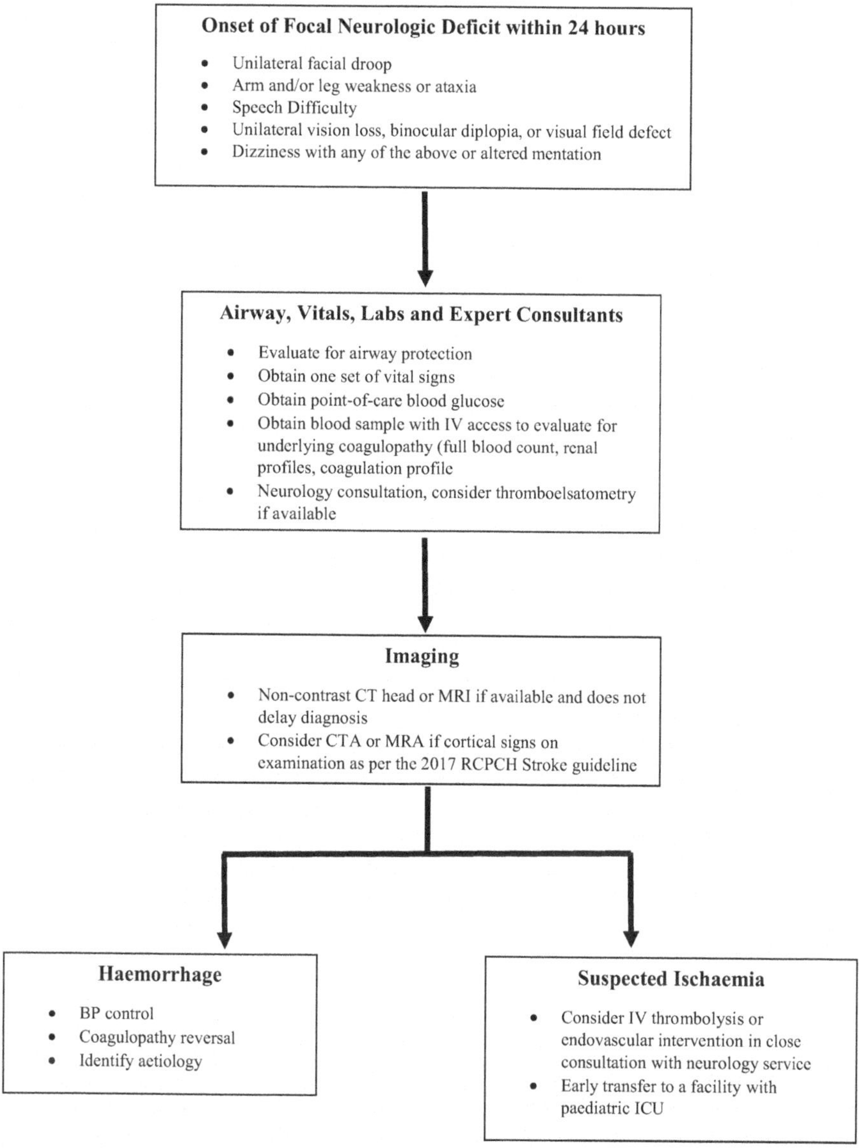

Fig. 6.1 Management of paediatric stroke flowchart [52]

electrolyte levels should be maintained, and hypotension should be promptly addressed. It is essential to maintain oxygen saturation above 92%.

6.3.1.9 Specific Therapy for AIS

For paediatric patients under 18 years old with acute ischaemic stroke (AIS) may be considered for treatment with either intravenous tissue plasminogen activator (tPA) or mechanical thrombectomy [56]. The efficacy of these treatments is directly related to the time delay with earlier initiation associated with better outcomes, and they should be administered within predefined time windows [57, 58]. However, these treatment may not be widely available in all countries as they are considered highly specialised interventions.

Nonetheless, the decision to administer these treatment options in paediatric patients should be individualised and carefully considered for each case. This is because up to half of patients with AIS may recover without neurological deficits even without treatment, and in some cases, the risks associated with intervention may outweigh the potential benefits [51, 57]. While the overall risk of symptomatic intracranial haemorrhage after intravenous tPA is low (approximately 2.6%), it tends to be slightly higher in older children and young adults. Additionally, the risk of intracranial haemorrhage after tPA is approximately 3.48 times higher compared to no treatment, although this does not significantly affect in-hospital mortality [59, 60].

6.3.1.10 Prevention, Identification and Management of Complications

There are potential complications following acute ischaemic stroke and haemorrhagic stroke. The 2017 RCPCH guidelines outline these complications in Diagram 6.3 [34].

6.3.1.11 Rehabilitation Model

The rehabilitation process begins with early intervention strategies that have proven effective in older children. For example, the Goals Activity Motor Enrichment model, which has been assessed in infants, demonstrated improved motor outcomes compared to standard care [61]. Similarly, another study investigated the combination of baby constraint-induced movement therapy and baby massage in infants under 12 months with unilateral cerebral palsy, with the goal of enhancing manual ability [62].

The rehabilitation model should encompass a irrefutable assessment of the child's body structure, functions and activities by a multidisciplinary team (MDT) comprising paediatricians, rehabilitation physicians, occupational therapists, physiotherapists, speech and language therapists, community care providers, social workers, and dieticians, conducted as early as possible during hospital admission. A collaborative approach involving children, young people, and families is crucial for recognising precedence areas for assessment and intervention.

Involving parents/carers, family members, and the child/young person themselves as active participants in the assessment of activities such as dressing, feeding, mobilising, and toileting, when clinically appropriate, aids in identifying early rehabilitation priorities. Various clinical and instrumental methods exist to assess body structures, functions, and activity limitations, ensuring a smooth transition to community discharge.

School-aged children should undergo neuropsychological assessment to assist with their return to school during the initial post-stroke period and at regular

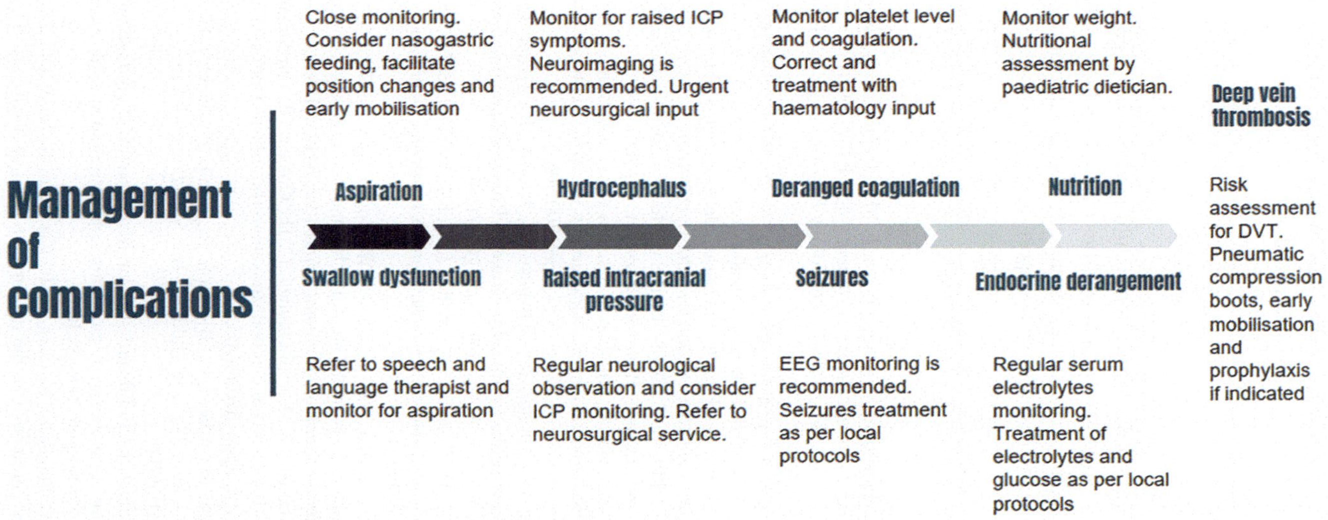

Diagram 6.3 List of complication and management recommendation by the 2017 RCPCH stroke guideline

intervals as their educational needs evolve and new challenges emerge. Childhood stroke poses considerable psychosocial stressors for patients and families. Thus, early support from a coordinated care team, which often includes mental health counselling, is essential for optimising outcomes.

6.4 Conclusion

Perinatal and neonatal stroke, along with childhood stroke, significantly impact childhood morbidity. Strategies to address these issues include early recognition, prevention of early and late recurrence, management of complications, and comprehensive rehabilitation and psychosocial support.

References

1. Rutherford, M. A., Ramenghi, L. A., & Cowan, F. M. (2012). Neonatal stroke. *Archives of Disease in Childhood. Fetal and Neonatal Edition, 97*(5), F377–F384.
2. Lynch, J. K., & Nelson, K. B. (2001). Epidemiology of perinatal stroke. *Current Opinion in Pediatrics, 13*(6), 499–505.
3. Raju, T. N., Nelson, K. B., Ferriero, D., Lynch, J. K., & NICHD-NINDS Perinatal Stroke Workshop Participants. (2007). Ischemic perinatal stroke: Summary of a workshop sponsored by the national institute of child health and human development and the national institute of neurological disorders and stroke. *Pediatrics, 120*(3), 609–616.
4. Lehman, L. L., & Rivkin, M. J. (2014). Perinatal arterial ischemic stroke: Presentation, risk factors, evaluation, and outcome. *Pediatric Neurology, 51*, 760–768.
5. Benders, M. J., Groenendaal, F., Uiterwaal, C. S., Nikkels, P. G., Bruinse, H. W., Nievelstein, R. A., et al. (2007). Maternal and infant characteristics associated with perinatal arterial stroke in the preterm infant. *Stroke, 38*, 1759–1765. Epub 2007/05/15.
6. Gunther, G., Junker, R., Strater, R., Schobess, R., Kurnik, K., Heller, C., et al. (2000). Symptomatic ischemic stroke in full-term neonates: Role of acquired and genetic prothrombotic risk factors. *Stroke, 31*, 2437–2441. Epub 2000/10/07.
7. Wu, Y. W., March, W. M., Croen, L. A., Grether, J. K., Escobar, G. J., & Newman, T. B. (2004). Perinatal stroke in children with motor impairment: A population-based study. *Pediatrics, 114*, 612–619. Epub 2004/09/03.
8. Harteman, J. C., Groenendaal, F., Kwee, A., Welsing, P. M., Benders, M. J., & de Vries, L. S. (2012). Risk factors for perinatal arterial ischaemic stroke: A case–control study. *Archives of Disease in Childhood. Fetal and Neonatal Edition, 97*, F411–F416.
9. Berfelo, F. J., Kersbergen, K. J., van Ommen, C. H., Govaert, P., van Straaten, H. L., Poll-The, B. T., et al. (2010). Neonatal cerebral sinovenous thrombosis from symptom to outcome. *Stroke, 41*, 1382–1388. Epub 2010/06/05.
10. deVeber, G., Andrew, M., Adams, C., Bjornson, B., Booth, F., Buckley, D. J., et al. (2001). Cerebral sinovenous thrombosis in children. *The New England Journal of Medicine, 345*, 417–423. Epub 2001/08/11.
11. Nwosu, M. E., Williams, L. S., Edwards-Brown, M., Eckert, G. J., & Golomb, M. R. (2008). Neonatal sinovenous thrombosis: Presentation and association with imaging. *Pediatric Neurology, 39*, 155–161. Epub 2008/08/30.
12. Fitzgerald, K. C., Williams, L. S., Garg, B. P., Carvalho, K. S., & Golomb, M. R. (2006). Cerebral sinovenous thrombosis in the neonate. *Archives of Neurology, 63*, 405–409. Epub 2006/03/15.

13. Wu, Y. W., Miller, S. P., Chin, K., Collins, A. E., Lomeli, S. C., Chuang, N. A., et al. (2002). Multiple risk factors in neonatal sinovenous thrombosis. *Neurology, 59*, 438–440. Epub 2002/08/15.
14. Yeo, K. T., Thomas, R., Chow, S. S., Bolisetty, S., Haslam, R., Tarnow-Mordi, W., et al. (2020). Improving incidence trends of severe intraventricular haemorrhages in preterm infants <32 weeks gestation: A cohort study. *Archives of Disease in Childhood. Fetal and Neonatal Edition, 105*(2), 145–150.
15. Cizmeci, M. N., de Vries, L. S., Ly, L. G., van Haastert, I. C., Groenendaal, F., Kelly, E. N., et al. (2020). Periventricular hemorrhagic infarction in very preterm infants: Characteristic sonographic findings and association with neurodevelopmental outcome at age 2 years. *The Journal of Pediatrics, 217*, 79–85.e1.
16. Gupta, S. N., Kechli, A. M., & Kanamalla, U. S. (2009). Intracranial hemorrhage in term newborns: Management and outcomes. *Pediatric Neurology, 40*, 1–12. [PubMed: 19068247].
17. Takenouchi, T., Kasdorf, E., Engel, M., et al. (2012). Changing pattern of perinatal brain injury in term infants in recent years. *Pediatric Neurology, 46*, 106–110. [PubMed: 22264705].
18. Towner, D., Castro, M. A., Eby-Wilkens, E., et al. (1999). Effect of mode of delivery in nulliparous women on neonatal intracranial injury. *The New England Journal of Medicine, 341*, 1709–1714. [PubMed: 10580069].
19. Benedetti, T. J. (1999). Birth injury and method of delivery. *The New England Journal of Medicine, 341*, 1758–1759. [PubMed: 10580076].
20. Armstrong-Wells, J., Johnston, S. C., Wu, Y. W., et al. (2009). Prevalence and predictors of perinatal hemorrhagic stroke: Results from the Kaiser pediatric stroke study. *Pediatrics, 123*, 823–828. [PubMed: 19255009].
21. Clive, B., Vincer, M., Ahmad, T., Khan, N., Afifi, J., & El-Naggar, W. (2020). Epidemiology of neonatal stroke: A population-based study. *Paediatrics & Child Health, 25*(1), 20–25. https://doi.org/10.1093/pch/pxy194
22. Elgendy, M. M., Puthuraya, S., LoPiccolo, C., Liu, W., Aly, H., & Karnati, S. (2022). Neonatal stroke: Clinical characteristics and neurodevelopmental outcomes. *Pediatrics and Neonatology, 63*(1), 41–47. https://doi.org/10.1016/j.pedneo.2021.06.017
23. Benders, M. J., Groenendaal, F., & De Vries, L. S. (2009). Preterm arterial ischemic stroke. *Seminars in Fetal & Neonatal Medicine, 14*(5), 272–277.
24. Golomb, M. R., Garg, B. P., Edwards-Brown, M., & Williams, L. S. (2008). Very early arterial ischemic stroke in premature infants. *Pediatric Neurology, 38*(5), 329–334.
25. Nowak-Gottl, U., Duering, C., Kempf-Bielack, B., & Strater, R. (2003). Thromboembolic diseases in neonates and children. *Pathophysiology of Haemostasis and Thrombosis, 33*, 269–274. Epub 2005/ 02/05.
26. Cowan, F., Mercuri, E., Groenendaal, F., Bassi, L., Ricci, D., Rutherford, M., et al. (2005). Does cranial ultrasound imaging identify arterial cerebral infarction in term neonates? *Archives of Disease in Childhood. Fetal and Neonatal Edition, 90*, F252–F256. Epub 2005/04/23.
27. Roze, E., Harris, P. A., Ball, G., Elorza, L. Z., Braga, R. M., Allsop, J. M., et al. (2012). Tractography of the corticospinal tracts in infants with focal perinatal injury: Comparison with normal controls and to motor development. *Neuroradiology, 54*, 507–516. Epub 2011/10/19.
28. van der Aa, N. E., Leemans, A., Northington, F. J., van Straaten, H. L., van Haastert, I. C., Groenendaal, F., et al. (2011). Does diffusion tensor imaging-based tractography at 3 months of age contribute to the prediction of motor outcome after perinatal arterial ischemic stroke? *Stroke, 42*, 3410–3414. Epub 2011/10/25.
29. Mallick, A. A., & O'Callaghan, F. J. (2010). The epidemiology of childhood stroke. *European Journal of Paediatric Neurology, 14*, 197–205.
30. Tsze, D. S., & Valente, J. H. (2011). Pediatric stroke: A review. *Emergency Medicine International, 2011*, 734506. https://doi.org/10.1155/2011/734506
31. Chung, B., & Wong, V. (2004). Pediatric stroke among Hong Kong Chinese subjects. *Pediatrics, 114*, e206–e212. https://doi.org/10.1542/peds.114.2.e206

32. Lee, K. L., Tseng, Y. C., Yang, S. T., & Kuo, Y. T. (2018). Uncommon pediatric stroke caused by MCA dissection presenting as initial loss of consciousness. *Pediatrics and Neonatology, 59*, 634–635. https://doi.org/10.1016/j.pedneo.2018.04.012
33. deVeber, G. A., Kirton, A., Booth, F. A., Yager, J. Y., Wirrell, E. C., Wood, E., et al. (2017). Epidemiology and outcomes of arterial ischemic stroke in children: The Canadian Pediatric Ischemic Stroke Registry. *Pediatric Neurology, 69*, 58–70.
34. RCPCH. (2017). *Stroke in childhood—Clinical guideline for diagnosis, management and rehabilitation*. RCPCH: Royal College of Paediatrics and Child Health.
35. Catenaccio, E., Riggs, B. J., Sun, L. R., Urrutia, V. C., Johnson, B., Torriente, A. G., et al. (2020). Performance of a pediatric stroke alert team within a comprehensive stroke center. *Journal of Child Neurology, 35*, 571–577. https://doi.org/10.1177/0883073820920111
36. Lauzier, D. C., Galardi, M. M., Guilliams, K. P., Goyal, M. S., Amlie-Lefond, C., Hallam, D. K., et al. (2021). Pediatric thrombectomy. *Stroke, 52*, 1511–1519. https://doi.org/10.1161/STROKEAHA.120.032268
37. Mackay, M. T., Yock-Corrales, A., Churilov, L., Monagle, P., Donnan, G. A., & Babl, F. E. (2016). Differentiating childhood stroke from mimics in the emergency department. *Stroke, 47*, 2476–2481. https://doi.org/10.1161/STROKEAHA.116.014179
38. Elbers, J., Wainwright, M. S., & Amlie-Lefond, C. (2015). The pediatric stroke code: Early management of the child with stroke. *The Journal of Pediatrics, 167*, 19–24.e4. https://doi.org/10.1016/j.jpeds.2015.03.051
39. Harrar, D. B., Salussolia, C. L., Kapur, K., Danehy, A., Kleinman, M. E., Mannix, R., et al. (2020). A stroke alert protocol decreases the time to diagnosis of brain attack symptoms in a pediatric emergency department. *The Journal of Pediatrics, 216*, 136–141.e6. https://doi.org/10.1016/j.jpeds.2019.09.027
40. Wintermark, M., Hills, N. K., DeVeber, G. A., Barkovich, A. J., Bernard, T. J., Friedman, N. R., Mackay, M. T., Kirton, A., Zhu, G., Leiva-Salinas, C., Hou, Q., Fullerton, H. J., & Investigators, V. I. P. S. (2017). Clinical and imaging characteristics of arteriopathy subtypes in children with arterial ischemic stroke: Results of the VIPS study. *AJNR. American Journal of Neuroradiology, 38*, 2172–2179. https://doi.org/10.3174/ajnr.A5376
41. Tuckuviene, R., Christensen, A. L., Helgestad, J., Johnsen, S. P., & Kristensen, S. R. (2011). Paediatric arterial ischaemic stroke and cerebral sinovenous thrombosis in Denmark 1994–2006: A nationwide population-based study. *Acta Paediatrica, 100*, 543–549. https://doi.org/10.1111/j.1651-2227.2010.02100.x
42. Hartman, A. L., Lunney, K. M., & Serena, J. E. (2009). Pediatric stroke: Do clinical factors predict delays in presentation? *The Journal of Pediatrics, 154*, 727–732. https://doi.org/10.1016/j.jpeds.2008.11.011
43. McGlennan, C., & Ganesan, V. (2008). Delays in investigation and management of acute arterial ischaemic stroke in children. *Developmental Medicine and Child Neurology, 50*, 537–540. https://doi.org/10.1111/j.1469-8749.2008.03012.x
44. Mackay, M. T., Chua, Z. K., Lee, M., Yock-Corrales, A., Churilov, L., Monagle, P., Donnan, G. A., & Babl, F. E. (2014). Stroke and nonstroke brain attacks in children. *Neurology, 82*, 1434–1440.
45. Braun, K. P., Kappelle, L. J., Kirkham, F. J., & Deveber, G. (2006). Diagnostic pitfalls in paediatric ischaemic stroke. *Developmental Medicine and Child Neurology, 48*, 985–990.
46. Mallick, A. A., Ganesan, V., Kirkham, F. J., Fallon, P., Hedderly, T., McShane, T., Parker, A. P., Wassmer, E., Wraige, E., Amin, S., Edwards, H. B., Tilling, K., & O'Callaghan, F. J. (2014). Childhood arterial ischaemic stroke incidence, presenting features, and risk factors: A prospective population-based study. *Lancet Neurology, 13*, 35–43.
47. Fox, C. K., Sidney, S., & Fullerton, H. J. (2015). Community-based casecontrol study of childhood stroke risk associated with congenital heart disease. *Stroke, 46*, 336–340.
48. Uohara, M. Y., Beslow, L. A., Billinghurst, L., et al. (2017). Incidence of recurrence in posterior circulation childhood arterial ischemic stroke. *JAMA Neurology, 74*(3), 316–323. https://doi.org/10.1001/jamaneurol.2016.5166

49. Steinlin, M., Pfister, I., Pavlovic, J., Everts, R., Boltshauser, E., Capone Mori, A., Gubser Mercati, D., Hänggeli, C. A., Keller, E., Luetschg, J., Marcoz, J., Ramelli, G. P., Roulet Perez, E., Schmitt-Mechelke, T., & Weissert, M. (2005). Swiss Societies of Paediatric Neurology and Neonatology. The first three years of the Swiss Neuropaediatric Stroke Registry (SNPSR): A population-based study of incidence, symptoms and risk factors. *Neuropediatrics, 36*(2), 90–97. https://doi.org/10.1055/s-2005-837658
50. Mackay, M. T., Prabhu, S. P., & Coleman, L. (2010). Childhood posterior circulation arterial ischemic stroke. *Stroke, 41*, 2201–2209.
51. Ferriero, H. J., Fullerton, et al. (2019). Management of stroke in neonates and children a scientific statement from the American Heart Association/American Stroke Association. On behalf of the American Heart Association Stroke Council and Council on Cardiovascular and Stroke Nursing. *Stroke, 50*, e51–e96.
52. Hollist, M., Au, K., Morgan, L., Shetty, P. A., Rane, R., Hollist, A., Amaniampong, A., & Kirmani, B. F. (2021). Pediatric stroke: Overview and recent updates. *Aging and Disease, 12*(4), 1043. https://doi.org/10.14336/ad.2021.0219
53. Mackay, M. T., Monagle, P., & Babl, F. E. (2017). Improving diagnosis of childhood arterial ischaemic stroke. *Expert Review of Neurotherapeutics, 17*, 1157–1165. https://doi.org/10.1080/14737175.2017.1395699
54. Mackay, M. T., Churilov, L., Donnan, G. A., Babl, F. E., & Monagle, P. (2016). Performance of bedside stroke recognition tools in discriminating childhood stroke from mimics. *Neurology, 86*, 2154–2161. https://doi.org/10.1212/WNL.0000000000002736
55. Rafay, M. F., Pontigon, A. M., Chiang, J., Adams, M., Jarvis, D. A., Silver, F., et al. (2009). Delay to diagnosis in acute pediatric arterial ischemic stroke. *Stroke, 40*, 58–64.
56. Goyal, M., Menon, B. K., van Zwam, W. H., Dippel, D. W. J., Mitchell, P. J., Demchuk, A. M., Dávalos, A., Majoie, C. B. L. M., van der Lugt, A., de Miquel, M., et al. (2016). Endovascular thrombectomy after large-vessel ischaemic stroke: A meta-analysis of individual patient data from five randomised trials. *Lancet, 387*, 1723–1731.
57. Albers, G. W., Marks, M. P., Kemp, S., Christensen, S., Tsai, J. P., Ortega-Gutierrez, S., McTaggart, R. A., Torbey, M. T., Kim-Tenser, M., Leslie-Mazwi, T., et al. (2018). Thrombectomy for stroke at 6 to 16 hours with selection by perfusion imaging. *The New England Journal of Medicine, 378*, 708–718.
58. Nogueira, R. G., Jadhav, A. P., Haussen, D. C., Bonafe, A., Budzik, R. F., Bhuva, P., Yavagal, D. R., Ribo, M., Cognard, C., Hanel, R. A., et al. (2018). Thrombectomy 6 to 24 hours after stroke with a mismatch between deficit and infarct. *The New England Journal of Medicine, 378*, 11–21.
59. Pacheco, J. T., Siepmann, T., Barlinn, J., Winzer, S., Penzlin, A. I., Puetz, V., von der Hagen, M., & Barlinn, K. (2018). Safety and efficacy of recanalization therapy in pediatric stroke: A systematic review and meta-analysis. *European Journal of Paediatric Neurology, 22*, 1035–1041.
60. Satti, S., Chen, J., Sivapatham, T., Jayaraman, M., & Orbach, D. (2016). Mechanical thrombectomy for pediatric acute ischemic stroke: Review of the literature. *Journal of NeuroInterventional Surgery, 9*, 732–737.
61. Morgan, C., Novak, I., Dale, R. C., Guzzetta, A., & Badawi, N. (2016). Single blind randomised controlled trial of GAME (Goals—Activity—Motor Enrichment) in infants at high risk of cerebral palsy. *Research in Developmental Disabilities, 55*, 256–267. https://doi.org/10.1016/j.ridd.2016.04.005
62. Wang, J. J., Shi, K. L., Li, J. W., Jiang, L. Q., Caspi, O., Fang, F., Xiao, J., Jing, H., & Zou, L. P. (2009). Risk factors for arterial ischemic and hemorrhagic stroke in childhood. *Pediatric Neurology, 40*, 277–281. https://doi.org/10.1016/j.pediatrneurol.2008.11.002

Type 2 Diabetes Mellitus and Stroke: Understanding the Connection

7

Fa'iza Abdullah

Abstract

Stroke or cerebrovascular disease is one of the main causes of death globally. The incidence of stroke is linked to the persistent rise in the prevalence of risk factors like obesity, diabetes, hypertension, and hyperlipidemia. As a significant risk factor for stroke, this chapter discussed the impact of diabetes on stroke events. It summarises the pathophysiology of hyperglycaemia and additional other co-existing common risk factors leading to stroke events. Likewise, this chapter discusses the clinical implications, management approach to diabetes with stroke events and preventive strategies to reduce stroke risk. The comprehensive medical intervention of diabetes, alongside other associated risk factors management, together with effective healthy lifestyle changes, is an essential and effective strategy to lower the risk of stroke in patients with T2DM.

Keywords

Type-2 diabetes mellitus · Stroke · Hyperglycaemia · Preventive measures

7.1 Introduction to Type 2 Diabetes Mellitus and Stroke

Diabetes mellitus (DM) encompasses a spectrum of metabolic disorders characterized by persistently elevated blood sugar levels. This spectrum includes type 1 diabetes mellitus (T1DM), type 2 diabetes mellitus (T2DM), maturity-onset diabetes of the young (MODY), gestational diabetes, and various secondary forms. T1DM,

F. Abdullah (✉)
Family Medicine Department, International Islamic University Malaysia,
Kuantan, Pahang, Malaysia
e-mail: drfaiza@iium.edu.my

W. Mohamed, M. B. Mat-Nor (eds.), *Nutrition and Stroke*, Nutritional Neurosciences, https://doi.org/10.1007/978-981-97-6818-9_7

predominantly affecting children and adolescents, results from autoimmune-mediated destruction of pancreatic beta cells, leading to absolute insulin deficiency. Conversely, T2DM which typically manifests in middle-aged and older adults initially characterized by insulin resistance, a disorder in which cells do not react to insulin as they should and potentially progressing to relative insulin deficiency as the disease advances. While the precise aetiology of each diabetes type remains incompletely elucidated, both genetic factors, poor lifestyle and dietary intake are thought to play significant roles. T2DM risk factors are predominantly associated with suboptimal lifestyle choices and dietary patterns. The distinct pathophysiological mechanisms underlying T1DM and T2DM contribute to their divergent aetiologies, clinical presentations, and treatments [1, 2].

T2DM is characterized by a gradual onset featuring a functional insulin deficit from an imbalance between insulin levels and insulin sensitivity. The development of Insulin resistance is multifactorial primarily stemming from obesity and ageing, reflecting the integration between genetic predisposition and lifestyle factors. There is evidence indicates that T2DM has a stronger genetic component compared to T1DM [3]. A majority of T2DM patients have a least one parent with the condition. Formerly known as adult-onset diabetes, T2DM comprises 90% to 95% of all diagnosed diabetes cases and is characterized by hyperglycaemia, insulin resistance, and relative insulin insufficiency. Typical manifestations include polydipsia, polyuria, increased hunger leading to polyphagia, fatigue and inexplicable weight loss [4]. Prolonged hyperglycaemia can also cause other symptoms like unhealing wounds and a pins and needles sensation [4]. Life-threatening acute complications of hyperglycaemia can be caused by a hyperosmolar hyperglycaemic state (HHS) and diabetic ketoacidosis (DKA) conditions. Chronic vasculopathy and neuropathy associated with DM can lead to severe long-term sequelae. These complications may include cardiovascular disease, cerebrovascular events, diabetic retinopathy, renal failure, and peripheral vascular disease, potentially resulting in limb amputation due to compromised circulation [5]. It is crucial for diabetes patients who experience these symptoms to consult healthcare professionals for appropriate immediate management.

The prevalence of T2DM has shown a consistent upward trend over recent decades, with a more pronounced increase observed in low- and middle-income countries (LMICs) compared to their high-income counterparts. According to the International Diabetes Federation (IDF) Diabetes Atlas published in 2021, diabetes affects 10.5% of the adult population aged 20–79 years worldwide. Notably, approximately half of these individuals are undiagnosed. T2DM accounts for over 90% of all diabetes cases globally [6]. The rise of T2DM is mainly due to the rise of an ageing population and urbanisation, coupled with reduced physical activity and rising rates of overweight and obesity. These trends are influenced by socioeconomic, demographic, environmental and genetic factors. Asia, being the world's most populous region, is home to over 60% of individuals with diabetes. Live in Asia, the world's most populous region [6]. This situation presents significant social and economic challenges to many countries in the region and has the potential to hinder both national and global development efforts. Malaysia, one of the Asia countries, showed an increased trend in diabetes prevalence in adults ≥18 years, from 13.4% in 2015 to 18.3% in 2019. What is worrying is the 24.1% increase in overall diabetes prevalence among adults aged 30 years and above in 2019. This figure includes both previously diagnosed cases and undiagnosed individuals with elevated fasting blood glucose levels (≥7.0 mmol/L) detected during surveys [7, 8]. A recent systematic review and meta-analysis published in 2021 further corroborates this trend, demonstrating that T2DM prevalence has risen over the years and increases with age [9]. As the most prevalent form of diabetes mellitus, T2DM frequently coexists with other non-communicable diseases, particularly hypertension, dyslipidaemia and obesity [7]. The highest prevalence of documented T2DM was between 55 and 59 years old, with the mean age at diagnosis being 53, and females were slightly higher in percentage. Data indicate a slight decrease in the proportion of patients with Ischemic Heart Disease (IHD), from 5.86% in 2019 to 5.64% in 2020. However, during the same period, the proportion of patients with stroke showed a small increase, rising from 1.79% to 1.88% [8].

One of the significant long-term sequelae of T2DM is cerebrovascular accident, also known as stroke or brain attack. This condition results from compromised blood flow to cerebral tissues, leading to neuronal death. Depending on the type of stroke, poor oxygen supply to the brain leading to stroke can be due to rupture of blood vessels causing bleeding known as haemorrhagic stroke or building up plaque causing blockage of bold vessels, mainly the artery, inhibiting adequate blood flow to the brain leading to brain cells ischaemia, known as ischemic stroke. Both stroke types ultimately cause localized neuronal death and loss of function in affected brain regions [10]. Approximately 87% of all strokes are ischemic, while remaining 13% are hemorrhagic. The clinical manifestations of stroke typically emerge rapidly following the event and may include hemiparesis or hemiplegia, hemianesthesia, aphasia or dysphasia, facial asymmetry, vertigo, and homonymous hemianopsia. These neurological deficits reflect the sudden onset of cerebral dysfunction characteristic of stroke syndromes [10].

Diabetes patients exhibit a 1.5 to 2-fold increased stroke risk compared to non-diabetics, with this risk escalating as the disease progresses. Several diabetes-associated mechanisms contribute to stroke risk, including arterial atherosclerosis, cerebral small vessel disease, endothelial dysfunction, and cardiac embolism. These pathophysiological processes, coupled with hyperglycemia, not only elevate the likelihood of stroke occurrence but also lead to poorer outcomes in acute ischemic stroke patients compared to those with normal blood glucose levels [11]. The pathophysiology of hyperglycaemia leading to stroke events is explained further in the sub-chapter T2DM and Stroke. In brief, Prolonged exposure to elevated blood glucose levels causes injury to blood vessels that supply and drain the heart. Over time, this damage promotes the formation of atherosclerotic plaques, ultimately resulting in impaired blood flow. To describe succinctly, chronic hyperglycemia leads to progressive damage of the cardiovascular system. To the blood vessels that carry blood to and from the heart. Over time, this damage promotes the formation of atherosclerotic plaques and ultimately resulting in impaired blood flow.

Patients with stroke and uncontrolled blood sugar levels have an increased mortality risk and worse recovery outcomes [12]. Meta-analysis and a literature review found diabetes affects about 28% of all types stroke patients. This percentage is higher in cases with ischemic stroke (33%) compared to haemorrhagic stroke (26%) [13]. Furthermore, those with T2DM who have previously experienced an ischemic stroke have a 50% higher likelihood of having another stroke compared to those without diabetes, as indicated by a hazard ratio of 1.50 (95% confidence interval: 1.36–1.65 [14]. Persistent uncontrol plasma glucose increases stroke risk in T2DM patients. When comparing a hyperglycemic group to one with normal blood sugar, a systematic review and meta-analysis study found a dose-dependent association that increased the risk of stroke by a significant margin, with a risk ratio (RR) of 1.12 for the caused-specific mortality [15]. This finding underscores the critical importance of limiting sugar intake for patients with T2DM as a means to reduce their chances of experiencing a stroke and to lower the risk of death from related complications.

7.2 Pathophysiology of Type 2 Diabetes Mellitus

Diabetes mellitus is a chronic hyperglycemic state resulting from various metabolic disorders. It is characterized by either impaired insulin production, reduced effectiveness of insulin on target tissues, or frequently a combination of both. The condition's manifestations can vary, spanning from initial resistance to insulin action with a relative shortage of insulin, to a more pronounced deficiency in insulin secretion coupled with insulin resistance.

Chronic high blood glucose levels characterize T2DM, which arises from various metabolic dysfunctions that affect the hormone insulin function, which is crucial for regulating blood sugar levels. The condition can manifest as either a resistance to the insulin produced by the body, a deficiency in insulin production, or a combination of both [16]. The pancreas's beta cells produce insulin, a peptide hormone, which is crucial for managing blood sugar. Its main function is to facilitate the uptake of glucose into cells, especially in the liver, muscles, and fat tissues. Once inside these insulin-responsive cells, the glucose can be stored or converted into energy. This process is crucial for maintaining homeostasis, ensuring that glucose levels in the bloodstream remain within a narrow, healthy range. Disruptions in insulin function or production can lead to high blood sugar levels.

The etiology of diabetes mellitus is rooted in the dysregulation of glucose homeostasis mechanisms. Insulin resistance, a hallmark of the disease, precipitates a triad of metabolic aberrations: hepatic glucose overproduction, impaired glycogen synthesis, and diminished glucose uptake in skeletal muscle tissue. These perturbations collectively contribute to the hyperglycemic state characteristic of diabetes mellitus. Concurrently, the dysfunction of pancreatic β-cells, which are responsible for insulin production, exacerbates the problem by failing to compensate for the increased demand. The pathogenesis T2DM is predicated on the disruption of glucose homeostasis, primarily driven by insulin resistance and β-cell dysfunction.

These metabolic perturbations, either independently or synergistically, constitute the fundamental etiological factors underlying the onset of T2DM. Over time, these persistent metabolic disturbances of the intricate glucose regulatory mechanisms culminates in the hyperglycemic state characteristic of T2DM, potentially precipitating a cascade of associated complications [17]. In a state of nutritional excess, for instance, an obese person, when consuming high sweet beverages or high carbohydrate-rich diets, induces persistent hyperglycaemia and hyperlipidaemia condition, predisposes to oxidative stress and a chronic inflammatory state that compromising pancreatic islet integrity through a complex cascade of molecular alterations and dysfunction. The damage to β-cells in T2DM has always been ascribed to cellular apoptosis, which eventually reduces insulin production [18].

It is well-known that T2DM is a major cause of various organ complications due to micro- and macro neuropathy as well as vasculopathy, and these include diabetic retinopathy, diabetic kidney diseases, cardiovascular disease (CVD), cerebrovascular accidents (CVA) or stroke, diabetic carbuncle and diabetic foot. Vascular diseases represent the principal source of morbidity and mortality in diabetic patients, significantly increasing the risk of coronary artery disease, cerebrovascular events, and peripheral arterial disease. The pathophysiological mechanisms underlying diabetic vasculopathy are multifactorial, encompassing dysfunction of endothelial cells, vascular smooth muscle cells, and platelets. This triad of cellular aberrations collectively contributes to the accelerated atherogenesis and thrombotic propensity characteristic of diabetes-associated vascular disorders. Elevated glucose levels can lead to changes in platelet function and fibrinogen levels, which may increase the risk of thrombosis and accelerate the degeneration of diabetic atherosclerotic plaques. It triggered molecular mechanisms that lead to vascular dysfunction by elevating free fatty acids and insulin resistance [19].

A critical pathway in the development of diabetic vasculopathy induced by hyperglycemia involves the non-enzymatic interaction between glucose and vascular proteins, particularly lipoproteins. In the hyperglycemic state, glucose initially binds to plasma or vessel wall proteins, forming early glycosylation products (Amadori products). These intermediates subsequently undergo complex molecular rearrangements, culminating in the formation of advanced glycosylation end products (AGEs). The accumulation of these products within the extracellular matrix (ECM) leads to structural and functional alterations in blood vessels, significantly contributing to the pathogenesis of diabetic vasculopathy [20].

7.3 Pathophysiology of Hyperglycaemia Leads to Stroke Event

Chronic excessive blood glucose can increase fatty deposits or blood vessel clots. These clots narrow or block blood vessels in various parts of the body, especially in the brain, neck and heart, occlude the blood flows, preventing the brain from oxygen supply, causing cerebrovascular (CVA) accidents or strokes and also cardiovascular disease (CVD) to the heart or myocardial infarction (MI). Apart from primary stroke events among T2DM patients with stroke, the other primary risk factor contributing to the recurrence of stroke was the presence of concomitant ischaemic heart disease [21]. It demonstrated that cardiovascular disease (IHD) not only increases cardiovascular risk but also increases the risk of recurrent stroke among post-stroke T2DM patients.

Persistent high blood sugar levels in the blood, known as hyperglycemia, intensify oxidative stress, which leads to several pathological processes, eventually leading to microvascular complications [22]. These several mechanisms in the pathological process include arterial stiffness, vascular endothelial dysfunction, chronic systemic inflammation and capillary basal membrane thickening [12, 22]. Vascular endothelial function plays a crucial role in preserving the structural and functional integrity of vessel walls and regulating vasomotor tone. Endothelial dysfunction refers to the impaired capacity of the endothelium to maintain and modulate vascular tone. This pathophysiological state is primarily characterized by a prothrombotic and proinflammatory milieu, coupled with attenuated vasodilatory responses. The dysregulation of these endothelial-dependent processes significantly contributes to the progression of vascular pathologies [23].

Nitric oxide (NO), essential for vasodilatation and maintaining vascular endothelial function, is also impaired in T2DM patients. This dysfunction may arise from two primary mechanisms: enhanced NO inactivation or reduced smooth muscle responsiveness to NO signaling. Decreasing nitric oxide availability to facilitate vasodilation can cause endothelial dysfunction, making arteries stiffer and less elastic compared to those with normal glucose levels. The increased systemic inflammatory response seen in T2DM patients can also contributes the development of atherosclerotic plaque. This inflammatory milieu is characterized by elevated levels of several biomarkers, including C-reactive protein, pro-inflammatory cytokines, and altered adiponectin profiles. The cascade of atherosclerosis, which is initiated by arterial stiffness and chronic inflammation, occurs in small cerebral vessels, meaning more platelets aggregate and increased vascular stasis, which results in ischemia and thrombogenic risk [12]. The vascular damage in diabetic patients causing vascular relaxation and constriction defects alters the vascular resistance and increases the risk for hypertensive disease, too [24]. Atherosclerotic plaques in diabetic patients exhibit heightened vulnerability to rupture, primarily due to their characteristic lipid-rich core and attenuated fibrous cap [25], thus increasing the risk of stroke events and worse outcomes when these diabetic atherosclerotic plaque eruptions.

It can be concluded that persistent high blood sugar and insulin resistance are believed to be the essential triggers for the pathophysiology process of diabetes-related cerebral vascular endothelial dysfunction associated with accelerated atherogenesis, cerebral small vessel disease, and an increased propensity for cardiac embolism leading to the stroke event.

7.4 Common Risk Factors and Shared Pathways

Demographic studies have shown that T2DM co-exist with other comorbidities, and the prevalence of concurrent chronic diseases can be quite high. A diabetes clinical audit in 2019 in one of the countries in the Pacific Asia region documented the prevalence of dyslipidemia, hypertension and overweight (either pre-obese or obese) among T2DM patients were high, 74.3%, 80.4% and 84.0%, respectively. Hence, it increases the risk of stroke [8]. These remarkably high percentages of concomitant chronic diseases with T2DM show that other comorbidities contribute to the pathophysiology processes that will further aggravate cerebral vascular dysfunction and lead to metabolic complications, including stroke [22]. The pathophysiology of other comorbidity, including dyslipidemia, hypertension and obesity as risk factors for stroke, are discussed in other chapters.

Both T1DM and T2DM are recognized as independent risk factors for accelerated atherogenesis [26]. The coexistence of dyslipidemia with hyperglycemia and insulin resistance engenders a cascade of physiological alterations. This metabolic milieu promotes the formation of atherogenic low-density lipoprotein (LDL) particles, accelerates the production AGEs, and activates pro-inflammatory signaling pathways. These processes synergistically compromise arterial wall integrity, facilitating the development and progression of atherosclerotic lesions [26].

The coexistence of diabetes, dyslipidemia, and hypertension significantly amplifies the risk of cerebrovascular events, encompassing both thromboembolic ischemic incidents and non-thromboembolic occurrences such as aneurysmal ruptures and hemorrhagic episodes [16]. The rising prevalence of obesity in conjunction with diabetes creates a metabolic milieu characterized by hyperglycemia, hypertension, and dyslipidemia. This pathophysiological triad exacerbates vascular pathology, promoting both generalized vasculopathy and the formation of diabetes-specific atherosclerotic plaques. Consequently, this complex interplay of metabolic derangements substantially elevates the risk of stroke events [17] as illustrated in Fig. 7.1.

7.5 Epidemiological Link Between Type 2 Diabetes Mellitus and Stroke

Cerebrovascular accidents or strokes generally occur either through the development of diabetic atherosclerotic plaque within the cerebral vessels or rupture of the atherosclerotic plaque elsewhere, e.g. coronary artery disease that blocks the blood flow to the brain. According to the World Health Organization (WHO) key facts on diabetes, age-standardised mortality rates attributable to diabetes exhibited a 3% increase over the period of 2000 and 2019. Diabetes, as an independent risk factor, conferred a rise of approximately two-fold to develop a wide range of vascular diseases, including stroke, when compared with people without diabetes [2]. An increasing trend of global and local prevalence of T2DM patients and cerebrovascular accidents (stroke) among T2DM patients accelerates morbidity and mortality proportion, subsequently leading to a substantial socio-economic burden. In terms

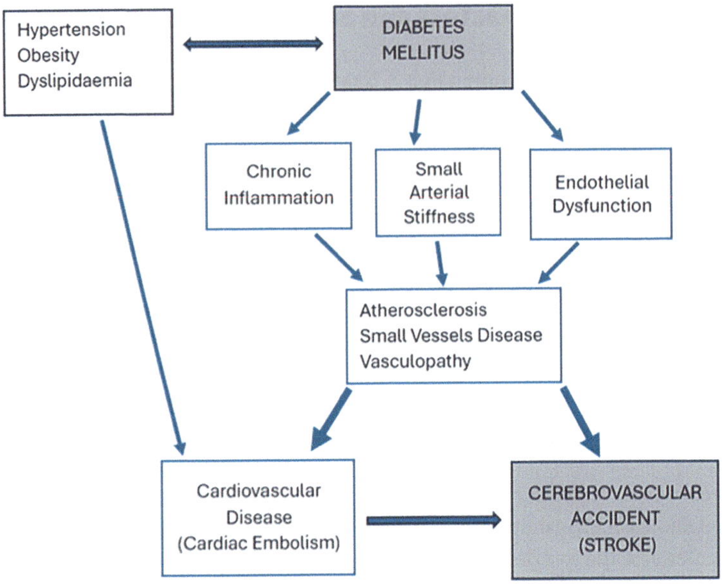

Fig. 7.1 T2DM as vascular risk factor in stroke pathogenesis

of stroke patterns among T2DM patients, a higher proportion of ischemic stroke was documented compared to hemorrhagic stroke. In addition, lacunar infarcts (a small infarct measuring 0.2 to 15 mm or noncortical infarcts) are the most common type, especially among those with hypertension [27–29].

Uncontrolled diabetes significantly elevates the risk of both ischemic and hemorrhagic stroke events. However, the magnitude of this risk varies by stroke subtype. In patients with T2DM, the adjusted hazard ratio (HR) for ischemic stroke is notably higher at 2.27 (95% CI: 1.95–2.65), compared to an HR of 1.56 (95% CI: 1.19–2.05) for hemorrhagic stroke, when contrasted with non-diabetic individuals [30].

Another risk factor of stroke in T2DM patients is cardiac embolism, a known risk of the ischemic stroke event. The underlying cause of cardiac embolism leading to stroke event is atrial fibrillation, which has a 35% higher risk of occurring among T2DM patients compared with the general population [31]. The current mechanisms of stroke in atrial fibrillation are complex, but it may be due to the increase of oxidative stress and production of advanced glycation end products (AGEs) in T2DM, which trigger the structural, electrical as well as autonomic remodelling in the cardiac atrium [32]. Other contributing factors for cardioembolic stroke are the presence of comorbidities like hypertension, hyperlipidaemia, and obesity, which aggravate the macro and microvascular complications such as diabetic nephropathy and neuropathy, and ischaemic heart disease, leading to heart failure, and subsequently will also increase the risk of stroke among T2DM patients shown in Diagram 1.

7.6 Impact of Glycemic Control on Stroke Risk

T2DM represents a significant independent risk factor for cerebrovascular events. Moreover, T2DM often occurs in conjunction with other cardiometabolic disorders, each of which independently contributes to stroke risk [33]. Studies have shown that T2DM patients tend to experience worse outcomes following a stroke event and a higher risk for stroke to recur than those without diabetes [14, 34]. Glycaemia contributes to the underlying vascular and neurodegeneration, which also can lead to cognitive decline compared to those without diabetes [34]. Although the pathophysiology interaction between stroke and diabetes on cognitive function is not clear, the study showed that T2DM is associated with poorer cognitive function 3–6 months post-ischemic stroke than those with normal blood glucose levels [35]. Thus, improving blood sugar control undeniably decreases the risk of stroke events and poor post-stroke outcomes in patients with T2DM.

A comprehensive systematic review and meta-analysis revealed that T2DM patients with suboptimal glycemic control, as indicated by elevated HbA1c levels, face a significantly increased risk of primary stroke events. The analysis demonstrated a hazard ratio (HR) of 2.15 (95% CI: 1.76, 2.63) for first-ever stroke in this cohort compared to individuals with normal HbA1c values [36]. HbA1c, a critical biomarker in diabetes management, provides insight into average glycemic levels over a 2–3 month period. Its utility extends beyond mere glycemic assessment, serving as a predictor of both micro- and macrovascular complications associated with diabetes. Moreover, HbA1c plays a pivotal role in evaluating the efficacy of glycemic management strategies, thereby informing therapeutic adjustments aimed at optimizing patient outcomes.

Diabetic management should be discussed as a shared-care intervention approach according to patient considerations. Blood glucose levels or HbA1c targets should be individualized based on age, coexisting morbidities, duration of diabetes, and hypoglycaemia events. In general, T2DM patients should aim for fasting plasma glucose or pre-prandial plasma glucose of 4.0 to 7.0 mmol/L. [34] to achieve a target of A1c $\leq$7.0%. There are recent studies that have published and documented new antidiabetic drug classes that have shown evidence to reduce the risk of stroke [11, 37]. The new practices in T2DM management in stroke patients have recommended to consider other new antihyperglycemic agents with proven benefits on major cardiovascular outcomes and stroke for those with a history of stroke whose blood glucose levels do not achieve targets with standard oral antihyperglycemic medications [38].

7.7 Preventive Strategies and Management

Limited clinical trials have evaluated specific stroke prevention strategies in T2DM patients individually. Yet, the clinical management of T2DM for primary and secondary prevention of stroke remains an important intervention to be accomplished. Nevertheless, anti-platelet agents like aspirin are prescribed as secondary prevention of cardiovascular disease in T2DM patients. Primary prevention is only recommended if there is an increase in cardiovascular risk, for example, an estimated 20% or higher of a 10-year predicted coronary heart disease (CHD) risk calculated, and it should only be started after a discussion with the patient on the benefits of the treatment and the risk of bleeding [7]. As the prevalence of diabetes increases, the overall burden of cardiovascular and cerebrovascular complications also increases. Therefore, the prevention of stroke in T2DM patients should include strategies that co-manage co-existing chronic disease as well as lifestyle modification, and it should be individualised.

Non-pharmacotherapy advice for healthy lifestyle changes is the mainstay of first-line intervention in T2DM patients. This is especially important concerning the diet consumed commonly, which was not healthy. The persistent intake of sugary beverages like soft drinks/sodas, flavoured juice drinks, sweetened tea and coffee, energy drinks, and electrolyte replacement drinks results in high plasma sugar. To reduce the risk of T2DM, it is recommended to replace sugary beverages with equal water intake [5].

A healthy and balanced diet generally means consuming a diet high in vegetables, fruit, whole grains, fish, nuts and olive oil, focusing on whole, natural and low red meat and minimally intake of processed foods [38]. Dietitian advice helps achieve the target blood sugar level and control co-existing chronic diseases. The dietary counselling will recommend less consumption of confectionaries, sugary drinks, processed or pre-prepared meats and foods. Dietary recommendations for patients with Type 2 Diabetes Mellitus (T2DM) should emphasize the consumption of fruits in their natural state, either fresh or frozen, without added sugars. When selecting canned fruits, options without supplementary sweeteners are preferable. Similarly, frozen or canned vegetables should be chosen without additional sauces or sodium. Protein intake should be optimized by prioritizing plant-based sources such as legumes, nuts, and seeds. When consuming animal proteins, patients should be advised to select options lower in saturated fats, including fish, poultry, and lean cuts of meat [38].

A healthy diet not only controls diabetes, hypertension, dyslipidaemia and weight gain but also lowers the risk of having heart disease and stroke, particularly among T2DM patients. Nonetheless, there is still no recommended consensus on the best nutritional therapy for an ideal carbohydrate, protein and fat requirement for T2DM patients after comprehensive nutritional research was carried out [39].

A new recommendation in the clinical practice guideline for the management of ischaemic stroke has included a Mediterranean diet rich in olive oil and canola oil, with less meat and more vegetables, as a healthy diet found to be beneficial in stroke prevention [40, 41]. It is recommended to counsel T2DM patients on the Mediterranean diet, which comprises 40–50% carbohydrates and 25–35% fat (including <7% saturated, 10% polyunsaturated, 10% monosaturated) and 15–25% protein within the total energy value. T2DM patients are also advised to increase their fibre intake and reduce their sodium consumption to <2300 mg daily. This nutritional therapy with low levels of carbohydrates, fats, and sodium is perhaps the most recommended strategy in the management of T2DM and other coexisting chronic diseases like hypertension and dyslipidaemia to be well controlled [39]. Almost certainly, it reduces the risk of primary and recurrent stroke events among T2DM patients.

Individualising dietary counselling based on the patient's socio-demographic background, disease condition, co-existing morbidities, and individual patient's goal and preference is recommended in achieving an effective preventive strategy. Hence, referring to the dietitian is obligatory for all newly diagnosed T2DM patients so that they can receive optimal therapeutic dietary counselling. Other non-pharmacotherapy preventive strategies are exercise, weight loss and medical adherence.

T2DM patients without stroke events or medically stable post-stroke T2DM patients should be encouraged to exercise regularly and avoid sedentary behaviours. The intensity of the exercise should be personalised and aims towards upgrading activity as tolerated. For example, exercising four to seven days weekly, regularly accumulating to at least 150 min weekly. Patients should be reassessed episodically at any transition points of increased exercise intensity or exercise duration alongside medical care [38].

A multi-discipline team approach should be considered in T2DM in managing the overweight condition (PreObese and Obese) to achieve sustainable weight loss with assessable weight loss goals. Counselling and education on weight reduction should be followed by behavioural interventions such as increased physical activity in terms of duration and intensity, appropriate calorie consumption, and modification of diet. Ideally, a waist circumference of <80 centimetres for women and <100 cm for men and a body mass index (BMI) of 17.5 to 23.5 kg/m^2 to be targeted, achieved and maintained (refer to values for Asian ethnic groups) [40].

Adherence to prescribed medical regimens is paramount for patients with T2DM in both primary and secondary prevention of stroke and other cardiovascular complications. According to the WHO, medical adherence refers to the extent to which an individual's action align with mutually agreed-upon healthcare recommendations, including taking medication, following dietary advice, and performing healthy lifestyle changes [42]. Whereas, compliance is defined as the extent to which a patient's obedience to comply with the prescriber's advice [42]. Medical adherence infers that the patient and physician work together by integrating the medical intervention, taking into consideration the patient's lifestyle, values and preferences to improve the patient's well-being. Suboptimal adherence to medical treatments is widely acknowledged to negatively impact patient outcomes and increase mortality rates. The WHO asserts that enhancing adherence to therapies for hypertension, hyperlipidemia, and diabetes will improve health and subsequently will lead to personal, family and economic benefits [42].

Comprehensive co-existing risk factors management plays an important role in preventing stroke events among T2DM patients by directing and accomplishing well-controlled blood pressure and target-achieved lipid parameters. These co-morbidities can be addressed with intensive medical therapy and therapeutic, healthy lifestyle modification strategies, including dietary control, regular exercise, smoking cessation, and alcohol reduction. Management of hypertension, dyslipidaemia, obesity, alcohol and nutrition in regard to stroke is expansively discussed in other chapters.

7.8 Clinical Implications and Management Approaches

Immediate T2DM management after a stroke brings significant challenges to healthcare providers to ensure favourable post-stroke outcomes, long-term monitoring of the disease controlled and follow-up, and prevent recurrent stroke. The inappropriate intervention of post-stroke T2DM patients increases short and long-term sickness and higher-risk mortality, significantly elevating the risk for recurrent strokes [43]. Majority of T2DM patients have already taken glucose-lowering agents (GLA) either orally or through injection during the stroke event. Thus, it is imperative to look into the patient's diabetes condition, its treatment during the acute post-stroke phase, and whether diabetes treatment can be further intensified.

Post-stroke outcomes and hyperglycaemia show a bidirectional relationship. The uncontrolled blood glucose contributes to unfavourable stroke outcomes; meanwhile, an extensive ischemic stroke can give rise to a post-stroke hyperglycaemia state. This may be due to a generalized stress reaction during the occurrence of stroke events stimulating the hypothalamic-pituitary-adrenal axis mechanism, leading to insulin resistance and hyperglycaemia in the post-stroke phase [44, 45]. In the post-stroke acute phase, hyperglycaemia causing additional reperfusion injury by the increasing oxidative stress and barrier permeability, as well as stimulating systemic inflammation, aggravates the stroke consequences [44, 45]. Severe hyperglycaemia in the acute stroke condition may need to be treated by using a sliding scale of subcutaneous insulin. However, caution should be taken to avoid hypoglycaemic conditions. A systematic review that recruited over 1500 patients with ischemic stroke in the eleven randomized controlled trials did not advocate the benefits of intensive glycaemic control. Additionally, a study has shown that the intervention group had a higher symptomatic hypoglycaemia rate compared to the control group [46].

In the initial management of patients with acute ischemic stroke, The American Heart Association (AHA)/American Stroke Association guidelines have recommended achieving a target glucose level ranging from 140–180 mg/dL (7.8–10.0 mmol/L) during the first 24 h after stroke in all hospitalized patients [47]. Hence, an appropriate intervention during the acute phase of post-stroke is by closely monitoring blood sugar levels while actively treating hyperglycaemia condition in order to avoid overtreatment and hypoglycaemic complications. It is recommended to avoid very tight targets of glucose control in the first few hours of ischaemic strokes to avoid hypoglycaemic complications, and the blood glucose levels should be kept between 6.0 and 10.0 mmol/L post-acute stroke condition [42, 48, 49].

Patients with T2DM who have experienced a cerebrovascular accident (CVA) or stroke should undergo a comprehensive diabetes assessment, with individualized glycemic control targets established for optimal management. Nonetheless, the long-term benefits of stringent glycemic control on cardiovascular outcomes, including stroke, remain uncertain [48–50]. A decade-long follow-up study of the United Kingdom Prospective Diabetes Study (UKPDS) demonstrated that newly diagnosed T2DM patients in the intensive treatment group experienced significant reductions in microvascular complications, myocardial infarction incidence, and all-cause mortality. However, stroke incidence did not decline [48]. Another two studies on T2DM patients with well-known cardiovascular risk and longstanding diabetes (median 10-year duration) did not demonstrate significantly reduced cardiovascular events, including stroke, in the first few years in the intensive glycaemic control (HbA1c $\leq$ 6.5%) group [48, 50]. One of the studies was terminated early when it demonstrated a rise in mortality of 22% in the intensive- therapy group. In addition, more frequent hypoglycaemia events and more than 10 kg weight increase were observed in this intensive treatment group [48].

The Steno-2 study demonstrated that prolonged intensive management (average follow-up of 7.8 years) targeting multiple comorbidities including hyperglycemia, hypertension, dyslipidemia, overweight/obesity, and smoking, resulted in a decrease in cardiovascular events among T2DM patients with microalbuminuria. Showed a long-term intensified intervention (mean follow-up of 7.8 years) for multiple co-morbidities (hyperglycaemia, hypertension, dyslipidaemia, pre-obese and obesity, smoking) has led to a reduction of cardiovascular events among T2DM patients and microalbuminuria [51]. This study highlighted that managing multiple risk factors rather than treating a single factor can positively impact cardiovascular outcomes, including stroke. In addition, the extension of this steno-2 study to 21 years of follow-up demonstrated a median of 7.9 years of gain of life. Hence, managing other multiple risk factors alongside intensified diabetes treatment concurrently increases lifespan and positively influences stroke outcomes [51]. For T2DM patients, the most vital disease intervention is to ensure effective glycaemic control and monitoring to prevent the progression of diabetes neuropathy and macro and micro vasculopathy, including stroke or recurrent stroke. Intensifying the treatment of diabetes, managing other co-existing morbidities to achieve a well-controlled state, screening target organ complications, and therapeutic lifestyle changes are highly recommended.

A meta-analysis of diabetic patients revealed that aspirin therapy did not significantly mitigate cardiovascular event risk, including stroke, without elevating the risk of major hemorrhage [52]. For primary stroke prevention in T2DM patients with moderate risk or in geriatric populations, daily antiplatelet aspirin is not advised due to the bleeding risk outweighing potential benefits [53, 54]. Conversely, robust evidence demonstrates a 25% reduction in stroke risk among all patients with prior stroke who received aspirin, supporting the use of aspirin as secondary prevention [55]. Hence, prescribing oral aspirin 75–325 mg daily to post-stroke patients, including those with T2DM, is recommended [40]. Secondary prevention of stroke or recurrent stroke, in general, involves pharmacotherapy interventions like antiplatelet therapy, antihypertensive treatment, lipid-lowering agents, and glycaemic-lowering agents. At times, it may include prevention of cardioembolism and revascularisation procedures in selected cases [40].

7.9 Challenges and Future Directions

In addition to intensive diabetes intervention, post-stroke patients who present either at the primary care or hospital settings should receive a thorough clinical assessment to evaluate their risk of subsequent stroke events and to implement timely and suitable preventive measures [37]. According to the national stroke registry in Malaysia, one of the countries in the Asia Pacific rim, diabetes mellitus (41.4%) was the second top modifiable risk factor associated with first-ever strokes among the population after hypertension (69.9%), followed by smoking (26.3%), hyperlipidaemia (24.4%), ischaemic heart disease (IHD) and atrial fibrillation (3.4%) [41]. Thus, upkeep of these existing co-morbidities under well-controlled, as well as practising a healthy lifestyle, are vital strategies but challenging to follow and undertake.

Non-adherence to medication and healthy lifestyle changes will cause the blood glucose, blood pressure and lipid parameters not to achieve target control poses a great challenge to healthcare providers. Multiple factors contribute to poor adherence, for example: 1. Patient-related factors, such as lack of understanding, poor insight, and other psychosocial factors; 2. Physician-related factors, such as ineffective and inadequate communication; and 3. Health system-related factors, such as insufficient coordination between patients, their families and medical personnel, and patients' inaccessibility to health care and expensive drugs. There are no effective, straightforward measures to improve long-term medical adherence and health outcomes, as documented in one Cochrane review of 78 randomized trials, which has proven the difficulty of cultivating medication adherence [56]. Principally, effective, coordinated, multifactorial strategies are needed to improve medical intervention, including medication adherence.

Behavioural changes leading to a healthy lifestyle, such as appropriate reduction of calorie intake, modification of diet, graded exercise with appropriate intensity, smoking cessation, reduction of alcohol consumption as advised, and compliance to the medication require more than just modest instruction or a brochure from their physician. A comprehensive awareness program integrated with conducive models of cognitive behaviour counselling instilled by multidisciplinary support is very much needed to successfully change toward successful therapeutic lifestyle changes for T2DM patients to prevent stroke or recurrent stroke [57]. Patients with stroke are usually associated with a sedentary life and prolonged sitting behaviours. The motivation towards self-empowerment alongside good family support for stroke patients, especially those with coexisting diabetes, is essential to performing physical activity in a supervised and safe manner.

In general, prevention of stroke in T2DM patients should also address coexisting chronic disease intervention, including dyslipidaemia and hypertension, to be well-controlled; weight reduction strategy in obese T2DM, as well as lifestyle modification, including smoking cessation, balanced diet and maintaining physical activity. To address these challenges, healthcare providers may well implement personalised medicine approaches, which include discussing and documenting the patient's adherence to their prescribed treatment plans (pharmacotherapy and non-pharmacotherapy), exploring and addressing non-adherence, providing counselling and engaging the patients in achieving therapeutic goals at each medical visit. The approach should be integrated care between physicians, diabetic educators, dieticians, physical therapists, pharmacists and other healthcare providers in providing medical care to T2DM patients. This encourages effective diabetes and other co-existing disease treatment outcomes, especially in the prevention of stroke and recurrent stroke.

The metabolic syndrome screening by identifying the risk of cardiovascular disease and stroke in those who are at risk of diabetes can be an effective primordial intervention in the prevention of T2DM and its complications. The documented metabolic risk factors are dyslipidaemia, elevated blood pressure and plasma glucose, and central obesity give rise to prothrombotic and proinflammatory conditions shown by elevated concentration of inflammatory marker, C-reactive protein (CRP) [58]. Treatment of the metabolic syndrome is aimed at improving insulin resistance and weight loss through healthy lifestyle changes as a first-line in management. In patients with abnormal parameters for glucose, lipid and blood pressure, treatment of the individual components may yield a good result and positive impact on reducing overall cardiovascular disease risks, including stroke [58].

Identifying patients at risk and trying to prevent diabetes, not only lowering glucose levels once T2DM is diagnosed, is imperative to be emphasized. Genetic screening plays a crucial role in predicting both diabetes susceptibility and treatment response. Individual susceptibility to diabetes is significantly influenced by genetic variability. T2DM is recognized as a condition arising from an interplay of genetic predisposition and environmental factors, collectively increasing vulnerability to the disease [59]. Genetic variations in relevant genes have been identified as influencing insulin secretion and glucose metabolism, potentially compromising glycemic control [59]. Further studies for a better comprehensive understanding of the genetic factors as one of the underlying causes which may lead to diabetes can have significant implications in various aspects of T2DM intervention, including cardiovascular and stroke risk prediction, early detection, and the development of individualized intervention strategies [60, 61].

Telemedicine and remote patient monitoring are also being considered as part of advanced T2DM future management to enhance the accessibility of care and medication adherence [62]. Despite its progress, a few obstacles need to be addressed. These include providing adequate training, handling data ownership issues, and making the system applicable to larger populations to ensure that telemedicine and remote patient monitoring systems can work as a global, efficient, and sustainable health service. Hence, further research can be undertaken in this new direction of the therapeutic approach.

7.10 Conclusion

T2DM is a well-established independent risk factor for cerebrovascular accident or stroke, often coexisting with other cardiometabolic comorbidities, which further increase the risk of stroke. Higher plasma glucose levels predicted a poorer prognosis for stroke outcomes. Hence, establishing good glycaemic control is imperative in reducing stroke risk. Blood glucose and HbA1c targets should be tailored to each patient, taking into account factors such as age, existing comorbidities, duration of diabetes, and history of hypoglycemic events. An integrated and multidisciplinary approach to effectively managing diabetes and other concurrent cardiovascular risk factors has been demonstrated to reduce the risk of stroke and improve outcomes in post-stroke patients with T2DM. In addition to the medical intervention, healthy lifestyle changes, particularly focusing on individual nutritional management, are equally important in addressing the dual burden of these conditions.

References

1. Sapra, A., & Bhandari, P. (2024). *Diabetes*. StatPearls Publishing. [Updated 2023 Jun 21]. StatPearls [Internet]. https://www.ncbi.nlm.nih.gov/books/NBK551501/
2. Emerging Risk Factors Collaboration, Sarwar, N., Gao, P., Seshasai, S. R., Gobin, R., et al. (2010). Diabetes mellitus, fasting blood glucose concentration, and risk of vascular disease: A collaborative meta-analysis of 102 prospective studies. *Lancet (London, England), 375*(9733), 2215–2222. https://doi.org/10.1016/S0140-6736(10)60484-9
3. Ali, O. (2013). Genetics of type 2 diabetes. *World Journal of Diabetes, 4*(4), 114–123. https://doi.org/10.4239/wjd.v4.i4.114
4. Diagnosis of Diabetes and Prediabetes. National Institute of Diabetes and Digestive and Kidney Diseases (2014). Retrieved from https://web.archive.org/web/20160306174631/http://www.niddk.nih.gov/health-information/health-topics/Diabetes/diagnosis-diabetes-prediabetes/Pages/index.aspx
5. Deshpande, A. D., Harris-Hayes, M., & Schootman, M. (2008). Epidemiology of diabetes and diabetes-related complications. *Physical Therapy, 88*(11), 1254–1264. https://doi.org/10.2522/ptj.20080020
6. International Diabetes Federation (IDF). (2021). *IDF diabetes atlas* (10th ed.) Retrieved at: https://diabetesatlas.org/
7. Malaysian, C. P. G. (2020). *Management of type 2 diabetes mellitus* (6th ed.). Retrieved at: https://www.moh.gov.my/moh/resources/Penerbitan/CPG/Endocrine/CPG_T2DM_6th_Edition_2020_13042021.pdf
8. Malaysian National Diabetes Registry Report 2020. Retrieved at: https://www.moh.gov.my/moh/resources/Penerbitan/Rujukan/NCD/Diabetes/National_Diabetes_Registry_Report_2020.pdf
9. Akhtar, S., Nasir, J. A., Ali, A., et al. (2022). Prevalence of type-2 diabetes and prediabetes in Malaysia: A systematic review and meta-analysis. *PLoS One, 17*(1), e0263139. https://doi.org/10.1371/journal.pone.0263139
10. National Heart Lung and Blood Institute, USA (2014) What is a stroke?. Retrieved at: https://www.nhlbi.nih.gov/health/stroke
11. Mosenzon, O., Cheng, A. Y., Rabinstein, A. A., & Sacco, S. (2023). Diabetes and stroke: What are the connections? *Journal of Stroke, 25*(1), 26–38. https://doi.org/10.5853/jos.2022.02306

12. Chen, R., Ovbiagele, B., & Feng, W. (2016). Diabetes and stroke: Epidemiology, pathophysiology, pharmaceuticals and outcomes. *The American Journal of the Medical Sciences, 351*(4), 380–386. https://doi.org/10.1016/j.amjms.2016.01.011
13. Lau, L. H., Lew, J., Borschmann, K., et al. (2019). Prevalence of diabetes and its effects on stroke outcomes: A meta-analysis and literature review. *Journal of Diabetes Investigation, 10*, 780–792.
14. Zhang, L., Li, X., Wolfe, C. D. A., et al. (2021). Diabetes as an independent risk factor for stroke recurrence in ischemic stroke patients: An updated meta-analysis. *Neuroepidemiology, 55*, 427–435.
15. Wang, Y., Zhao, R., Wang, B., et al. (2022). The dose-response associations of sugar-sweetened beverage intake with the risk of stroke, depression, cancer, and cause-specific mortality: A systematic review and meta-analysis of prospective studies. *Nutrients, 14*(4), 777. https://doi.org/10.3390/nu14040777
16. Petersmann, A., Müller-Wieland, D., Müller, U. A., et al. (2019). Definition, classification and diagnosis of diabetes mellitus. *Experimental and Clinical Endocrinology & Diabetes, 127*, S1–S7.
17. Cerf, M. E. (2013). Beta cell dysfunction and insulin resistance. *Frontiers in Endocrinology, 4*, 37. https://doi.org/10.3389/fendo.2013.00037
18. Christensen, A. A., & Gannon, M. (2019). The beta cell in type 2 diabetes. *Current Diabetes Reports, 19*(9), 81. https://doi.org/10.1007/s11892-019-1196-4
19. Maida, C. D., Daidone, M., Pacinella, G., Norrito, R. L., Pinto, A., & Tuttolomondo, A. (2022). Diabetes and ischemic stroke: An old and new relationship an overview of the close interaction between these diseases. *International Journal of Molecular Sciences, 23*(4), 2397.
20. Jia, G., Bai, H., Mather, B., Hill, M. A., Jia, G., & Sowers, J. R. (2024). Diabetic vasculopathy: Molecular mechanisms and clinical insights. *International Journal of Molecular Sciences, 25, no. 2*, 804. https://doi.org/10.3390/ijms25020804
21. Aziz, S., Sheikh Ghadzi, S. M., Abidin, N. E., et al. (2019). Gender differences and risk factors of recurrent stroke in type 2 diabetic Malaysian population with history of stroke: The observation from Malaysian National Neurology Registry. *Journal of Diabetes Research, 2019*, 1794267. https://www.hindawi.com/journals/jdr/2019/1794267/
22. Brownlee, M. (2005). The pathobiology of diabetic complications: A unifying mechanism. *Diabetes, 54*, 1615–1625. https://doi.org/10.2337/diabetes.54.6.1615
23. Endemann, D. H., & Schiffrin, E. L. (2004). Endothelial dysfunction. *Journal of the American Society of Nephrology, 15*(8), 1983–1992.
24. Mota, R. I., Morgan, S. E., & Bahnson, E. M. (2020). Diabetic vasculopathy: Macro and microvascular injury. *Current Pathobiology Reports, 8*(1), 1–14. https://doi.org/10.1007/s40139-020-00205-x
25. Menezes, L. J., et al. (2011). Investigating vulnerable atheroma using combined (18)F-FDG PET/ET angiography of carotid plaque with immunohistochemical validation. *Journal of Nuclear Medicine, 52*(11), 1698–1703.
26. Poznyak, A., Grechko, A. V., Poggio, P., Myasoedova, V. A., Alfieri, V., & Orekhov, A. N. (2020). The diabetes mellitus–atherosclerosis connection: The role of lipid and glucose metabolism and chronic inflammation. *International Journal of Molecular Sciences, 21*, 1835.
27. Vaidya, V., Gangan, N., & Sheehan, J. (2015). Impact of cardiovascular complications among patients with type 2 diabetes mellitus: A systematic review. *Expert Review of Pharmacoeconomics & Outcomes Research, 15*, 487–497. https://doi.org/10.1586/14737167.2015.1024661
28. Hata, J., Arima, H., Rothwell, P. M., et al. (2013). Visit variability in systolic blood pressure on macrovascular and microvascular complications in patients with type 2 diabetes mellitus: The ADVANCE trial. *Circulation, 128*, 1325–1334. https://doi.org/10.1161/CIRCULATIONAHA.113.002717
29. Tuttolomondo, A., Pinto, A., Salemi, G., et al. (2008). Diabetic and non-diabetic subjects with ischemic stroke: Differences, subtype distribution and outcome. *Nutrition, Metabolism, and Cardiovascular Diseases, 18*, 152–157. https://doi.org/10.1016/j.numecd.2007.02.003

30. Emerging Risk Factors Collaboration Diabetes mellitus. (2010). Fasting blood glucose concentration, and risk of vascular disease: A collaborative meta-analysis of 102 prospective studies. *Lancet, 375*, 2215–2222.
31. Seyed Ahmadi, S., Svensson, A. M., Pivodic, A., et al. (2020). Risk of atrial fibrillation in persons with type 2 diabetes and the excess risk in relation to glycaemic control and renal function: A Swedish cohort study. *Cardiovascular Diabetology, 19*(1), 9. https://doi.org/10.1186/s12933-019-0983-1
32. Bohne, L. J., Johnson, D., Rose, R. A., et al. (2019). The association between diabetes mellitus and atrial fibrillation: Clinical and mechanistic insights. *Frontiers in Physiology, 26*(10), 135. https://doi.org/10.3389/fphys.2019.00135
33. Visseren, F. L. J., Mach, F., Smulders, Y. M., et al. (2021). ESC National Cardiac Societies; ESC Scientific Document Group. 2021 ESC Guidelines on cardiovascular disease prevention in clinical practice. *European Heart Journal, 42*(34), 3227–3337. https://doi.org/10.1093/eurheartj/ehab484. Erratum in: Eur Heart J. 2022;43(42):4468.
34. Van Sloten, T. T., Sedaghat, S., Carnethon, M. R., et al. (2020). Cerebral microvascular complications of type 2 diabetes: Stroke, cognitive dysfunction, and depression. *The Lancet Diabetes and Endocrinology, 8*(4), 325–336. https://doi.org/10.1016/S2213-8587(19)30405-X
35. Lo, J. W., Crawford, J. D., Samaras, K., et al. (2020). Association of prediabetes and type 2 diabetes with cognitive function after stroke: A STROKOG collaboration study. *Stroke, 51*(6), 1640–1646. https://doi.org/10.1161/STROKEAHA.119.028428
36. Mitsios, J. P., Ekinci, E. I., Mitsios, G. P., et al. (2018). Relationship between glycated hemoglobin and stroke risk: A systematic review and meta-analysis. *Journal of the American Heart Association, 7*(11), e007858.
37. Diabetes Management in Stroke_ Canadian Stroke Best Practices. 7e édition—2020 MISE À JOUR. Retrieved at: https://www.strokebestpractices.ca/recommendations/secondary-prevention-of-stroke/diabetes-management-in-stroke#:~:text=Lifestyle%20changes%2C%20tight%20glycemic%20control,can%20all%20have%20beneficial%20effects.
38. Lifestyle Behaviours and Risk Factor Management_ Canadian Stroke Best Practices. 7th Edition. 2020 UPDATED. Retrieved at: https://www.strokebestpractices.ca/recommendations/secondary-prevention-of-stroke/lifestyle-behaviours-and-risk-factor-management
39. Minari, T. P., Tácito, L. H. B., Yugar, L. B. T., Ferreira-Melo, S. E., Manzano, C. F., Pires, A. C., Moreno, H., Vilela-Martin, J. F., Cosenso-Martin, L. N., & Yugar-Toledo, J. C. (2023). Nutritional strategies for the management of type 2 diabetes mellitus: A narrative review. *Nutrients, 15*(24), 5096. https://doi.org/10.3390/nu15245096
40. Malaysian CPG. (2020). *Management of ischaemic stroke* (3rd ed.). Retrieved at: https://www.moh.gov.my/moh/resources/Penerbitan/CPG/CARDIOVASCULAR/CPG_Management_of_Ischaemic_Stroke_3rd_Edition_2020_28.02_.2021_.pdf
41. Estruch, R., Ros, E., Salas-Salvadó, J., et al. (2013). Primary prevention of cardiovascular disease with a Mediterranean diet. *The New England Journal of Medicine, 368*, 1279–1290.
42. Sabaté, E. (Ed.). (2003). *Adherence to long-term therapies: Evidence for action*. World Health Organization. Retrieved at: https://iris.who.int/bitstream/handle/10665/42682/9?sequence=1
43. Maahs, D. M., Daniels, S. R., de Ferranti, S. D., et al. (2014). Cardiovascular disease risk factors in youth with diabetes mellitus: A scientific statement from the American Heart Association. *Circulation, 130*, 1532–1558.
44. Kruyt, N. D., Biessels, G. J., Devries, J. H., et al. (2010). Hyperglycemia in acute ischemic stroke: Pathophysiology and clinical management. *Nature Reviews. Neurology, 6*(3), 145–155. https://doi.org/10.1038/nrneurol.2009.231
45. Maida, C. D., Daidone, M., Pacinella, G., et al. (2022). Diabetes and ischemic stroke: An old and new relationship an overview of the close interaction between these diseases. *International Journal of Molecular Sciences, 23*(4), 2397. https://doi.org/10.3390/ijms23042397
46. Bellolio, M. F., Gilmore, R. M., & Ganti, L. (2014). Insulin for glycaemic control in acute ischaemic stroke. *Cochrane Database of Systematic Reviews, 2014*(1), CD005346. https://doi.org/10.1002/14651858.CD005346.pub4

47. Jauch, E. C., Saver, J. L., Adams, H. P., Jr., et al. (2013). American Heart Association Stroke Council; Guidelines for the early management of patients with acute ischemic stroke: A guideline for healthcare professionals from the American Heart Association/American Stroke Association. *Stroke, 44*(3), 870–947. https://doi.org/10.1161/STR.0b013e318284056a
48. Gerstein, H. C., Miller, M. E., Byington, R. P., et al. (2008). Effects of intensive glucose lowering in type 2 diabetes. *The New England Journal of Medicine, 358*, 2545–2559. https://doi.org/10.1056/NEJMoa0802743
49. Patel, A., MacMahon, S., Chalmers, J., et al. (2008). Intensive blood glucose control and vascular outcomes in patients with type 2 diabetes. *The New England Journal of Medicine, 358*, 2560–2572. https://doi.org/10.1056/NEJMoa0802987
50. Gaede, P., Vedel, P., Larsen, N., et al. (2003). Multifactorial intervention and cardiovascular disease in patients with type 2 diabetes. *The New England Journal of Medicine, 348*, 383–393. https://doi.org/10.1056/NEJMoa021778
51. Gaede, P., Oellgaard, J., Carstensen, B., et al. (2016). Years of life gained by multifactorial intervention in patients with type 2 diabetes mellitus and microalbuminuria: 21 years follow-up on the Steno-2 randomised trial. *Diabetologia, 59*(11), 2298–2307. https://doi.org/10.1007/s00125-016-4065-6
52. Zhang, C., Sun, A., Zhang, P., et al. (2010). Aspirin for primary prevention of cardiovascular events in patients with diabetes: A meta-analysis. *Diabetes Research and Clinical Practice, 87*(2), 211–218. https://doi.org/10.1016/j.diabres.2009.09.029
53. Gaziano, M., Brotons, C., Coppolecchia, R., et al. (2018). Use of aspirin to reduce risk of initial vascular events in patients at moderate risk of cardiovascular disease (ARRIVE): A randomised, double-blind, placebo-controlled trial. *Lancet, 392*, 1036–1046.
54. The ASCEND Study Collaborative Group. (2018). Effects of aspirin for primary prevention in persons with diabetes mellitus. *The New England Journal of Medicine, 379*, 1529–1539.
55. Antithrombotic Trialists' Collaboration. (2002). Collaborative meta-analysis of randomized clinical trials of anti-platelet therapy for prevention of death, myocardial infarction and stroke in high-risk patients. *BMJ, 324*, 71–86.
56. Brown, M. T., & Bussell, J. K. (2011). Medication adherence: WHO cares? *Mayo Clinic Proceedings, 86*(4), 304–314. https://doi.org/10.4065/mcp.2010.0575
57. Kleindorfer, D. O., Towfighi, A., Chaturvedi, S., et al. (2021). Guideline for the prevention of stroke in patients with stroke and transient ischemic attack: A guideline from the American Heart Association/American Stroke Association. *Stroke, 52*(7), e364–e467. https://doi.org/10.1161/STR.0000000000000375. Epub 2021 May 24. Erratum in: Stroke. 2021;52(7):e483–e484.
58. Lau, D. C., Yan, H., & Dhillon, B. (2006). Metabolic syndrome: A marker of patients at high cardiovascular risk. *The Canadian Journal of Cardiology, 22*(Suppl B), 85B–90B. https://doi.org/10.1016/s0828-282x(06)70992-8
59. Bhattamisra, S. K., Siang, T. C., Rong, C. Y., et al. (2019). Type-3c diabetes mellitus, diabetes of exocrine pancreas—An update. *Current Diabetes Reviews, 15*(5), 382–394. https://doi.org/10.2174/1573399815666190115145702
60. Rescalli, A., Varoni, E. M., Cellesi, F., et al. (2022). Analytical challenges in diabetes management: Towards glycated albumin point-of-care detection. *Biosensors, 12*(9), 687. https://doi.org/10.3390/bios12090687
61. Pandit, S., Ho, A. C., & Yonekawa, Y. (2023). Recent advances in the management of proliferative diabetic retinopathy. *Current Opinion in Ophthalmology, 34*(3), 232–236. https://doi.org/10.1097/ICU.0000000000000946. Epub 2023 Mar 3.
62. Sugandh, F., Chandio, M., Raveena, F., et al. (2023). Advances in the management of diabetes mellitus: A focus on personalized medicine. *Cureus, 15*(8), e43697. https://doi.org/10.7759/cureus.43697

Exploring the Relationship Among Stroke, Hypertension, and Hyperlipidemia: A 10-Year Bibliometric Analysis

8

Wesam Ibrahim Abo-Elenien and Wael Mohamed

Abstract

Background: Stroke is a neurological disorder characterized by blockage of blood vessels. Clots form in the brain and interrupt blood flow, clogging arteries and causing blood vessels to break, leading to bleeding. Rupture of the arteries leading to the brain during stroke results in the sudden death of brain cells owing to a lack of oxygen. Stroke can also lead to depression and dementia. Hypertension and hyperlipidemia are the major risk factors of stroke. In this regard, knowledge mapping of the research linking stroke, hypertension and hyperlipidemia is necessary.

Objective: We perform a bibliometric analysis to explore the link between stroke, hypertension and hyperlipidemia in the last 10 years.

Methods: All publications related to this topic were retrieved from Scopus data base using the search terms (stroke AND hypertension AND hyperlipidemia). Using filters AND PUBYEAR > 2012 AND PUBYEAR < 2024 AND (LIMIT-TO (SUBJAREA , "PHAR") OR LIMIT-TO (SUBJAREA , "NEUR") OR LIMIT-TO (SUBJAREA , "MEDI")) AND (LIMIT-TO (DOCTYPE , "ar") OR LIMIT-TO (DOCTYPE , "re")) AND (LIMIT-TO (LANGUAGE , "English")). Microsoft office excel 2021 and VOSviewer (v.1.6.20) were used to conduct the bibliometric analysis.

Results: The filtered search identified 3496 articles published between 2013 to 2023. In terms of research and scientific collaboration in this field, the United States is in the lead, followed by China, Taiwan, Japan, south Korea and Germany.

W. I. Abo-Elenien
Clinical Department, Faculty of Pharmacy, Alexandria University, Alexandria, Egypt
e-mail: dc06202110129@alexu.edu.eg

W. Mohamed (✉)
Basic Medical Science Department, Kulliyyah of Medicine, International Islamic University Malaysia (IIUM), Kuantan, Pahang, Malaysia

W. Mohamed, M. B. Mat-Nor (eds.), *Nutrition and Stroke*, Nutritional Neurosciences, https://doi.org/10.1007/978-981-97-6818-9_8

Conclusion: The link between stroke, hypertension and hyperlipidemia was realized a decade ago and becomes increasingly important over time. Our findings can aid scholars in comprehending the global context of stroke, hypertension and hyperlipidemia relationship and provide significant insights for future researches.

Keywords

Bibliometric · Stroke · Hypertension · Hyperlipidemia

8.1 Introduction

Stroke is the third most common cause of disability and the second most common cause of death worldwide [1]. Stroke is a risk factor for one in four persons throughout their lives. Physical inactivity, alcohol abuse, tobacco use, and being overweight or obese are lifestyle risk factors for stroke. High blood pressure, high cholesterol, diabetes, and a personal or family history of heart attack or stroke are examples of medical risk factors. Low- and middle-income nations account for 87% of stroke-related fatalities and disability-adjusted life years, and an estimated 70% of strokes occur in these countries [2]. Also, the mortality and morbidity rates linked to stroke provide a clear picture of its significant consequences. Stroke not only has a significant impact on long-term impairment but also permanently alters one's physical, mental, and emotional health. The financial consequences are tangible and include direct medical expenditures, rehabilitation costs, as well as the indirect costs of missed production and caretaker load. Stroke is a permanent illness that requires lifetime medical care, rehabilitation, and lifestyle modifications. The fact that stroke is a chronic condition emphasizes the significance of taking preventative action to address modifiable risk factors, raising awareness, and encouraging lifestyle changes. In light of these considerations, stroke emerges not only as a health challenge but as a global health priority. Acknowledging its multifaceted impact is pivotal for mobilizing resources, fostering international collaboration, and implementing evidence-based strategies to curtail the incidence and repercussions of stroke on a global scale.

Bibliometrics is a branch of informatics that takes the literature system and bibliometric characteristics as the research object and conducts quantitative and qualitative analyses of literature. This method can quantitatively measure the contour distribution, relationship, and clustering of the research field [3] and has become one of the prevalent techniques for assessing academic work credibility, quality, and impact [4, 5]. In particular, the assessment might include the contributions and effect of various writers, nations/regions, organizations, fields, and journals; it can also evaluate the state, directions, and boundaries of research endeavors [6, 7].

VOSviewer is the most common used bibliometric visualization tool for data analysis and visualization [8].

8.2 Methodology

8.2.1 Data Collection and Search Strategy

Relevant publications were collected from Scopus database. The following search terms were employed to retrieve literature from Scopus database (TITLE-ABS-KEY (stroke AND hypertension AND hyperlipidemia) AND PUBYEAR > 2012 AND PUBYEAR < 2024 AND (LIMIT-TO (SUBJAREA , "PHAR") OR LIMIT-TO (SUBJAREA , "NEUR") OR LIMIT-TO (SUBJAREA , "MEDI")) AND (LIMIT-TO (DOCTYPE , "ar") OR LIMIT-TO (DOCTYPE , "re")) AND (LIMIT-TO (LANGUAGE , "English"))). A total of 3496 records were obtained from total 6200 after applied filters on them. The retrieved data were collected on January 6, 2024, to avoid any potential deviation due to daily updates. The metadata were downloaded and saved as .csv format.

8.2.2 Data Analysis and Visualization

The Microsoft Office Excel 2021, VOSviewer (v.1.6.20) analyzed all 3496 documents. VOSviewer (a bibliometric software) is a Java-based free software developed by Van Eck and Waltman of the center for science and technology studies (CWTS) of Leiden University in the Netherlands in 2009 (https://www.vosviewer.com/). It has a strong graphic ability and is suitable for processing large-scale data [8]. Figure 8.1 shows the flowchart for search strategy and selection process in this study.

8.3 Results

8.3.1 Temporal Distribution Map of the Literatures

The number of publications over a period can reflect the research speed and trend in this field [9]. As depicted in Fig. 8.2, there have been publications in this field since 2013, demonstrating that it has garnered interest. From 2013 to 2017 the number of publications was almost constant by average 187. In the recent 5 years, the number of publications has been increased to 462 in average, which reflect the extend of scientists' interest in this topic.

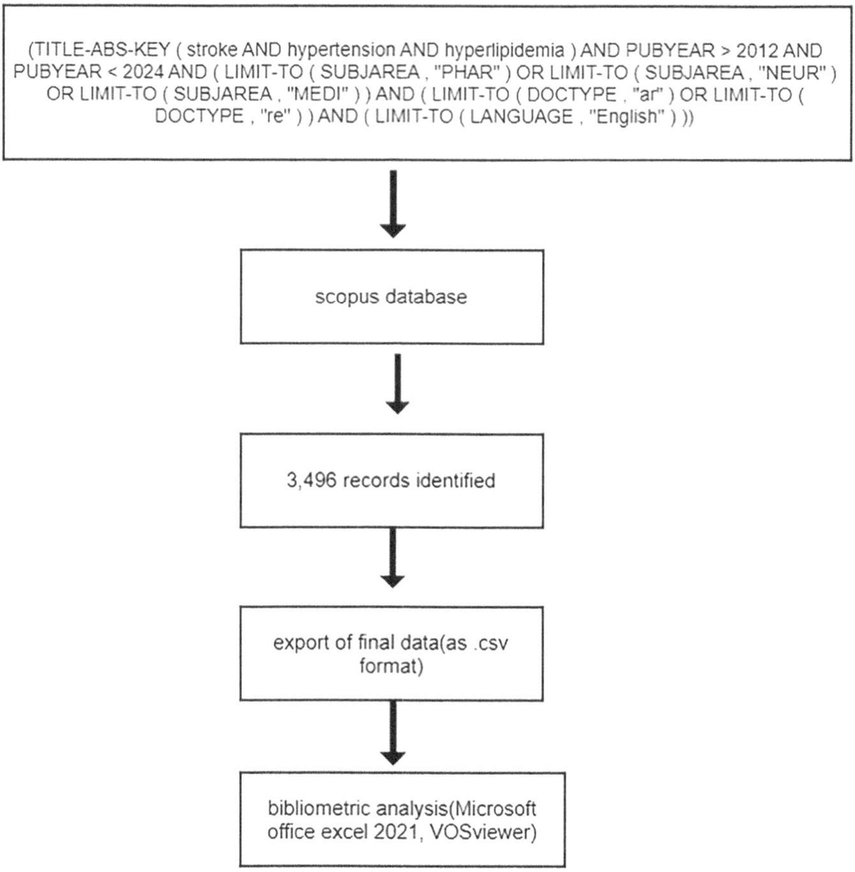

Fig. 8.1 The flowchart illustrating the search strategy and selection process

8.3.2 Distribution of Countries

As revealed in Fig. 8.3, largest number of publications was from USA, China and Taiwan. The total number of studies from the three countries accounted for more than half of the total reports.

VOSviewer parameters were set as follows: Method (Linlog/modularity) and a minimum number of country documents: 5. The obtained results were retrieved from 159 countries, with 56 meeting the thresholds. Based on co-authorship analysis, VOSviewer classifies countries into different clusters. Nodes with different colors in the figure represent different clusters. The size of nodes represents the number of documents, and the thickness of lines represents the number of connections between nodes. Figures 8.4 and 8.5 display the cooperation map of countries/regions.

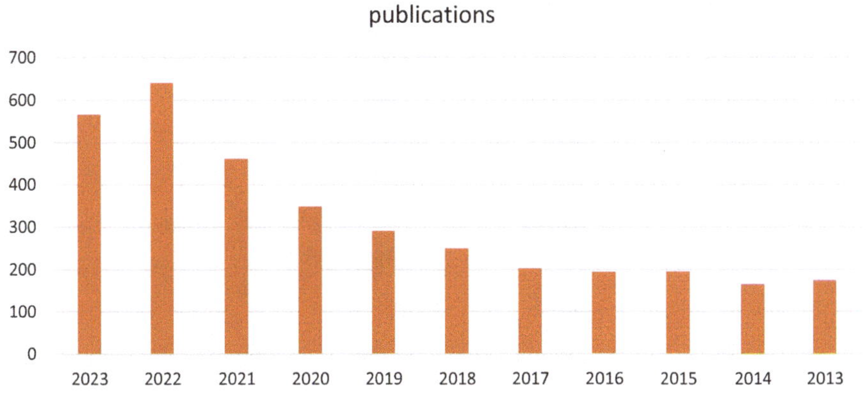

Fig. 8.2 Total number of annual publications

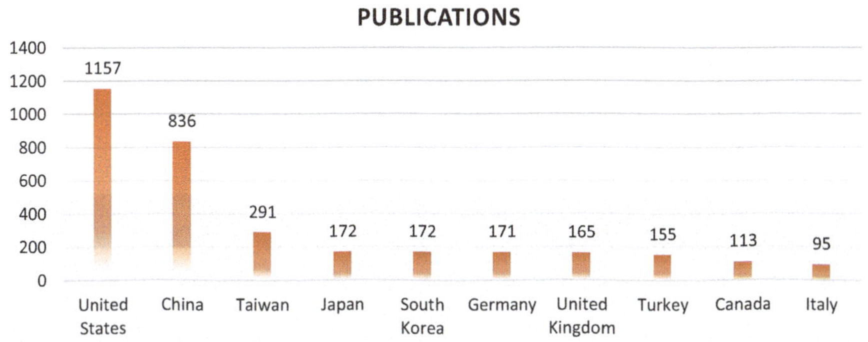

Fig. 8.3 Top 10 countries with higher number of publications

8.3.3 Distribution of Authors and Research Institutions

As revealed in Fig. 8.6, Kao, C.H. was the author with the most published articles (36 papers), followed by Wang, Y (27 papers).

VOSviewer parameters were set as follows: Method (Linlog/modularity) and a minimum number of documents of an author: 5. The obtained results were retrieved for 20,580 authors, and 236 met the thresholds. Based on co-authorship analysis, VOSviewer classifies authors into different clusters and colors them according to the time course in which they appear, superimposing time on the network of co-occurrence authors. Each node in Fig. 8.7 represents each author, the size of the circle reflects the number of articles published by researchers, and the line connecting the circles represents the co-occurrence relationship between authors.

As presented in Fig. 8.8, capital medical university (139 papers) was the institution that published the most research papers, followed by China medical university (100 papers) and Harvard medical school (98 papers).

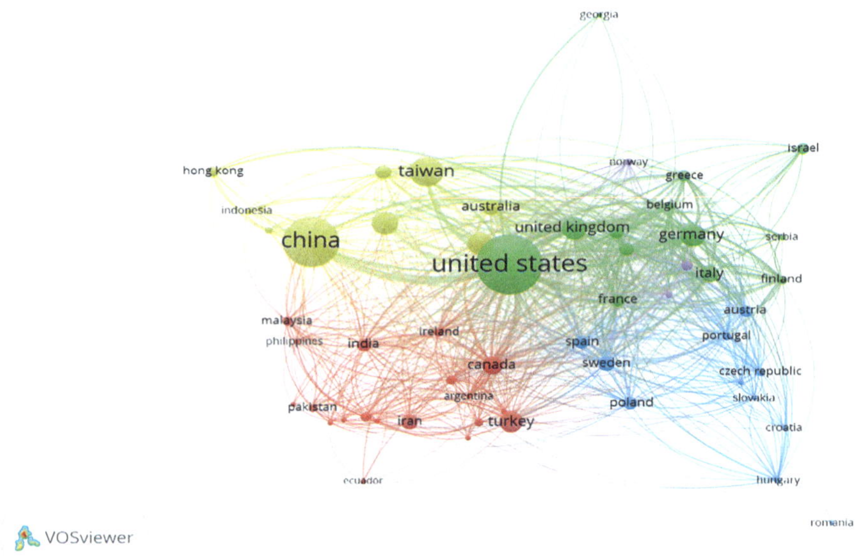

Fig. 8.4 A visual map for VOSviewer network

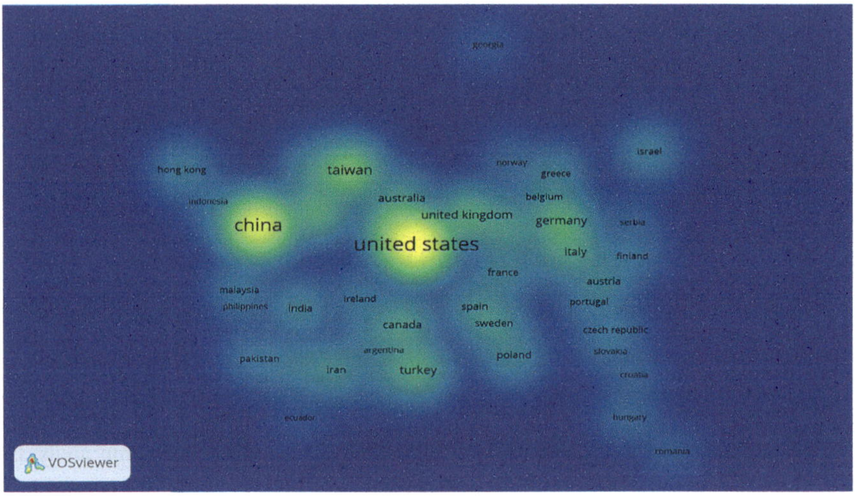

Fig. 8.5 Density visualization map of VOSviewer

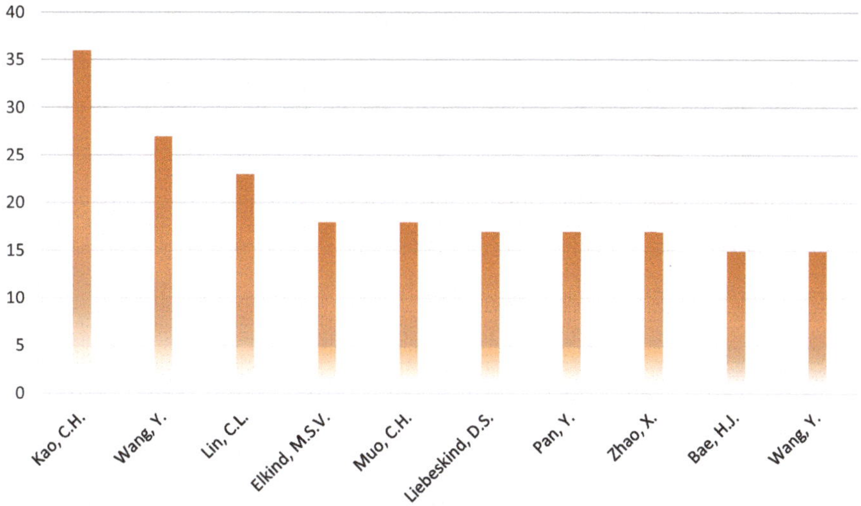

Fig. 8.6 Top 10 authors with higher number of publications

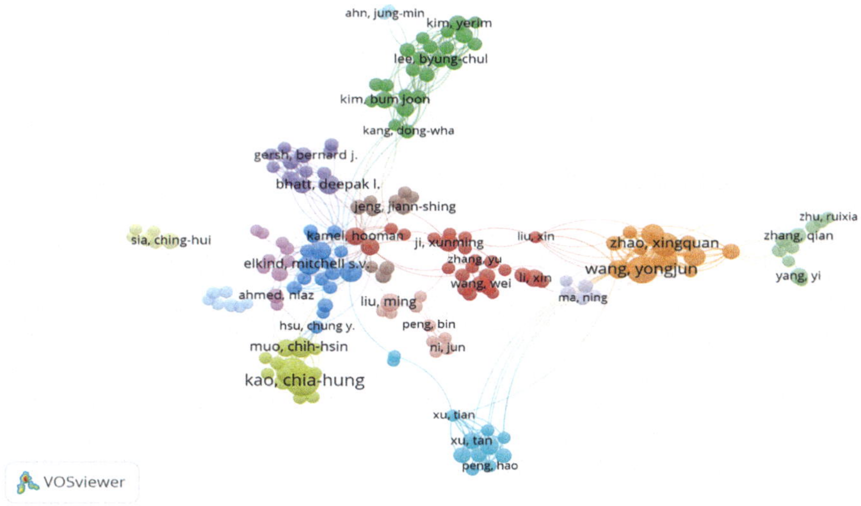

Fig. 8.7 A visual map for VOSviewer network among authors

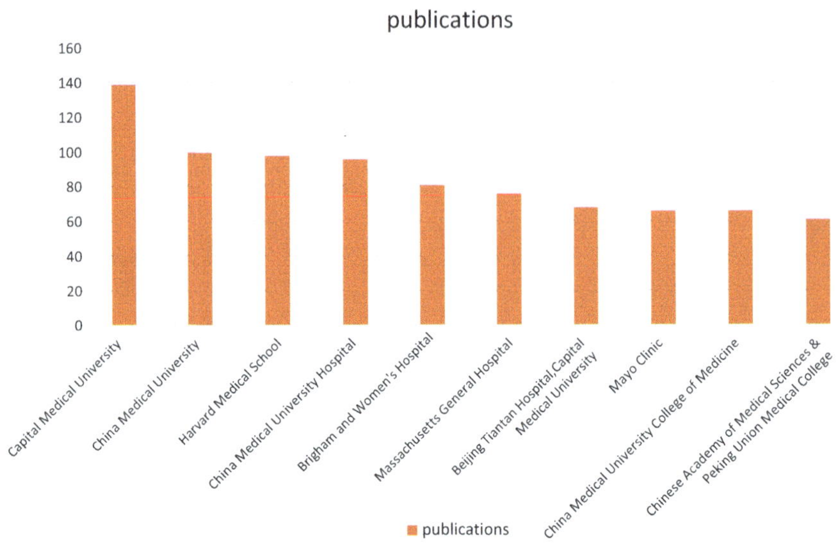

Fig. 8.8 Top 10 institutions that score the higher publications score

As listed in Table 8.1, *Journal of Stroke and Cerebrovascular Diseases* was the journal with the highest number of publications, followed by stroke.

8.3.4 Analysis of Key Words

VOSviewer parameters were as follows: Method (Linlog/modularity) and a minimum number of occurrences of a keyword: 50. There were 17,647 keywords, with 457 keywords meeting the thresholds. For each of 457 keywords, the total strength of co-occurrence links with other keywords was calculated. Keywords with the greatest total link strength were selected. The keyword co-occurrence network graph (Fig. 8.9) displays that the thicker the connection between nodes, the higher the frequency of two keywords appearing together.

8.4 Discussion

In this study, we analyzed stroke, hypertension and hyperlipidemia documents and reviewed the research results and progress using VOSviewer quantitative analysis software. The basic information of annual publication quantity, country, author, institution, and journal is quantitatively analyzed. Based on the statistical analysis of the number of papers published by various countries/regions and institutions, it is possible to identify key countries/regions and research institutions that have published many documents related to this topic and have greater influence and determine their cooperative relationship. USA, China, and Taiwan are the three top

Table 8.1 Top 5 journals and most cited article for 2023

Journal	TP	TC	Cite score	The most cited article (2023)	Times cited	Publisher
Journal of Stroke and Cerebrovascular Diseases	2357	10,216	4.3	Effects of virtual reality-based telerehabilitation for stroke patients: A systematic review and meta-analysis of randomized controlled trials	7	Elsevier
Stroke	1605	20,769	12.9	2023 guideline for the Management of Patients with Aneurysmal Subarachnoid Hemorrhage: A guideline from the American Heart Association/American Stroke Association	45	American Heart Association
Frontiers in neurology	8544	40,985	4.8	Myalgic Encephalomyelitis/Chronic Fatigue Syndrome is common in post-acute sequelae of SARS-CoV-2 infection (PASC): Results from a post-COVID-19 multidisciplinary clinic	13	Frontiers Media S.A.
Medicine (United States)	17,904	52,631	2.9	Non-small cell lung cancer (NSCLC): A review of risk factors, diagnosis, and treatment	12	Wolters Kluwer Health
Frontiers in Cardiovascular Medicine	4356	11,178	2.6	Sex and gender differences in myocarditis and dilated cardiomyopathy: An update	15	Frontiers Media S.A.

countries with higher number of publications related to this topic as displayed in Fig. 8.3. Among the top ten institutions, Capital Medical University is the institution with the most publications. From the top 10 authors, Kao, C.H. was the author with the most published articles (36 papers), followed by Wang, Y (27 papers) and Lin, C.L. (23 papers). According to journal distribution in Table 8.1, Journal of Stroke and Cerebrovascular Diseases is the journal with the highest number of publications about this topic followed by Stroke.

This review explores the intricate relationship between stroke, hypertension, and hyperlipidemia, highlighting the synergistic impact of these interconnected factors on cardiovascular health. Stroke, a leading cause of morbidity and mortality globally, is often influenced by modifiable risk factors such as hypertension and

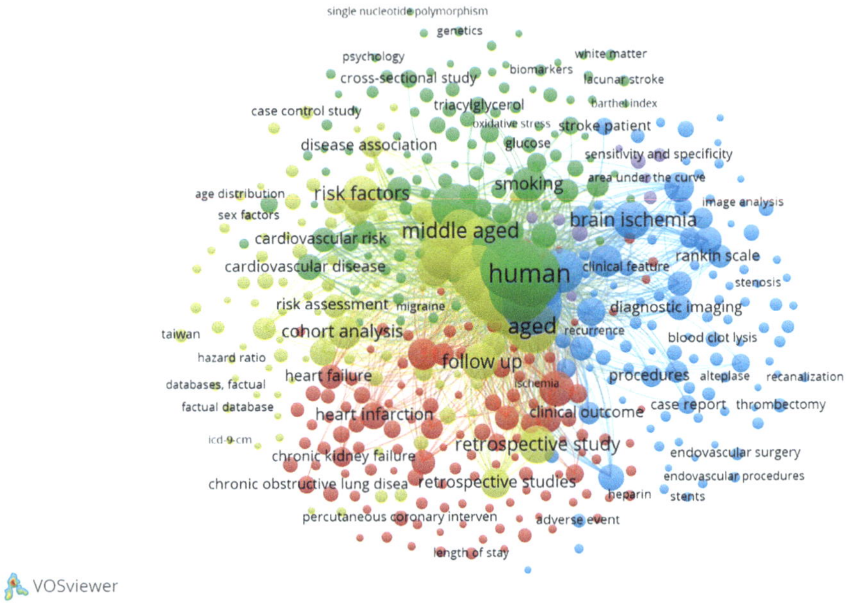

Fig. 8.9 The cluster of key words in the studies

hyperlipidemia. Understanding the complex interplay among these conditions is crucial for effective prevention and management strategies. The review delves into the mechanisms through which hypertension and hyperlipidemia contribute to stroke risk, emphasizing the importance of comprehensive approaches for mitigating these interconnected health challenges.

8.4.1 Stroke

Stroke is a neurological disorder characterized by blockage of blood vessels. Clots form in the brain and interrupt blood flow, clogging arteries and causing blood vessels to break, leading to bleeding. Rupture of the arteries leading to the brain during stroke results in the sudden death of brain cells owing to a lack of oxygen. Stroke can also lead to depression and dementia.

8.4.1.1 Types of Stroke

The bulk of stroke cases are divided into two categories. Both ischemic strokes and hemorrhagic strokes result in local hypoxia that destroys brain tissue. Ischemic strokes are triggered by the rupture of an internal blood vessel in the brain. Even though they are both dangerous and common, ischemic strokes account for 87% of all strokes in the United States [10].

Ischemic stroke: the more prevalent of the two types, occurs when an artery supplying blood to the brain is blocked, typically by a blood clot or plaque buildup. This type of stroke accounts for a majority of stroke cases.

- Subcategory: Thrombotic Strokes: These strokes manifest when a blood clot forms within one of the brain's arteries, often due to atherosclerosis, a condition marked by plaque accumulation in the arteries.
- Subcategory: Embolic Strokes: Embolic strokes occur when a blood clot or plaque fragment travels from another part of the body, primarily the heart, to the brain, blocking a cerebral artery.

Hemorrhagic Stroke: The less common but equally severe hemorrhagic stroke arises when a weakened blood vessel within the brain ruptures, leading to bleeding within the brain.

- Subcategory: Intracerebral Hemorrhage: This type occurs when a blood vessel ruptures, causing blood to leak into the surrounding brain tissue.
- Subcategory: Subarachnoid Hemorrhage: Subarachnoid hemorrhages involve bleeding into the space between the brain and its protective tissues. Blockages in ischemic stroke is blood clots that become stuck in one of the brain's arteries.

8.4.1.2 Stroke Risk Factors

Examining the risk factors linked to strokes reveals a complex interaction of circumstances that lead to this illness. These variables may be generally divided into two categories: modifiable and non-modifiable. Each has an important role in determining a person's risk of having a stroke.

Non-Modifiable Factors: The following are examples of nonmodifiable risk variables (also known as risk indicators) for stroke: sex, age, race/ethnicity, and genetics. Stroke is often considered an aging illness. After the age of 55, the incidence of stroke doubles every ten years as the patient ages [11].

Age determines the link between sex and stroke risk. Women are just as or more likely than men to suffer a stroke while they are younger, but as they get older, men's relative risk is slightly higher [12]. Younger women are more prone to have strokes because of hormonal variables including using hormonal contraceptives, as well as hazards associated with pregnancy and the postpartum period. Overall, more strokes occur in women than men, because of the longer life span of women compared with men [11, 13].

It is also well recognized that genetic variables are non-modifiable risk factors for stroke, with family and paternal history raising the risk of stroke. The hereditary risks of stroke differ by age, sex, and ethnicity, much like other stroke risk factors [14–16].

Modifiable Factors: Because stroke risk may be decreased by implementing intervention techniques targeted at lowering modifiable risk factors, these variables are extremely important. It is essential to identify and modify risk factors as soon as possible [11]. Behavioral risk factors and medical disorders are two more categories

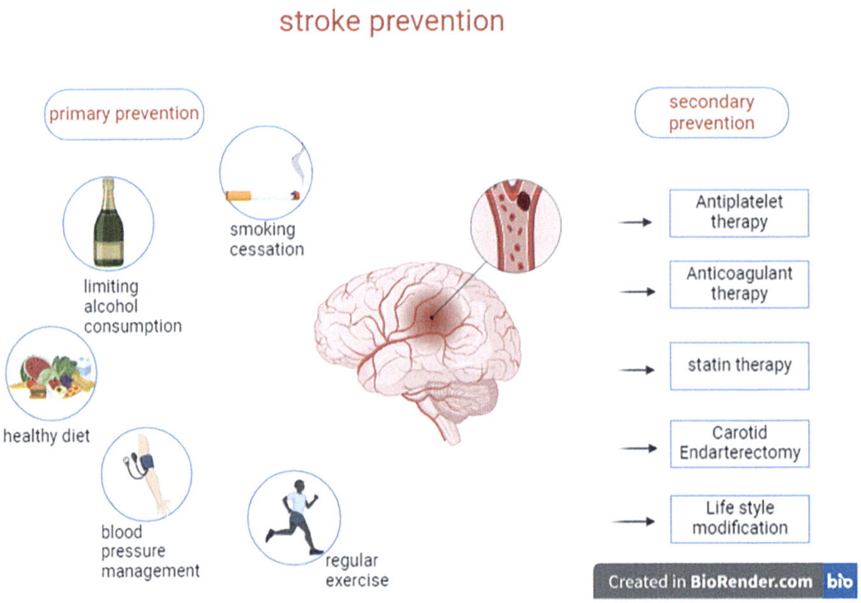

Fig. 8.10 Some of primary and secondary stroke prevention

of modifiable risk factors. Numerous conventional risk factors, including smoking, diabetes mellitus, hypertension, and hyperlipidemia, are known to contribute to stroke. Research is still ongoing in the field of novel or emerging risk factors.

8.4.1.3 Stroke Prevention

Stroke prevention is a crucial component of the complex web of healthcare that combines preventive measures with alert treatments to protect people from the catastrophic consequences of cerebrovascular accidents. Stroke is a major global cause of death and disability; thus, preventative programs must be comprehensive and include both primary and secondary prevention. The field of stroke prevention is broad and ever-changing, ranging from specialized therapies and early diagnosis to lifestyle adjustments and health promotion programs. This path toward stroke prevention is more than just a medical undertaking; it is a group effort to promote healthy communities, provide people with information, and accept a comprehensive concept of wellbeing. Let's dive into the many approaches, interventions, and shared accountability that support the goal of preventing strokes and preserving lives as we set out on our investigation.

Stroke prevention falls into two categories: primary prevention, which tries to keep people from having a stroke in the first place, and secondary prevention, which concentrates on keeping those who have already had a stroke from having another one (Fig. 8.10). Each point will be discussed later in section of prevention and management in detail.

8.4.2 Hypertension

Adults with a systolic blood pressure of 130 mmHg or higher and a diastolic blood pressure of 80 mmHg or higher are considered to have hypertension [17].

Range of reference: less than 120 mmHg at the systolic and 80 mmHg at the diastolic levels.

Elevated: diastolic less than 80 mmHg and systolic between 120- and 129-mmHg.

Systolic 130 to 139 mmHg or diastolic 80 to 89 mmHg is considered stage 1 hypertension.

Systolic or diastolic blood pressure of 140 mmHg or above indicates stage 2 hypertension.

Systolic pressure less than 130 mmHg and diastolic pressure of 80 mmHg or more is known as isolated diastolic hypertension [18].

8.4.2.1 Classification of Hypertension According to ACC/AHA HTN (Table 8.2) [17]

Causes and Risk Factors

Essential hypertension is considered idiopathic.

Secondary hypertension occurs as a result of the following (Fig. 8.11):

8.4.3 Hyperlipidemia

An imbalance in blood cholesterol levels, specifically between low-density lipoprotein cholesterol (LDL-C) and high-density lipoprotein cholesterol (HDL-C), is known as hyperlipidemia. An imbalance in the levels of LDL and HDL cholesterol can raise the risk of cardiovascular events, such as myocardial infarction and stroke. Other types of hyperlipidemias include mixed hyperlipidemia, which is characterized by increased levels of both triglycerides and cholesterol, and hypertriglyceridemia. Elevated LDL-C is linked to an increased risk of atherosclerotic cardiovascular disease (ASCVD), which includes coronary artery disease and stroke, and can cause plaques to accumulate in the arteries. Elevated HDL-C levels (≥60 mg/dL) can help lower the risk of ASCVD because HDL-C helps the body eliminate cholesterol [19].

Table 8.2 Blood pressure classifications

BP category	SBP		DBP
Normal	<120 mmHg	And	<80 mmHg
Elevated	120–129 mmHg	And	<80 mmHg
Hypertension			
Stage 1	130–139 mmHg	Or	80–89 mmHg
Stage 2	≥140 mmHg	Or	≥90 mmHg

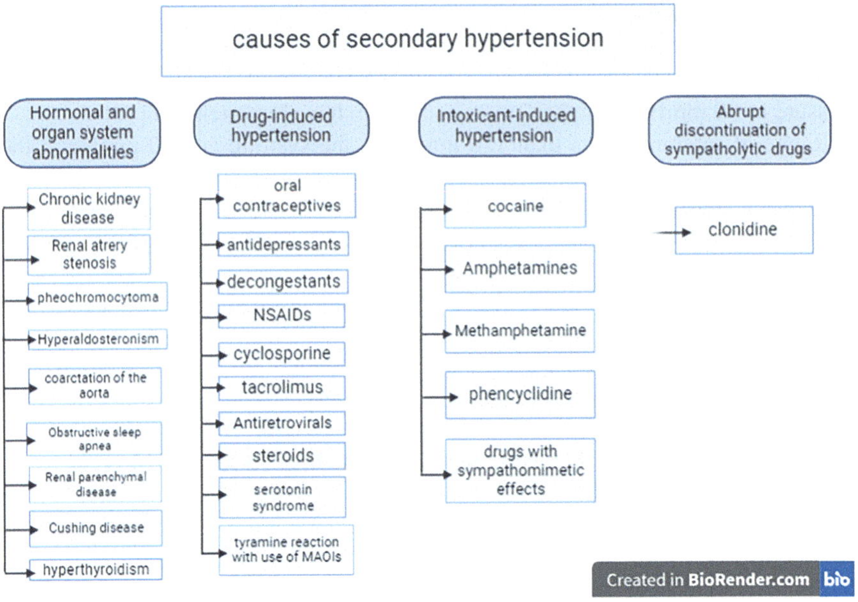

Fig. 8.11 Some causes of secondary hypertension

8.4.3.1 Risk Factors of Hyperlipidemia

Several factors increase risk of hyperlipidemia as shown in the next figure (Fig. 8.12):

8.4.4 The Link Between Hypertension and Stroke

Elevated blood pressure, or hypertension, is a significant risk factor for stroke. The relationship between elevated blood pressure and stroke risk is well-established, and controlling blood pressure is a key strategy for stroke prevention. Here are some key points to understand the connection between elevated blood pressure and stroke:

Increased Stress on Arteries: Elevated blood pressure puts increased stress on the walls of arteries. Over time, this can lead to damage and weakening of the arteries, making them more susceptible to the formation of blood clots or the development of aneurysms [20].

Atherosclerosis Formation: Hypertension is a major contributor to the development of atherosclerosis, a condition characterized by the buildup of fatty deposits (plaque) on the inner walls of arteries. Atherosclerosis can narrow and block arteries, reducing blood flow to the brain and increasing the risk of stroke [21].

Risk of Thrombosis: High blood pressure is associated with an increased risk of blood clot formation. Clots can travel to the brain and cause a blockage in a blood vessel, leading to an ischemic stroke. Ischemic strokes are the most common type of stroke and occur when blood flow to the brain is blocked.

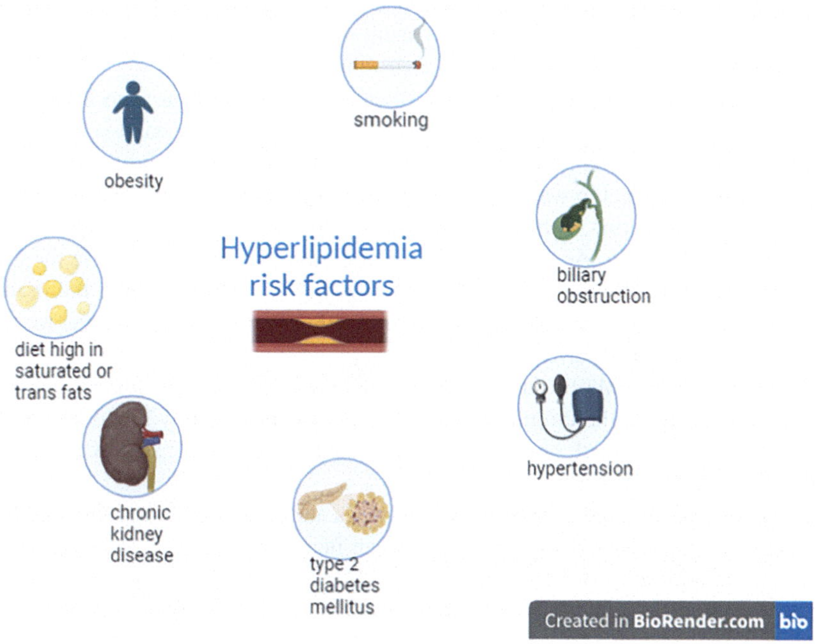

Fig. 8.12 Some examples of risk factors of hyperlipidemia

Hemorrhagic Stroke Risk: Elevated blood pressure can also contribute to the risk of hemorrhagic stroke, which occurs when a weakened blood vessel ruptures and causes bleeding in the brain. Hypertension can lead to the weakening of blood vessel walls, increasing the likelihood of a rupture [22].

Target Organ Damage: Chronic hypertension can cause damage to various organs, including the heart, kidneys, and blood vessels. The damage to blood vessels can further contribute to the risk of stroke [23].

8.4.5 The Role of Hyperlipidemia in Stroke Risk

High blood lipid levels, especially high LDL (low-density lipoprotein) cholesterol levels, are critical for the development of atherosclerosis, a significant risk factor for stroke. An analysis of the relationship between high lipid levels and atherosclerosis is provided below:

Lipid Accumulation: The build-up of lipids, especially low-density lipoprotein (LDL) cholesterol, in the arterial walls is known as atherosclerosis. Excessive cholesterol buildup in the artery walls, known as fatty plaques, can result from elevated low-density lipoprotein (LDL) cholesterol levels [24].

Fatty Plaque Formation: An inflammatory reaction is triggered by the buildup of low-density lipoprotein (LDL) cholesterol in the artery walls. Macrophages and other inflammatory cells move to the area, eat the cholesterol that has collected, and become foam cells. Fatty plaques are a result of the combination of these foam cells and other elements such as smooth muscle cells [25].

Plaque Growth and Rupture: Fatty plaques have the potential to enlarge over time, hardening and thickening the artery walls. Plaque progression is the term for this process. Additionally, the plaques may become brittle and prone to breaking. Blood clots can develop when a broken plaque exposes its contents, such as cholesterol and tissue factors, to blood components [26].

Artery Narrowing: As plaques develop and multiply, they may cause the artery lumen to narrow, lowering blood flow to important organs including the brain. A stroke is more likely when there is inadequate blood flow to the brain due to stenosis, a narrowing of the arteries [27].

Formation of Emboli: Atherosclerotic plaques occasionally release pieces or clots that can enter the circulation and lodge in the brain's tiny blood arteries. An embolic stroke, in which blood flow is obstructed to a particular region of the brain, may result from this [28].

Other Lipid-Related variables: Triglycerides and lipoprotein (a), two other lipid-related variables, have also been linked to an increased risk of atherosclerosis and stroke, in addition to LDL cholesterol. Accelerated plaque development and enhanced thrombogenicity have been associated with elevated triglyceride and high lipoprotein (a) levels [29].

8.4.6 Synergistic Effects

Individuals with hypertension and hyperlipidemia frequently coexist, and their combined detrimental effects on cardiovascular health might be amplified. Let's examine this connection and the ways in which they work together to raise the risk of cardiovascular issues.

Coexistence of Hypertension and Hyperlipidemia: Obesity, an unhealthy diet, a sedentary lifestyle, and a genetic predisposition are major risk factors for both conditions, which often coexist. Many people have both disorders coexisting because of these risk factors, which also have a role in the development of both conditions [21].

Synergistic Effects on Atherosclerosis: The accumulation of fatty deposits, or plaques, in the arteries is the hallmark of atherosclerosis, a disease for which both hypertension and hyperlipidemia are essential. Chronic hypertension damages the artery walls, increasing their vulnerability to the deleterious effects of lipids. Atherosclerotic plaques are created when triglycerides and cholesterol build up in the artery walls due to hyperlipidemia [30].

Vascular Damage: Blood vessels can sustain damage from both hypertension and hyperlipidemia on their own. Excessive strain applied to the artery walls by hypertension results in inflammation, oxidative stress, and endothelial damage. Because hyperlipidemia inhibits the synthesis of nitric oxide, a vital chemical for the upkeep of healthy blood vessels, it can lead to endothelial dysfunction. Combining these illnesses might worsen vascular damage and raise the possibility of problems [31].

Risk of Cardiovascular Events: Heart attacks and strokes are among the cardiovascular events that are greatly increased when hypertension and hyperlipidemia coexist. Atherosclerotic plaques are more likely to burst due to hypertension, which can result in blood clots that obstruct blood flow to the heart or brain. Unstable plaques that are more likely to burst are a result of hyperlipidemia. The chance of cardiovascular events is increased when these diseases coexist [32].

The total risk of stroke is considerably increased by the combined effects of hypertension, hyperlipidemia, and the risk factors that are linked with each condition. Let's examine the interactions between these risk factors and how they raise the likelihood of a stroke.

Synergistic Effect on Atherosclerosis: Atherosclerosis is a key underlying cause of stroke, and both hypertension and hyperlipidemia contribute to its development and progression. Because of the damage that hypertension causes to the artery walls, lipids, particularly cholesterol, are more likely to accumulate inside the vessel walls. Elevated levels of triglycerides and low-density lipoprotein cholesterol (LDL) are indicative of hyperlipidemia, which encourages the buildup of fatty deposits in the arteries. These risk factors work together to hasten the development of atherosclerotic plaques, which restrict the arteries and raise the risk of stroke [21].

Increased Blood Clot Formation: Blood clots can cause strokes and are also made more likely by hypertension and hyperlipidemia. Blood clots occur as a result of endothelium, the lining of blood vessels, being damaged by hypertension. Particularly when linked to elevated LDL cholesterol levels, hyperlipidemia plays a role in the development of unstable plaques that have the potential to burst and initiate the clotting process. These risk factors work together to strengthen the prothrombotic condition and raise the possibility of strokes caused by blood clots.

Effect on Small Blood Vessels: A condition known as cerebral small vessel disease is caused by both hypertension and hyperlipidemia, which can have a negative impact on the brain's tiny blood vessels. Chronic hypertension damages these arteries, causing them to thicken, constrict, and eventually stop blood flow. Small vascular disease is facilitated by hyperlipidemia, which also increases endothelial dysfunction and inflammation. When these risk factors combine, there may be a decrease in blood flow to vital brain areas, which raises the possibility of a stroke [33].

Exacerbation of Other Risk Factors: Hypertension and hyperlipidemia have the potential to aggravate additional risk factors for stroke. For instance, atrial fibrillation, a disorder marked by an erratic heartbeat and a separate risk factor for stroke, can be made more likely by hypertension. The development of diabetes and metabolic syndrome, both of which raise the risk of stroke, can be attributed to hyperlipidemia. The total risk of stroke increases when several risk factors overlap [34].

8.4.7 Advances in Understanding the Link

8.4.7.1 Free Radicals, Oxidative Stress, and Atherosclerosis [35]

In Atherosclerosis, Oxidative Stress: An imbalance between the body's antioxidant defenses and reactive oxygen species, or free radicals, is what defines oxidative stress. Within the framework of atherosclerosis, oxidative stress is essential for the start and development of plaque development. Low-density lipoprotein (LDL) cholesterol is oxidized by free radicals, which aids in the development of atherosclerotic plaques. The oxidative changes that follow increase the likelihood that immune cells may absorb LDL cholesterol and cause inflammation in the walls of blood vessels.

Function of NADPH Oxidase: Hypertension is frequently associated with an upregulation of NADPH oxidase, which is a key generator of reactive oxygen species in blood vessels. This particular enzyme system plays a major role in the oxidative stress linked to hypertension and hyperlipidemia. The vicious cycle of oxidative damage, endothelial dysfunction, and atherosclerosis is sustained by elevated NADPH oxidase activity.

Antioxidant Defense Mechanisms: The body's natural defenses against free radicals include enzymes that neutralize them, such as glutathione peroxidase and superoxide dismutase. On the other hand, prolonged exposure to high blood pressure and dyslipidemia might override these protective mechanisms, resulting in persistent oxidative stress.

8.4.7.2 Vascular Regulation and MicroRNAs [36]

Hypertension and Hyperlipidemia with MicroRNAs: Tiny RNA molecules called microRNAs control gene expression post-transcriptionally. The pathogenesis of hypertension and hyperlipidemia has been linked to certain miRNAs by recent study. Certain miRNAs are involved in the regulation of lipid metabolism, inflammation, and vascular function. The complex biochemical network connecting various diseases may be influenced by dysregulation of these miRNAs.

Prospective Therapeutic Targets: Gaining insight into the function of miRNAs provides opportunities for prospective therapeutic measures. One potential tactic to affect vascular function and lessen the incidence of atherosclerosis caused by hypertension and hyperlipidemia is to modify the expression of miRNA. To find miRNAs that may be used as therapeutic targets to prevent strokes, research in this field is still underway.

8.4.7.3 Customized Approaches to Therapy [37]

Genetic and Molecular Profiling: As precision medicine progresses, genetic and molecular profiling is used to customize treatment plans based on each patient's distinct features. In order to improve the selection of antihypertensive and lipid-lowering drugs, this strategy takes into account genetic differences, biomarker profiles, and other relevant parameters.

Pharmacogenomics: Research into genetic variants that affect a person's reaction to a particular medicine is the goal of pharmacogenomic investigations. Knowing a person's genetic composition can help choose drugs with less side effects and increased efficacy for treating hypertension and hyperlipidemia.

Tailored Lifestyle Interventions: Tailored treatment plans also include advice for a tailored lifestyle in addition to medication interventions. Stroke prevention may be approached holistically by taking into account variables including stress management practices, physical activity levels, and dietary choices.

8.4.7.4 Disruption of the Blood-Brain Barrier [38]

The semipermeable blood-brain barrier (BBB) is a highly selective barrier that divides the extracellular fluid of the brain from the blood that is circulated. It is essential for preserving the microenvironment of the brain and shielding it from potentially dangerous chemicals in the blood.

BBB Disruption and Hypertension: Hypertension compromises the integrity of the BBB by contributing to its disruption. High blood pressure causes the brain's fragile blood vessels to experience mechanical stress, which alters the BBB's structural makeup. This perturbation permits the entry of molecules that are ordinarily prohibited, such as immune cells and inflammatory mediators, into the brain tissue.

Effect on Stroke Risk: A weakened blood-brain barrier raises the possibility of brain injury and is linked to the etiology of strokes. It further contributes to the chain of events that might result in ischemia or hemorrhagic strokes by fostering an environment that is favorable for neuroinflammation and oxidative stress.

8.4.7.5 The Correlation Between Non-Alcoholic Fatty Liver Disease (NAFLD) and Cardiovascular Risk [39]

The buildup of fat in the liver in people who do not drink excessive amounts of alcohol is a defining feature of non-alcoholic fatty liver disease, or NAFLD. NAFLD is becoming more well acknowledged as a cardiovascular risk factor in addition to its effects on the liver.

Relationship between Hyperlipidemia and Stroke Risk: NAFLD and hyperlipidemia frequently coexist, which adds to the inflammatory milieu throughout the body. This inflammatory condition can raise the risk of atherosclerosis and, in turn, stroke, when combined with related metabolic disorders including insulin resistance and dyslipidemia.

Clinical Considerations: Reducing body weight and changing one's diet are part of managing non-alcoholic fatty liver disease (NAFLD). Not only is treating NAFLD essential for liver health, but it also reduces the risk of cardiovascular events, such as stroke.

8.4.7.6 Inflammatory Markers' Function [40]

Hyper-Inflammation and Heart Disease: One common factor in the intricate interactions among hypertension, hyperlipidemia, and stroke is inflammation. High concentrations of inflammatory markers, such as C-reactive protein (CRP), indicate that the vascular system is experiencing an inflammatory reaction.

Association with Stroke Risk: There is a correlation between elevated inflammatory markers and a higher risk of stroke. Within the framework of atherosclerosis, inflammation plays a role in the development of plaques and their propensity to burst, resulting in embolism or thrombosis, both of which can result in strokes.

Clinical Implications: By keeping an eye on inflammatory indicators, medical professionals can learn important details about a patient's cardiovascular risk profile. In those with increased inflammatory state, reducing inflammation with lifestyle changes and, occasionally, medication treatments, becomes a method for avoiding stroke.

8.4.8 Preventive Strategies and Management

The management and maintenance of hypertension, or high blood pressure, and hyperlipidemia, or high cholesterol, depend heavily on lifestyle changes and interventions. The main lifestyle changes and treatments for managing these disorders are summarized as follows:

8.4.8.1 Healthy Diet

Limit your salt consumption: Controlling hypertension requires limiting your sodium intake. The American Heart Association (AHA) states that people should aim to consume 1500 mg of salt daily, or less than 2300 mg per day [41].

Up your potassium intake: Research has shown that consuming more potassium lowers blood pressure. Consuming foods high in potassium, such as fruits, vegetables, and low-fat dairy products, is advised by the American Heart Association [42].

Place a strong emphasis on fruits and vegetables: Research has shown that the Dietary Approaches to Stop Hypertension (DASH) diet can lower blood pressure [43].

Opt for whole grains: They have been linked to better lipid profiles and lowered blood pressure due to their high fiber content [44].

Reduce your consumption of trans and saturated fats: Cutting back on these fats will help you control your hyperlipidemia. Intake of trans fats should be avoided, according to the American College of Cardiology (ACC) [45].

Eat lean proteins: As opposed to high-fat meats, lean protein sources including fish, poultry, lentils, and tofu are better [46].

Boost your intake of omega-3 fatty acids because they have been demonstrated to lower cholesterol and blood pressure. Eating walnuts, flaxseeds, and fatty fish can help raise your intake of omega-3 fatty acids [47].

8.4.8.2 Controlling Weight

It's critical to maintain a healthy weight in order to control hypertension and hyperlipidemia. Blood pressure can be lowered and lipid profiles can be improved with weight loss [17].

Caloric balance: Achieving and maintaining a healthy weight requires a balance between calorie consumption and physical activity [48].

8.4.8.3 Frequent Exercise

Regular aerobic exercise has been demonstrated to enhance lipid profiles and reduce blood pressure. AHA guidelines suggest engaging in moderate-intensity aerobic activity for at least 150 min every week [49]. Including strength training activities in your program can assist increase muscular strength and general health [50]. It's crucial to speak with a healthcare provider before beginning an exercise program, particularly if you have any underlying medical issues. They are able to offer tailored advice and suggestions.

8.4.8.4 Smoking Cessation

In order to control hypertension and lower the risk of cardiovascular illnesses, quitting smoking is imperative [51]. People can stop smoking with the use of counseling, medicine, and smoking cessation programs.

8.4.8.5 Limit Your Alcohol Consumption

For women, one drink per day and for males, two drinks per day is considered moderate alcohol intake. Drinking too much alcohol might impair lipid profiles and increase blood pressure [52].

8.4.8.6 Handling Stress

Deep breathing exercises, yoga, meditation, and mindfulness are a few stress-reduction practices that can help decrease blood pressure and enhance general well-being [53].

For the efficient management of hypertension and hyperlipidemia, a multidisciplinary strategy combining medical experts, patients, and public health efforts is essential. The following explains why every stakeholder is important:

8.4.8.7 Healthcare Professionals

Physicians that specialize in primary care, cardiology, and other medical fields are qualified to diagnose, treat, and keep an eye on hypertension and hyperlipidemia. They give individualized treatment programs, write prescriptions for drugs, and do routine follow-ups to evaluate the success of therapy. They can impart knowledge to patients on drug adherence, lifestyle changes, and the value of routine checkups. Medical practitioners are able to keep an eye on risk factors, modify treatment regimens, and handle any potential consequences.

8.4.8.8 Patients

It is the patients' responsibility to take charge of their own health. They must take an active role in their treatment program and alter their lifestyle as needed. Patients should take their medications on time, eat healthily, exercise frequently, give up smoking, and drink in moderation as directed. To have a better understanding of their disease and available treatments, they should ask questions, report any adverse effects or concerns, and communicate honestly with their healthcare professionals. Patients should schedule routine examinations and check-ups to keep an eye on their cholesterol, blood pressure, and general wellness.

8.4.8.9 Initiatives in Public Health

Initiatives related to public health are essential for raising awareness, encouraging preventative care, and putting policies in place to deal with hypertension and hyperlipidemia in the general population. Certain programs have the potential to inform the public on the signs, symptoms, and risk factors of certain illnesses. They can promote healthy lifestyle choices, support routine screenings, and offer tools for stress reduction, quitting smoking, and maintaining a nutritious diet. Campaigns for public health may focus on boosting physical activity, lowering salt consumption, and improving cardiovascular health in general. Collaboration and coordination among healthcare professionals, patients, and public health initiatives are necessary to achieve optimal outcomes in the prevention and management of hypertension and hyperlipidemia. By working together, they can improve awareness, provide comprehensive care, and implement strategies that address the broader societal impact of these conditions.

Patients with hypertension and hyperlipidemia can significantly reduce their risk of stroke by using pharmaceutical therapies. Here are some details on and their advantages:

8.4.9 Nutritional Strategies for Optimal Blood Pressure Management

Nutritional and dietary strategies are essential for controlling blood pressure. Eating a heart-healthy diet can improve cardiovascular health by lowering blood pressure. The following dietary guidelines are suggested:

Dietary Approaches to Stop Hypertension, or DASH Diet: Fruits, vegetables, healthy grains, lean meats, and low-fat dairy products are the mainstays of the DASH diet. It is linked to lowered blood pressure and promotes a decreased consumption of salt [54].

Rich in Magnesium Foods: The relaxing of blood vessels is aided by magnesium. Consume foods high in magnesium, such as whole grains, nuts, and seeds, and leafy green vegetables [55].

Foods High in Calcium: Calcium is good for your heart health in general. Include foods high in calcium, such as dairy, leafy greens, and fortified plant-based milk substitutes [56].

Limit Alcohol Intake: Excessive drinking can cause blood pressure to rise, while moderate drinking may offer some cardiovascular advantages. The key is moderation [57].

8.4.10 Lipid Lowering and Diet

Synergy between Diet and Statins: A family of drugs that is frequently recommended, statins, successfully decrease LDL cholesterol levels. Statin medication works better in controlling lipid profiles when combined with a heart-healthy diet [58].

Dietary Patterns and Sustainable Lipid Management: Adopting heart-healthy eating habits, such the DASH or Mediterranean diets, offers a long-term basis for controlling cholesterol levels and enhancing general cardiovascular health [59].

Red Yeast Rice as a Natural Alternative: The potential lipid-lowering benefits of red yeast rice, a supplement containing chemicals comparable to statins, are investigated. However, using it calls for close observation and thought [60].

Vitamin B3 (niacin) and HDL Boost: Although niacin may raise HDL cholesterol, its administration should be done so with caution because of possible negative effects. Niacin's ability to modulate lipid levels is enhanced when combined with dietary changes [61].

Plant Sterols and Stanols in Daily Nutrition: By preventing the absorption of LDL cholesterol, foods fortified with plant sterols and stanols help reduce LDL cholesterol levels. Including them in a well-balanced diet helps to control blood cholesterol levels generally [62].

Omega-3 Fatty Acids for Triglyceride Control: Adding omega-3 fatty acids to your diet—found in foods like walnuts, flaxseeds, and fatty fish—will help reduce your triglyceride levels and enhance the benefits of prescription lipid-lowering drugs [63].

8.4.11 The Interplay Between Antiplatelet Therapy and Nutritional Strategies for Cardiovascular Wellness

People who have been diagnosed with or are at risk for cardiovascular disorders are frequently recommended antiplatelet medication because it inhibits platelet aggregation, which helps prevent blood clots from forming. Although drugs such as aspirin are frequently employed for this goal, certain nutritional and dietary elements can also contribute to the maintenance of cardiovascular health. Here are some considerations:

Omega-3 Fatty Acids: Omega-3 fatty acids, found in fatty fish (such as salmon, mackerel, and trout), flaxseeds, chia seeds, and walnuts, have anti-inflammatory properties and may help reduce platelet aggregation [64].

Garlic: Garlic has antiplatelet and antithrombotic effects. It may help inhibit platelet aggregation and reduce the risk of blood clot formation [65].

Ginger: Ginger has antiplatelet and anti-inflammatory properties. It may help prevent platelet aggregation and improve cardiovascular health [66].

Turmeric/Curcumin: Curcumin, the active compound in turmeric, has antiplatelet and anti-inflammatory effects. It may help prevent the formation of blood clots [67].

Green Tea: mGreen tea contains catechins, which have antiplatelet and anticoagulant effects. Regular consumption may contribute to cardiovascular health [68].

Vitamin E: Vitamin E, found in nuts, seeds, and vegetable oils, may have antiplatelet effects. However, it's important to note that high doses of vitamin E supplements may not be advisable [69].

Fruits and Vegetables: A diet rich in fruits and vegetables provides antioxidants and anti-inflammatory compounds, which can contribute to overall cardiovascular health [70].

8.4.12 Close Monitoring and Adjustments

8.4.12.1 Monitoring of Blood Pressure

Blood pressure monitoring on a regular basis is crucial for those with hypertension. This entails taking blood pressure during routine medical examinations or at home with a blood pressure monitor. Although certain health factors may influence blood pressure objectives, the basic objective is to keep blood pressure below 130/80 mmHg for the majority of people. By monitoring, medical professionals may evaluate how antihypertensive drugs are working and make any necessary modifications [71].

8.4.12.2 Monitoring Lipid Profile

To monitor cholesterol, people with hyperlipidemia should have their lipid profile evaluated on a regular basis. This comprises triglycerides, total cholesterol, high-density lipoprotein (HDL) cholesterol, and low-density lipoprotein (LDL) cholesterol. Although certain risk factors may influence target cholesterol levels, decreasing LDL cholesterol is frequently the main goal. Healthcare professionals can assess and make required modifications to cholesterol-lowering drugs by keeping an eye on patient lipid levels [72].

8.4.12.3 Tests for Kidney Function

Certain antihypertensive drugs, especially ARBs and ACE inhibitors, may have an adverse effect on renal function. Consequently, it's critical to regularly assess kidney function with tests like serum creatinine and estimated glomerular filtration rate (eGFR). Modifications in renal function may need a change in medication dose or the investigation of substitute drugs [73].

8.4.12.4 Monitoring Electrolyte Levels

Certain antihypertensive drugs, particularly diuretics, can have an impact on potassium and other electrolyte levels. Monitoring potassium levels and other electrolytes on a regular basis helps guarantee that the balance is maintained. Changes to medication schedules or dietary guidelines may be necessary in response to electrolyte imbalances [17].

8.4.12.5 Compliance with Medication

One aspect of close observation is evaluating the patient's compliance with recommended drug regimens. Treatment efficacy can be severely impacted by non-adherence. During check-ups, medical professionals may question about medication adherence and investigate any difficulties or worries the patient may have [74].

8.4.12.6 Evaluation of Adverse Effects

Monitoring is keeping an eye out for any possible drug side effects or bad responses. Common adverse effects might be exhaustion, lightheadedness, or gastrointestinal problems. People are urged to notify their healthcare professionals right away if they have any strange symptoms.

8.4.12.7 Frequent Follow-Up Meetings

Patients are scheduled for routine follow-up sessions with healthcare professionals to go over monitoring findings, talk about any changes in their health state, and modify treatment plans as needed. These consultations offer the chance for further education, supporting lifestyle changes, and addressing any worries the person may have [75].

8.4.12.8 Working Together to Make Decisions

A cooperative connection between patients and healthcare professionals is fostered via close observation. Treatment regimens can be modified through shared decision-making in accordance with each patient's preferences, way of life, and reaction to medicine [76].

8.4.12.9 Education and Promotion of Health

Monitoring aims to promote general health and well-being in addition to evaluating clinical data. An essential component of the monitoring process is education on food choices, physical exercise, and lifestyle adjustments.

Close observation maximizes the treatment of hypertension and hyperlipidemia by ensuring that each patient receives individualized, high-quality therapy. Maintaining good communication between patients and medical professionals is essential to meeting treatment objectives, averting problems, and improving general cardiovascular health.

8.5 Conclusion

In summary, the complex story of the relationship between hypertension, hyperlipidemia, and stroke is played out within the context of cardiovascular health. The mutually reinforcing nature of these circumstances emphasizes the necessity of a strong defense against the epidemic of strokes. By means of strict blood pressure regulation, focused hyperlipidemia treatment, and all-encompassing lifestyle adjustments, people can strengthen their resistance and lessen the effects of these interrelated risk factors. The quest to prevent strokes is more than just a medical undertaking; it is a social effort to promote a culture of health and wellbeing. In order to create a better tomorrow, healthcare professionals, legislators, and the general public must work together to reduce strokes, save lives, and understand the complex relationship between hypertension, hyperlipidemia, and stroke.

8.6 Challenges and Perspectives

8.6.1 Challenges

1. Awareness and Screening: Lack of awareness and routine screening for hypertension and hyperlipidemia remains a significant challenge. Many individuals with these conditions may go undiagnosed or untreated, increasing their risk of stroke. Efforts are needed to raise public awareness about the importance of regular screenings, especially among high-risk populations, such as those with a family history of cardiovascular diseases or certain lifestyle factors.
2. Treatment Adherence: Adherence to medication and lifestyle modifications can be challenging for individuals with hypertension and hyperlipidemia. Long-term commitment to medication regimens, dietary changes, exercise routines, and lifestyle modifications can be difficult for some patients. Factors such as cost, side effects, forgetfulness, and lack of support or understanding from family and peers can contribute to poor treatment adherence. Healthcare providers should focus on patient education, counseling, and ongoing support to improve adherence rates.
3. Complex Interactions: The relationship between stroke, hypertension, and hyperlipidemia is complex, with multiple interrelated factors at play. These conditions often coexist and interact with each other, making it challenging to determine the precise contribution of each factor and establish optimal treatment strategies. The underlying mechanisms connecting these conditions require further research and understanding to develop targeted interventions.
4. Risk Stratification: Accurately stratifying the risk of stroke in individuals with hypertension and hyperlipidemia poses a challenge. While these conditions are known risk factors for stroke, not all individuals with elevated blood pressure or lipid levels will inevitably develop a stroke. Improved risk assessment tools that consider multiple factors, such as age, gender, smoking status, diabetes, and other comorbidities, are needed to better predict an individual's risk and guide treatment decisions effectively.
5. Socioeconomic Factors: Socioeconomic factors can significantly impact the management of stroke, hypertension, and hyperlipidemia. Disparities in access to healthcare services, including preventive screenings, medications, and specialized care, can hinder early detection and appropriate management. Additionally, the availability of affordable healthy food options and opportunities for physical activity may vary across different socioeconomic groups. Addressing these disparities is crucial to ensure equitable prevention and management strategies.
6. Multimorbidity: Individuals with stroke, hypertension, and hyperlipidemia often have other comorbid conditions, such as diabetes, obesity, or chronic kid-

ney disease. The presence of multiple medical conditions can complicate treatment regimens and increase the risk of adverse events. Coordinating care among different specialties and ensuring effective communication among healthcare providers is essential to provide comprehensive and coordinated care for individuals with multimorbidity.

7. Translating Research into Practice: Bridging the gap between research findings and clinical practice poses a challenge. While scientific advancements continue to enhance our understanding of the link between stroke, hypertension, and hyperlipidemia, it can take time for these findings to be incorporated into routine clinical care. Translating evidence-based guidelines and recommendations into actionable steps in real-world healthcare settings is crucial to improve patient outcomes.
8. Health Disparities: Disparities in healthcare access and outcomes among different populations present a significant challenge. Certain demographic groups, such as racial and ethnic minorities, individuals with lower socioeconomic status, and those living in rural areas, may face barriers to accessing adequate healthcare services. These disparities can contribute to disparities in the prevention, diagnosis, and management of stroke, hypertension, and hyperlipidemia, resulting in poorer health outcomes.
9. Behavioral and Lifestyle Factors: Modifying behaviors and adopting healthy lifestyle habits can be challenging for individuals with stroke, hypertension, and hyperlipidemia. Unhealthy behaviors such as poor diet, sedentary lifestyle, smoking, and excessive alcohol consumption can exacerbate these conditions and increase the risk of stroke. Encouraging behavior change and providing support for individuals to make sustainable lifestyle modifications is crucial but can be a complex and ongoing process.
10. Treatment Complexity: The management of stroke, hypertension, and hyperlipidemia often involves multiple medications, lifestyle modifications, and regular monitoring. The complexity of treatment regimens can pose challenges for patients, leading to medication errors, treatment non-compliance, and suboptimal control of risk factors. Simplifying treatment regimens, enhancing patient education, and providing clear instructions and support can help overcome these challenges.
11. Cost of Treatment: The cost of medications, diagnostic tests, and regular follow-up visits can be a significant burden for individuals with stroke, hypertension, and hyperlipidemia, especially for those without adequate health insurance coverage. High healthcare costs may limit access to essential medications and preventive services, leading to suboptimal management and increased risk of complications. Addressing affordability issues and expanding access to affordable healthcare services is crucial to ensure equitable care for all individuals.
12. Aging Population: As the global population continues to age, the prevalence of stroke, hypertension, and hyperlipidemia is expected to increase. Aging is a significant risk factor for these conditions, and the management of stroke and cardiovascular risk factors in older adults can be complex due to the presence of multiple comorbidities and age-related physiological changes. Tailoring

treatment strategies and guidelines to the unique needs of the aging population is a challenge that requires specific attention and research.

13. Patient and Caregiver Education: Education and health literacy play a vital role in the prevention and management of stroke, hypertension, and hyperlipidemia. However, limited health literacy among patients and caregivers can hinder their understanding of the conditions, treatment options, and self-management strategies. Providing accessible and culturally sensitive educational resources, promoting health literacy, and involving caregivers in the care process can help overcome these challenges.
14. Data Integration and Information Sharing: Effective management of stroke, hypertension, and hyperlipidemia requires the integration and sharing of health information across different healthcare settings and providers. However, fragmented healthcare systems and lack of interoperability between electronic health records can impede seamless data exchange. Implementing health information exchange systems and standardized protocols for data sharing can enhance care coordination and improve patient outcomes.
15. Prevention and Public Health Efforts: While there have been significant advancements in the treatment of stroke, hypertension, and hyperlipidemia, prevention efforts remain crucial. Public health strategies targeting the promotion of healthy lifestyles, early detection, and management of risk factors need to be strengthened. Implementing population-based interventions, such as community screenings, health education campaigns, and policy changes to create supportive environments, can have a substantial impact on reducing the burden of these conditions.

Addressing these challenges requires a comprehensive and multi-faceted approach, involving healthcare providers, policymakers, researchers, communities, and individuals themselves. By focusing on health equity, improving access to care, enhancing patient education and empowerment, and strengthening preventive efforts, we can overcome these challenges and improve the prevention, diagnosis, and management of stroke, hypertension, and hyperlipidemia.

Addressing these challenges requires collaborative efforts from healthcare providers, policymakers, researchers, and the community. By raising awareness, improving treatment adherence, fostering research and innovation, addressing socioeconomic disparities, and promoting integrated care models, we can overcome these challenges and make significant progress in preventing and managing stroke, hypertension, and hyperlipidemia effectively.

8.6.2 Perspectives

1. Integrated Approach: A comprehensive and integrated approach is crucial in managing the link between stroke, hypertension, and hyperlipidemia. Healthcare providers should consider the interplay between these conditions and develop personalized treatment plans tailored to each patient's needs. Collaborative care

involving primary care physicians, cardiologists, neurologists, and other specialists can optimize treatment outcomes.

2. Early Intervention: Early identification and intervention are key in preventing stroke. Timely diagnosis and treatment of hypertension and hyperlipidemia can help prevent or delay the onset of stroke and its complications. Routine screenings, especially in high-risk individuals, can aid in early detection and prompt management.
3. Patient Education: Educating patients about the relationship between hypertension, hyperlipidemia, and stroke is essential. Raising awareness about the importance of lifestyle modifications, such as adopting a heart-healthy diet, engaging in regular physical activity, and managing stress, can empower individuals to take an active role in their own health. Patient education should also address the significance of medication adherence and regular follow-up appointments.
4. Multidisciplinary Collaboration: Collaboration among healthcare professionals is crucial for managing stroke, hypertension, and hyperlipidemia effectively. A multidisciplinary approach ensures comprehensive care, shared decision-making, and optimal outcomes. Healthcare providers should work together to develop standardized protocols, guidelines, and referral systems to streamline care delivery.
5. Research and Innovation: Continued research is necessary to better understand the underlying mechanisms linking hypertension, hyperlipidemia, and stroke. Advancements in diagnostic tools, therapeutic strategies, and personalized medicine can further improve prevention and management approaches. Ongoing research should focus on identifying novel therapeutic targets, exploring lifestyle interventions, and evaluating the long-term outcomes of different treatment approaches.
6. Personalized Medicine: Advances in personalized medicine hold great potential in the management of stroke, hypertension, and hyperlipidemia. Genetic testing and biomarker profiling can aid in identifying individuals at higher risk and enable targeted interventions. Tailoring treatment plans based on an individual's genetic makeup, metabolic profile, and response to medications can optimize therapeutic outcomes.
7. Digital Health Technologies: The integration of digital health technologies, such as wearable devices and mobile applications, can revolutionize the management of stroke, hypertension, and hyperlipidemia. These technologies can facilitate real-time monitoring of blood pressure, lipid levels, and other relevant health parameters. They can also provide educational resources, reminders for medication adherence, and personalized feedback, empowering individuals to actively participate in their own care.
8. Lifestyle Modification Programs: Comprehensive lifestyle modification programs can play a significant role in preventing and managing stroke, hypertension, and hyperlipidemia. These programs should focus on promoting healthy eating habits, regular exercise, stress reduction techniques, and smoking cessation. Implementing community-based initiatives, workplace wellness programs,

and educational campaigns can support individuals in making sustainable lifestyle changes.

9. Collaborative Care Models: Collaborative care models involving healthcare professionals, patients, and caregivers can enhance the management of stroke, hypertension, and hyperlipidemia. Coordinated efforts among primary care physicians, specialists, pharmacists, and allied health professionals can ensure seamless transitions of care, effective medication management, patient education, and continuous support. This team-based approach can improve treatment outcomes and patient satisfaction.
10. Health Policy and Prevention Strategies: Health policies and prevention strategies play a crucial role in addressing the link between stroke, hypertension, and hyperlipidemia. Governments, healthcare organizations, and policymakers should prioritize the development and implementation of comprehensive public health initiatives. These initiatives should focus on raising awareness, improving access to healthcare services, promoting healthy lifestyles, and reducing risk factors. Additionally, policies that support affordable access to medications, cardiovascular screenings, and preventive care can contribute to better population health outcomes.
11. Patient Empowerment and Self-Management: Empowering patients to actively participate in their own care is vital. Patient education, self-management tools, and support networks can encourage individuals to take ownership of their health and make informed decisions. Promoting health literacy, shared decision-making, and self-monitoring can foster a sense of responsibility and improve treatment adherence.
12. Long-term Follow-up and Rehabilitation: Stroke survivors require long-term follow-up and rehabilitation to optimize their recovery and prevent recurrent events. Comprehensive post-stroke care should include regular assessments of blood pressure, lipid levels, and adherence to medication. Rehabilitation programs focusing on physical therapy, occupational therapy, and lifestyle modifications can improve functional outcomes and quality of life.

These perspectives highlight the importance of personalized, integrated, and patient-centered approaches to managing stroke, hypertension, and hyperlipidemia. By embracing advancements in technology, promoting healthy lifestyles, ensuring collaborative care, implementing effective policies, and empowering patients, healthcare systems can make significant strides in reducing the burden of these interconnected conditions and improving long-term outcomes.

Addressing the challenges and embracing these perspectives can enhance the link between stroke, hypertension, and hyperlipidemia, leading to improved outcomes and a reduced burden of these conditions on individuals and healthcare systems. It requires a concerted effort from healthcare professionals, policymakers, and the community to implement effective strategies for prevention, early detection, and comprehensive management.

References

1. Young, M. J., Regenhardt, R. W., Leslie-Mazwi, T. M., & Stein, M. A. (2020). Disabling stroke in persons already with a disability: Ethical dimensions and directives. *Neurology, 94*(7), 306–310. https://doi.org/10.1212/WNL.0000000000008964
2. World Stroke Day. (n.d.). Retrieved December 4, 2023, from https://www.who.int/southeastasia/news/detail/28-10-2021-world-stroke-day
3. Ma, D., Guan, B., Song, L., Liu, Q., Fan, Y., Zhao, L., Wang, T., Zhang, Z., Gao, Z., Li, S., & Xu, H. (2021). A bibliometric analysis of exosomes in cardiovascular diseases from 2001 to 2021. *Frontiers in Cardiovascular Medicine, 8*, 734514. https://doi.org/10.3389/fcvm.2021.734514
4. Ellegaard, O., & Wallin, J. A. (2015). The bibliometric analysis of scholarly production: How great is the impact? *Scientometrics, 105*(3), 1809–1831. https://doi.org/10.1007/s11192-015-1645-z
5. Rondanelli, M., Perna, S., Peroni, G., & Guido, D. (2016). A bibliometric study of scientific literature in Scopus on botanicals for treatment of androgenetic alopecia. *Journal of Cosmetic Dermatology, 15*(2), 120–130. https://doi.org/10.1111/jocd.12198
6. Pu, Q.-H., Lyu, Q.-J., & Su, H.-Y. (2016). Bibliometric analysis of scientific publications in transplantation journals from Mainland China, Japan, South Korea and Taiwan between 2006 and 2015. *BMJ Open, 6*(8), e011623. https://doi.org/10.1136/bmjopen-2016-011623
7. Smith, D. R. (2008). Bibliometrics, dermatology and contact dermatitis. *Contact Dermatitis, 59*(3), 133–136. https://doi.org/10.1111/j.1600-0536.2008.01405.x
8. van Eck, N. J., & Waltman, L. (2010). Software survey: VOSviewer, a computer program for bibliometric mapping. *Scientometrics, 84*(2), 523–538. https://doi.org/10.1007/s11192-009-0146-3
9. Qin, Y., Zhang, Q., & Liu, Y. (2020). Analysis of knowledge bases and research focuses of cerebral ischemia-reperfusion from the perspective of mapping knowledge domain. *Brain Research Bulletin, 156*, 15–24. https://doi.org/10.1016/j.brainresbull.2019.12.004
10. Talwalkar, A., & Uddin, S. (2015). Trends in emergency department visits for ischemic stroke and transient ischemic attack: United States, 2001–2011. NCHS Data Brief, 194, 1–8.
11. Roger, V. L., Go, A. S., Lloyd-Jones, D. M., Benjamin, E. J., Berry, J. D., Borden, W. B., Bravata, D. M., Dai, S., Ford, E. S., Fox, C. S., Fullerton, H. J., Gillespie, C., Hailpern, S. M., Heit, J. A., Howard, V. J., Kissela, B. M., Kittner, S. J., Lackland, D. T., Lichtman, J. H., et al. (2012). Executive summary: Heart disease and stroke statistics—2012 update: A report from the American Heart Association. *Circulation, 125*(1), 188–197. https://doi.org/10.1161/CIR.0b013e3182456d46
12. Kapral, M. K., Fang, J., Hill, M. D., Silver, F., Richards, J., Jaigobin, C., Cheung, A. M., & Investigators of the Registry of the Canadian Stroke Network. (2005). Sex differences in stroke care and outcomes: Results from the Registry of the Canadian Stroke Network. *Stroke, 36*(4), 809–814. https://doi.org/10.1161/01.STR.0000157662.09551.e5
13. Reeves, M. J., Fonarow, G. C., Zhao, X., Smith, E. E., Schwamm, L. H., Committee, G. W. T. G.-S. S., & Investigators. (2009). Quality of care in women with ischemic stroke in the GWTG program. *Stroke, 40*(4), 1127–1133. https://doi.org/10.1161/STROKEAHA.108.543157
14. Schulz, U. G. R., Flossmann, E., & Rothwell, P. M. (2004). Heritability of ischemic stroke in relation to age, vascular risk factors, and subtypes of incident stroke in population-based studies. *Stroke, 35*(4), 819–824. https://doi.org/10.1161/01.STR.0000121646.23955.0f
15. Seshadri, S., Beiser, A., Pikula, A., Himali, J. J., Kelly-Hayes, M., Debette, S., DeStefano, A. L., Romero, J. R., Kase, C. S., & Wolf, P. A. (2010). Parental occurrence of stroke and risk of stroke in their children: The Framingham study. *Circulation, 121*(11), 1304–1312. https://doi.org/10.1161/CIRCULATIONAHA.109.854240

16. Touzé, E., & Rothwell, P. M. (2008). Sex differences in heritability of ischemic stroke: A systematic review and meta-analysis. *Stroke, 39*(1), 16–23. https://doi.org/10.1161/STROKEAHA.107.484618
17. Whelton, P. K., Carey, R. M., Aronow, W. S., Casey, D. E., Collins, K. J., Dennison Himmelfarb, C., DePalma, S. M., Gidding, S., Jamerson, K. A., Jones, D. W., MacLaughlin, E. J., Muntner, P., Ovbiagele, B., Smith, S. C., Spencer, C. C., Stafford, R. S., Taler, S. J., Thomas, R. J., Williams, K. A., & Wright, J. T. (2018). 2017 ACC/AHA/AAPA/ABC/ACPM/AGS/APHA/ASH/ASPC/NMA/PCNA guideline for the prevention, detection, evaluation, and management of high blood pressure in adults: A report of the American College of Cardiology/American Heart Association Task Force on Clinical Practice Guidelines. *Hypertension, 71*(6), 1269–1324. https://doi.org/10.1161/HYP.0000000000000065
18. McEvoy, J. W., Daya, N., Rahman, F., Hoogeveen, R. C., Blumenthal, R. S., Shah, A. M., Ballantyne, C. M., Coresh, J., & Selvin, E. (2020). Association of isolated diastolic hypertension as defined by the 2017 ACC/AHA blood pressure guideline with incident cardiovascular outcomes. *The Journal of the American Medical Association, 323*(4), 329–338. https://doi.org/10.1001/jama.2019.21402
19. Karr, S. (2017). Epidemiology and management of hyperlipidemia. *The American Journal of Managed Care, 23*(9 Suppl), S139–S148.
20. Franklin, S. S., & Wong, N. D. (2013). Hypertension and cardiovascular disease: Contributions of the Framingham heart study. *Global Heart, 8*(1), 49–57. https://doi.org/10.1016/j.gheart.2012.12.004
21. Benjamin, E. J., Muntner, P., Alonso, A., Bittencourt, M. S., Callaway, C. W., Carson, A. P., Chamberlain, A. M., Chang, A. R., Cheng, S., Das, S. R., Delling, F. N., Djousse, L., Elkind, M. S. V., Ferguson, J. F., Fornage, M., Jordan, L. C., Khan, S. S., Kissela, B. M., Knutson, K. L., et al. (2019). Heart disease and stroke statistics-2019 update: A report from the American Heart Association. *Circulation, 139*(10), e56–e528. https://doi.org/10.1161/CIR.0000000000000659
22. Qureshi, A. I., Suri, M. F. K., Kirmani, J. F., & Divani, A. A. (2005). Prevalence and trends of prehypertension and hypertension in United States: National Health and Nutrition Examination Surveys 1976 to 2000. *Medical Science Monitor, 11*(9), CR403-9.
23. Williams, B., Mancia, G., Spiering, W., Agabiti Rosei, E., Azizi, M., Burnier, M., Clement, D. L., Coca, A., de Simone, G., Dominiczak, A., Kahan, T., Mahfoud, F., Redon, J., Ruilope, L., Zanchetti, A., Kerins, M., Kjeldsen, S. E., Kreutz, R., Laurent, S., & ESC Scientific Document Group. (2018). 2018 ESC/ESH guidelines for the management of arterial hypertension. *European Heart Journal, 39*(33), 3021–3104. https://doi.org/10.1093/eurheartj/ehy339
24. Tabas, I., & Bornfeldt, K. E. (2016). Macrophage phenotype and function in different stages of atherosclerosis. *Circulation Research, 118*(4), 653–667. https://doi.org/10.1161/CIRCRESAHA.115.306256
25. Weber, C., & Noels, H. (2011). Atherosclerosis: Current pathogenesis and therapeutic options. *Nature Medicine, 17*(11), 1410–1422. https://doi.org/10.1038/nm.2538
26. Lüscher, T. F. (2015). Prevention of non-communicable diseases and special causes of heart failure. *European Heart Journal, 36*(31), 2019–2022. https://doi.org/10.1093/eurheartj/ehv303
27. Libby, P., Ridker, P. M., & Hansson, G. K. (2011). Progress and challenges in translating the biology of atherosclerosis. *Nature, 473*(7347), 317–325. https://doi.org/10.1038/nature10146
28. Amarenco, P., Lavallée, P. C., Labreuche, J., Ducrocq, G., Juliard, J.-M., Feldman, L., Cabrejo, L., Meseguer, E., Guidoux, C., Adraï, V., Ratani, S., Kusmierek, J., Lapergue, B., Klein, I. F., Gongora-Rivera, F., Jaramillo, A., Mazighi, M., Touboul, P.-J., & Steg, P. G. (2011). Prevalence of coronary atherosclerosis in patients with cerebral infarction. *Stroke, 42*(1), 22–29. https://doi.org/10.1161/STROKEAHA.110.584086
29. Nordestgaard, B. G., Chapman, M. J., Ray, K., Borén, J., Andreotti, F., Watts, G. F., Ginsberg, H., Amarenco, P., Catapano, A., Descamps, O. S., Fisher, E., Kovanen, P. T., Kuivenhoven, J. A., Lesnik, P., Masana, L., Reiner, Z., Taskinen, M.-R., Tokgözoglu, L., Tybjærg-Hansen, A., & European Atherosclerosis Society Consensus Panel. (2010). Lipoprotein(a) as a cardio-

vascular risk factor: Current status. *European Heart Journal, 31*(23), 2844–2853. https://doi.org/10.1093/eurheartj/ehq386
30. European Association for Cardiovascular Prevention & Rehabilitation, Reiner, Z., Catapano, A. L., De Backer, G., Graham, I., Taskinen, M.-R., Wiklund, O., Agewall, S., Alegria, E., Chapman, M. J., Durrington, P., Erdine, S., Halcox, J., Hobbs, R., Kjekshus, J., Filardi, P. P., Riccardi, G., Storey, R. F., Wood, D., & ESC Committee for Practice Guidelines (CPG) 2008-2010 and 2010-2012 Committees. (2011). ESC/EAS guidelines for the management of dyslipidaemias: The Task Force for the management of dyslipidaemias of the European Society of Cardiology (ESC) and the European Atherosclerosis Society (EAS). *European Heart Journal, 32*(14), 1769–1818. https://doi.org/10.1093/eurheartj/ehr158
31. Chobanian, A. V., Bakris, G. L., Black, H. R., Cushman, W. C., Green, L. A., Izzo, J. L., Jones, D. W., Materson, B. J., Oparil, S., Wright, J. T., Roccella, E. J., Heart, N., Lung, and Blood Institute Joint National Committee on Prevention, Detection, Evaluation, and Treatment of High Blood Pressure, & National High Blood Pressure Education Program Coordinating Committee. (2003). The seventh report of the Joint National Committee on Prevention, Detection, Evaluation, and Treatment of High Blood Pressure: The JNC 7 report. *The Journal of the American Medical Association, 289*(19), 2560–2572. https://doi.org/10.1001/jama.289.19.2560
32. Yusuf, S., Hawken, S., Ounpuu, S., Dans, T., Avezum, A., Lanas, F., McQueen, M., Budaj, A., Pais, P., Varigos, J., Lisheng, L., & INTERHEART Study Investigators. (2004). Effect of potentially modifiable risk factors associated with myocardial infarction in 52 countries (the INTERHEART study): Case-control study. *The Lancet, 364*(9438), 937–952. https://doi.org/10.1016/S0140-6736(04)17018-9
33. Pantoni, L. (2010). Cerebral small vessel disease: From pathogenesis and clinical characteristics to therapeutic challenges. *Lancet Neurology, 9*(7), 689–701. https://doi.org/10.1016/S1474-4422(10)70104-6
34. O'Donnell, M. J., Xavier, D., Liu, L., Zhang, H., Chin, S. L., Rao-Melacini, P., Rangarajan, S., Islam, S., Pais, P., McQueen, M. J., Mondo, C., Damasceno, A., Lopez-Jaramillo, P., Hankey, G. J., Dans, A. L., Yusoff, K., Truelsen, T., Diener, H.-C., Sacco, R. L., & INTERSTROKE investigators. (2010). Risk factors for ischaemic and intracerebral haemorrhagic stroke in 22 countries (the INTERSTROKE study): A case-control study. *The Lancet, 376*(9735), 112–123. https://doi.org/10.1016/S0140-6736(10)60834-3
35. Madamanchi, N. R., & Runge, M. S. (2007). Mitochondrial dysfunction in atherosclerosis. *Circulation Research, 100*(4), 460–473. https://doi.org/10.1161/01.RES.0000258450.44413.96
36. Pan, Y., Hui, X., Hoo, R. L. C., Ye, D., Chan, C. Y. C., Feng, T., Wang, Y., Lam, K. S. L., & Xu, A. (2019). Adipocyte-secreted exosomal microRNA-34a inhibits M2 macrophage polarization to promote obesity-induced adipose inflammation. *The Journal of Clinical Investigation, 129*(2), 834–849. https://doi.org/10.1172/JCI123069
37. Strianese, O., Rizzo, F., Ciccarelli, M., Galasso, G., D'Agostino, Y., Salvati, A., Del Giudice, C., Tesorio, P., & Rusciano, M. R. (2020). Precision and personalized medicine: How genomic approach improves the management of cardiovascular and neurodegenerative disease. *Genes, 11*(7), 747. https://doi.org/10.3390/genes11070747
38. Iadecola, C. (2017). The neurovascular unit coming of age: A journey through neurovascular coupling in health and disease. *Neuron, 96*(1), 17–42. https://doi.org/10.1016/j.neuron.2017.07.030
39. Younossi, Z. M., Koenig, A. B., Abdelatif, D., Fazel, Y., Henry, L., & Wymer, M. (2016). Global epidemiology of nonalcoholic fatty liver disease-meta-analytic assessment of prevalence, incidence, and outcomes. *Hepatology, 64*(1), 73–84. https://doi.org/10.1002/hep.28431
40. Pearson, T. A., Mensah, G. A., Alexander, R. W., Anderson, J. L., Cannon, R. O., Criqui, M., Fadl, Y. Y., Fortmann, S. P., Hong, Y., Myers, G. L., Rifai, N., Smith, S. C., Taubert, K., Tracy, R. P., Vinicor, F., Centers for Disease Control and Prevention, & American Heart Association. (2003). Markers of inflammation and cardiovascular disease: Application to clinical and public health practice: A statement for healthcare professionals from the Centers for Disease Control

and Prevention and the American Heart Association. *Circulation, 107*(3), 499–511. https://doi.org/10.1161/01.cir.0000052939.59093.45
41. (2013). AHA/ACC Guideline on Lifestyle Management to Reduce Cardiovascular Risk: A Report of the American College of Cardiology/American Heart Association Task Force on Practice Guidelines. https://doi.org/10.1161/01.cir.0000437740.48606.d1
42. How Potassium Can Help Control High Blood Pressure | American Heart Association. (n.d.). Retrieved December 15, 2023, from https://www.heart.org/en/health-topics/high-blood-pressure/changes-you-can-make-to-manage-high-blood-pressure/how-potassium-can-help-control-high-blood-pressure
43. Soltani, S., Shirani, F., Chitsazi, M. J., & Salehi-Abargouei, A. (2016). The effect of dietary approaches to stop hypertension (DASH) diet on weight and body composition in adults: A systematic review and meta-analysis of randomized controlled clinical trials. *Obesity Reviews, 17*(5), 442–454. https://doi.org/10.1111/obr.12391
44. McRae, M. P. (2017). Health benefits of dietary whole grains: An umbrella review of meta-analyses. *Journal of Chiropractic Medicine, 16*(1), 10–18. https://doi.org/10.1016/j.jcm.2016.08.008
45. Arnett, D. K., Blumenthal, R. S., Albert, M. A., Buroker, A. B., Goldberger, Z. D., Hahn, E. J., Himmelfarb, C. D., Khera, A., Lloyd-Jones, D., McEvoy, J. W., Michos, E. D., Miedema, M. D., Muñoz, D., Smith, S. C., Virani, S. S., Williams, K. A., Yeboah, J., & Ziaeian, B. (2019). 2019 ACC/AHA guideline on the primary prevention of cardiovascular disease: A report of the American College of Cardiology/American Heart Association Task Force on Clinical Practice Guidelines. *Circulation, 140*(11), e596–e646. https://doi.org/10.1161/CIR.0000000000000678
46. Schwingshackl, L., Hoffmann, G., Lampousi, A.-M., Knüppel, S., Iqbal, K., Schwedhelm, C., Bechthold, A., Schlesinger, S., & Boeing, H. (2017). Food groups and risk of type 2 diabetes mellitus: A systematic review and meta-analysis of prospective studies. *European Journal of Epidemiology, 32*(5), 363–375. https://doi.org/10.1007/s10654-017-0246-y
47. Delgado-Lista, J., Perez-Martinez, P., Lopez-Miranda, J., & Perez-Jimenez, F. (2012). Long chain omega-3 fatty acids and cardiovascular disease: A systematic review. *The British Journal of Nutrition, 107*(Suppl 2), S201–S213. https://doi.org/10.1017/S0007114512001596
48. Rippe, J. M. (2018). Lifestyle medicine: The health promoting power of daily habits and practices. *American Journal of Lifestyle Medicine, 12*(6), 499–512. https://doi.org/10.1177/1559827618785554
49. Barone Gibbs, B., Hivert, M.-F., Jerome, G. J., Kraus, W. E., Rosenkranz, S. K., Schorr, E. N., Spartano, N. L., Lobelo, F., American Heart Association Council, & on Lifestyle and Cardiometabolic Health; Council on Cardiovascular and Stroke Nursing; and Council on Clinical Cardiology. (2021). Physical activity as a critical component of first-line treatment for elevated blood pressure or cholesterol: Who, what, and how?: A scientific statement from the American Heart Association. *Hypertension, 78*(2), e26–e37. https://doi.org/10.1161/HYP.0000000000000196
50. Strasser, B., Steindorf, K., Wiskemann, J., & Ulrich, C. M. (2013). Impact of resistance training in cancer survivors: A meta-analysis. *Medicine and Science in Sports and Exercise, 45*(11), 2080–2090. https://doi.org/10.1249/MSS.0b013e31829a3b63
51. Tsai, S.-Y., Huang, W.-H., Chan, H.-L., & Hwang, L.-C. (2021). The role of smoking cessation programs in lowering blood pressure: A retrospective cohort study. *Tobacco Induced Diseases, 19*, 82. https://doi.org/10.18332/tid/142664.
52. Piano, M. R. (2017). Alcohol's effects on the cardiovascular system. *Alcohol Research : Current Reviews, 38*(2), 219–241.
53. Brook, R. D., Appel, L. J., Rubenfire, M., Ogedegbe, G., Bisognano, J. D., Elliott, W. J., Fuchs, F. D., Hughes, J. W., Lackland, D. T., Staffileno, B. A., Townsend, R. R., Rajagopalan, S., American Heart Association Professional Education Committee of the Council for High Blood Pressure Research, Council on Cardiovascular and Stroke Nursing, Council on Epidemiology and Prevention, & Council on Nutrition, Physical Activity. (2013). Beyond medications and diet: Alternative approaches to lowering blood pressure: A scientific statement from

the American Heart Association. *Hypertension, 61*(6), 1360–1383. https://doi.org/10.1161/HYP.0b013e318293645f
54. Appel, L. J., Moore, T. J., Obarzanek, E., Vollmer, W. M., Svetkey, L. P., Sacks, F. M., Bray, G. A., Vogt, T. M., Cutler, J. A., Windhauser, M. M., Lin, P. H., & Karanja, N. (1997). A clinical trial of the effects of dietary patterns on blood pressure. DASH Collaborative Research Group. *The New England Journal of Medicine, 336*(16), 1117–1124. https://doi.org/10.1056/NEJM199704173361601
55. Magnesium - Health Professional Fact Sheet. (n.d.). Retrieved January 25, 2024, from https://ods.od.nih.gov/factsheets/Magnesium-HealthProfessional/
56. Calcium-Health Professional Fact Sheet. (n.d.). Retrieved January 25, 2024, from https://ods.od.nih.gov/factsheets/Calcium-HealthProfessional/
57. Pearson, T. A. (1996). Alcohol and heart disease. *Circulation, 94*(11), 3023–3025. https://doi.org/10.1161/01.cir.94.11.3023
58. Stone, N. J., Robinson, J. G., Lichtenstein, A. H., Bairey Merz, C. N., Blum, C. B., Eckel, R. H., Goldberg, A. C., Gordon, D., Levy, D., Lloyd-Jones, D. M., McBride, P., Schwartz, J. S., Shero, S. T., Smith, S. C., Watson, K., Wilson, P. W. F., & American College of Cardiology/American Heart Association Task Force on Practice Guidelines. (2014). 2013 ACC/AHA guideline on the treatment of blood cholesterol to reduce atherosclerotic cardiovascular risk in adults: A report of the American College of Cardiology/American Heart Association Task Force on Practice Guidelines. *Journal of the American College of Cardiology, 63*(25 Pt B), 2889–2934. https://doi.org/10.1016/j.jacc.2013.11.002
59. Estruch, R., Ros, E., Salas-Salvadó, J., Covas, M.-I., Corella, D., Arós, F., Gómez-Gracia, E., Ruiz-Gutiérrez, V., Fiol, M., Lapetra, J., Lamuela-Raventos, R. M., Serra-Majem, L., Pintó, X., Basora, J., Muñoz, M. A., Sorlí, J. V., Martínez, J. A., Fitó, M., Gea, A., & PREDIMED Study Investigators. (2018). Primary prevention of cardiovascular disease with a Mediterranean diet supplemented with extra-virgin olive oil or nuts. *The New England Journal of Medicine, 378*(25), e34. https://doi.org/10.1056/NEJMoa1800389
60. Becker, D. J., Gordon, R. Y., Morris, P. B., Yorko, J., Gordon, Y. J., Li, M., & Iqbal, N. (2008). Simvastatin vs therapeutic lifestyle changes and supplements: Randomized primary prevention trial. *Mayo Clinic Proceedings, 83*(7), 758–764. https://doi.org/10.4065/83.7.758
61. Taylor, A. J., Villines, T. C., Stanek, E. J., Devine, P. J., Griffen, L., Miller, M., Weissman, N. J., & Turco, M. (2009). Extended-release niacin or ezetimibe and carotid intima-media thickness. *The New England Journal of Medicine, 361*(22), 2113–2122. https://doi.org/10.1056/NEJMoa0907569
62. Demonty, I., Ras, R. T., van der Knaap, H. C. M., Duchateau, G. S. M. J. E., Meijer, L., Zock, P. L., Geleijnse, J. M., & Trautwein, E. A. (2009). Continuous dose-response relationship of the LDL-cholesterol-lowering effect of phytosterol intake. *The Journal of Nutrition, 139*(2), 271–284. https://doi.org/10.3945/jn.108.095125
63. Duda, M. K., O'Shea, K. M., & Stanley, W. C. (2009). Omega-3 polyunsaturated fatty acid supplementation for the treatment of heart failure: Mechanisms and clinical potential. *Cardiovascular Research, 84*(1), 33–41. https://doi.org/10.1093/cvr/cvp169
64. Mozaffarian, D., Geelen, A., Brouwer, I. A., Geleijnse, J. M., Zock, P. L., & Katan, M. B. (2005). Effect of fish oil on heart rate in humans: A meta-analysis of randomized controlled trials. *Circulation, 112*(13), 1945–1952. https://doi.org/10.1161/CIRCULATIONAHA.105.556886
65. Rahman, K. (2007). Effects of garlic on platelet biochemistry and physiology. *Molecular Nutrition & Food Research, 51*(11), 1335–1344. https://doi.org/10.1002/mnfr.200700058
66. Küster, L. J., & Frölich, J. C. (1986). Platelet aggregation and thromboxane release induced by arachidonic acid, collagen, ADP and platelet-activating factor following low dose acetylsalicylic acid in man. *Prostaglandins, 32*(3), 415–423. https://doi.org/10.1016/0090-6980(86)90009-2
67. Wongcharoen, W., & Phrommintikul, A. (2009). The protective role of curcumin in cardiovascular diseases. *International Journal of Cardiology, 133*(2), 145–151. https://doi.org/10.1016/j.ijcard.2009.01.073
68. Kuriyama, S., Shimazu, T., Ohmori, K., Kikuchi, N., Nakaya, N., Nishino, Y., Tsubono, Y., & Tsuji, I. (2006). Green tea consumption and mortality due to cardiovascular disease, cancer,

and all causes in Japan: The Ohsaki study. *The Journal of the American Medical Association, 296*(10), 1255–1265. https://doi.org/10.1001/jama.296.10.1255
69. Singh, U., & Devaraj, S. (2007). Vitamin E: Inflammation and atherosclerosis. *Vitamins and Hormones, 76*, 519–549. https://doi.org/10.1016/S0083-6729(07)76020-X
70. Dauchet, L., Amouyel, P., Hercberg, S., & Dallongeville, J. (2006). Fruit and vegetable consumption and risk of coronary heart disease: A meta-analysis of cohort studies. *The Journal of Nutrition, 136*(10), 2588–2593. https://doi.org/10.1093/jn/136.10.2588
71. Muntner, P., Carey, R. M., Gidding, S., Jones, D. W., Taler, S. J., Wright, J. T., & Whelton, P. K. (2018). Potential US population impact of the 2017 ACC/AHA high blood pressure guideline. *Circulation, 137*(2), 109–118. https://doi.org/10.1161/CIRCULATIONAHA.117.032582
72. Grundy, S. M., Stone, N. J., Bailey, A. L., Beam, C., Birtcher, K. K., Blumenthal, R. S., Braun, L. T., de Ferranti, S., Faiella-Tommasino, J., Forman, D. E., Goldberg, R., Heidenreich, P. A., Hlatky, M. A., Jones, D. W., Lloyd-Jones, D., Lopez-Pajares, N., Ndumele, C. E., Orringer, C. E., Peralta, C. A., & Yeboah, J. (2019). 2018 AHA/ACC/AACVPR/AAPA/ABC/ACPM/ADA/AGS/APHA/ASPC/NLA/PCNA guideline on the management of blood cholesterol: Executive summary: A report of the American College of Cardiology/American Heart Association Task Force on Clinical Practice Guidelines. *Journal of the American College of Cardiology, 73*(24), 3168–3209. https://doi.org/10.1016/j.jacc.2018.11.002
73. KDIGO 2012 clinical practice guideline for the evaluation and management of chronic kidney disease. (n.d.).
74. Osterberg, L., & Blaschke, T. (2005). Adherence to medication. *The New England Journal of Medicine, 353*(5), 487–497. https://doi.org/10.1056/NEJMra050100
75. Ettehad, D., Emdin, C. A., Kiran, A., Anderson, S. G., Callender, T., Emberson, J., Chalmers, J., Rodgers, A., & Rahimi, K. (2016). Blood pressure lowering for prevention of cardiovascular disease and death: A systematic review and meta-analysis. *The Lancet, 387*(10022), 957–967. https://doi.org/10.1016/S0140-6736(15)01225-8
76. Charles, C., Gafni, A., & Whelan, T. (1997). Shared decision-making in the medical encounter: What does it mean? (or it takes at least two to tango). *Social Science & Medicine, 44*(5), 681–692. https://doi.org/10.1016/s0277-9536(96)00221-3

Part II

Nutritional Intervention of Stroke

Role of Choline, Homocysteine and Branched-Chain Amino Acids in Stroke

9

Mahmoodullah Azimi and Muhammad Asif Atiq

Abstract

Stroke is a serious medical disorder that happens when the blood supply to the brain is impaired, leading to brain cell injury or death. Strokes rank as the second most prominent contributor to both mortality and disability on a global scale. Numerous clinical trials have been carried out to discover effective pharmacotherapy for stroke, as it continues to be considered a medical emergency due to the high risk of significant disability. The gold standard for treating acute ischemic stroke is the early administration of rtPA in a stroke unit. However, this approach can only be implemented in experienced and qualified centers located in developed countries. There is a pressing need for accessible therapeutic options for acute stroke. Numerous studies have delved into the intricate relationship between diet and stroke incidence. Certain micronutrients have been identified as beneficial, while others are deemed detrimental or associated with increased stroke risk. However, the findings across various observational and experimental studies have often been contradictory, yielding a complex landscape of research outcomes. In this chapter, specifically, we explore the role of choline, homocysteine and branched-chain amino acids in stroke.

Keywords

Branched-chain amino acids · CVA management · Choline · Homocysteine · Ischemic stroke

M. Azimi (✉) · M. A. Atiq
Department of Clinical Pharmacology, Kabul University of Medical Sciences, Kabul, Afghanistan

W. Mohamed, M. B. Mat-Nor (eds.), *Nutrition and Stroke*, Nutritional Neurosciences, https://doi.org/10.1007/978-981-97-6818-9_9

9.1 Introduction

Stroke is a serious medical disorder that happens when the blood supply to the brain is impaired, leading to brain cell injury or death. Two main types of stroke exist which include: Ischemic and hemorrhagic. Ischemic stroke is caused by an obstruction in a blood vessel supplying the brain, while hemorrhagic stroke is due to bleeding in the brain. Strokes rank as the second most prominent contributor to both mortality and disability on a global scale [1]. According to estimates, around 25% of the population is expected to experience a stroke at least once in their lifetime [2]. Every year, approximately 15 million individuals worldwide are affected by strokes. Out of this number, five million lose their lives, and an additional five million are left with permanent disabilities, which leads to a significant burden on both families and communities [3].

Roughly 87% of strokes are ischemic infarctions, and this proportion has significantly risen from 1990 to 2016. This increase is primarily linked to advancements in medical interventions and a decrease in mortality rates. The majority of strokes are primary hemorrhages, representing the first occurrence of bleeding in the brain. Secondary hemorrhages (second-time) make up approximately 10–25% of all strokes [4, 5]. Between 1990 and 2016, the incidence of strokes in low- and middle-income countries increased twofold, while high-income countries had an outstanding decline of 42% in stroke incidence during the same time frame.

Based on the study of Global Burden of Disease (GBD), while the total prevalence of stroke has dropped, factors like the age of individuals affected, their gender, and the geographical distribution indicates that the socio-economic burden associated with stroke has increased over time [5].

As individuals age, the incidence of stroke progressively increases, with a doubling rate after the age of 55 years [5–7].

Brain infarction and Intracerebral Hemorrhage (ICH) are more prevalent in male, while cardio-embolic stroke is more common among females. The fatality rate is also greater among women suffering stroke [7–9].

A significant inverse relationship exists between stroke and socioeconomic status, which can be attributed to inadequate hospital resources and post-stroke care within low-income populations [10].

The risk factor of stroke is divided to modifiable and non-modifiable groups. Age, gender, ethnicity, transient ischemic attacks and genetic factor are among non-modifiable risk factors, while hypertension, diabetes, lake of physical exercise, dyslipidemia, drug abuse, alcohol, and diet management are important modifiable risk factors [11]. Early and proper medical intervention can mitigate the risk of stroke in individuals who are prone to it.

9.2 Neurological Mechanisms Underlying Stroke

Stroke is characterized as a sudden neurological event resulting from impaired blood flow to the brain through the blood vessels. To understand the clinical presentation of a stroke, it is crucial to have knowledge of the neurovascular anatomy. The blood supply to the brain is provided by two internal carotid arteries in the front and two vertebral arteries in the back, forming a structure known as the circle of Willis. Ischemic stroke occurs when there is insufficient blood and oxygen supply to the brain, while hemorrhagic stroke is caused by bleeding or leakage from blood vessels. Ischemic occlusions, which lead to blockages in blood flow, account for approximately 85% of stroke cases, with the remaining cases attributed to intracerebral bleeding. Ischemic occlusion can result in thrombotic or embolic conditions within the brain [12].

Thrombotic stroke occurs when blood flow is impeded due to the narrowing of blood vessels caused by atherosclerosis. The accumulation of plaque gradually narrows the vascular chamber, leading to the formation of clots and resulting in a thrombotic stroke. On the other hand, embolic stroke is characterized by reduced blood flow to a specific region of the brain due to an embolism. This diminished blood flow causes substantial stress and untimely necrosis.

Following necrosis, breakdown of the plasma membrane, swelling of organelles, and leakage of cellular contents into the extracellular space occurs [13]. This process is accompanied by the loss of neuronal function. In addition, stroke pathology involves various other important events, including inflammation, energy depletion, disruption of homeostasis, acidosis, elevated levels of intracellular calcium, excitotoxicity, toxicity caused by free radicals, cytotoxicity mediated by cytokines, activation of the complement system, impairment of the blood-brain barrier, activation of glial cells, oxidative stress, and infiltration of white blood cells [14–18].

Hemorrhagic stroke constitutes approximately 10–15% of all strokes and is associated with a high mortality rate. This type of stroke occurs when stress on brain tissue and internal injury lead to the rupture of blood vessels. The rupture of these vessels results in lethal effects within the vascular system, leading to infarction [19]. Hemorrhagic stroke is further categorized into intracerebral hemorrhage (ICH) and subarachnoid hemorrhage. In ICH, blood vessels rupture, causing abnormal accumulation of blood within the brain. Hypertension, disrupted vasculature, excessive use of anticoagulants, and thrombolytic agents are among the main contributing factors to ICH. In subarachnoid hemorrhage, blood accumulates in the subarachnoid space of the brain, often due to a head injury or cerebral aneurysm.

Numerous clinical trials have been carried out to discover effective pharmacotherapy for stroke, as it continues to be considered a medical emergency due to the high risk of significant disability. In 1995, a significant development occurred in acute stroke care with the discovery that administering recombinant tissue plasminogen activator (rtPA) intravenously at an early stage could improve outcomes in a specific group of patient with acute ischemic stroke [20]. In addition to the utilization of rtPA, the management of acute ischemic stroke in stroke wards, early administration of aspirin, and decompressive surgery have been acknowledged as effective therapies in managing the condition. The gold standard for treatment of the acute ischemic stroke is the early administration of rtPA in a stroke unit. However, this approach can only be implemented in experienced and qualified centers located in developed countries [21, 22]. Despite its effectiveness, the utilization of rtPA in the management of the acute ischemic stroke is still limited worldwide [23]. Furthermore, there are significant disparities in the standard of care provided for this condition. Consequently, there is a pressing need for accessible therapeutic options for acute stroke. Certain micronutrients have been identified as beneficial, while others are deemed detrimental or associated with increased stroke risk. However, the findings across various observational and experimental studies have often been contradictory, yielding a complex landscape of research outcomes. In this chapter, we explore the role of choline, homocysteine and branched-chain amino acids in stroke.

9.3 Choline: Metabolism and Neurological Functions

Choline is a crucial nutrient that plays an important role in brain function. It is the precursor to acetylcholine, an important neurotransmitter that is involved in memory and learning. Choline also helps to maintain the integrity of the cell membranes and is involved in the synthesis of phospholipids, which are essential components of the cell membrane.

Phospholipids which contain choline, including Citicoline (Cytidine 5′-Diphosphocholine of CDP-choline) and Choline Alphoscerate (alpha—glyceryl—phosphorylcholine or α-GPC), which serve as precursors for acetylcholine, have been suggested as adjuncts in the management of acute strokes [24]. These compounds are utilized as substrates for the synthesis of acetylcholine, which can enhance cholinergic neurotransmission [24]. Following an ischemic stroke, the precursors of acetylcholine can also be employed in neuroprotective substances and contribute to neurogenesis [24, 25]. It has been observed that choline supplementation during various stages of brain development enhances brain performance, possibly due to its role in membrane synthesis during neuronal development. Consequently, an adequate supply of choline is necessary to preserve optimal brain function. Choline has a crucial role in various neurochemical pathways, contributing significantly to neurological functions [25].

9.4 Role of Choline in Stroke

Choline-containing phospholipids, specifically choline alphoscerate, have been suggested to possess anti-inflammatory effects. These properties may help to describe their neuroprotective effects after a stroke, leading to improved functional outcomes. Citicoline is a vital precursor of phosphatidylcholine, which serves as an important component of cell membrane structure. In the event of cerebral ischemia, phosphatidylcholine can undergo degradation, resulting in the formation of both free fatty acids and free radicals. Numerous studies have been conducted to explore the potential of choline alphoscerate in improving cognitive function and functional outcome in patients with acute cerebrovascular diseases including strokes and transient ischemic attacks (TIAs) [26, 27]. Earlier studies have demonstrated that the use of citicoline in the treatment of patients with acute ischemic stroke has shown encouraging results in terms of its safety profile [28, 29].

Clinical trials investigating the effects of citicoline on individuals with acute strokes have yielded conflicting results. For instance, a trail conducted in the United States of America involving 899 patients with acute ischemic stroke stated that citicoline did not demonstrate significant improvements in neurological and functional outcomes by the end of the designated follow-up period [30]. In a recent randomized controlled trail, the efficacy of citicoline treatment in improving neurological and functional recovery from acute ischemic stroke within a 90-day period was not demonstrated [31]. However, it is worth noting that in several European countries, citicoline is utilized for the treatment of cognitive impairment, specifically, in cases predominantly linked to cerebrovascular disease [32].

Citicoline, also known as CDP-choline, is a medication that exhibits both neurovascular protection and healing effects. It has been employed in the management of acute ischemic stroke and other neurological disorders, demonstrating a favorable safety profile [32, 33]. In 2002, a comprehensive meta-analysis was conducted on trials investigating the use of CDP-choline in acute and sub-acute strokes [34]. The meta-analysis advocated a significant and beneficial effect of citicoline treatment, with a reduction of 10–12% in the rate of long term disability and death. A pooled data analysis (PDA) revealed an odds ratio of 1.33 (95% confidence interval = 1.10–1.62) for complete functional and neurological recovery in patients with moderate to severe acute ischemic stroke who were received oral citicoline for a duration of 6 weeks, as compared to placebo [3].

Another meta-analysis of compiled data further reaffirmed the earlier findings, demonstrating a decrease in the risk of dependency or death [23]. Based on a systematic review and meta-analysis conducted in 2016, there is evidence supporting certain advantages of citicoline in the management of acute ischemic stroke. However, it should be noted that the citicoline's benefits appear to be limited when compared to the best available treatment, such as rtPA (recombinant tissue plasminogen activator) [35].

In a Cochrane systematic review conducted in 2020, the clinical advantages and potential adverse effects of citicoline were assessed in comparison to placebo or other standard managements for individuals with acute ischemic stroke. The review results indicate that there may be minimal or no significant difference observed between citicoline and its control groups in terms of disability or dependence in daily activities, all-cause mortality, neurological function, functional recovery, and sever adverse effects. However, it is important to note that the evidence supporting these conclusions is of low certainty [3].

On other hand, based on results of a recent systemic review and meta-analysis, citicoline does not demonstrate effectiveness in enhancement of functional recovery, neurological function, and independence in daily activities among individual with acute stroke [36]. Furthermore, the results indicate that the administration of citicoline does not demonstrate any observable improvements in functional and neurological outcomes among patients with hemorrhagic stroke who were treated with the drug [36]. On the contrary, this study indicates that choline alphoscerate is effective in improving neurological function, facilitating functional recovery, and yielding positive [36].

Based on aforementioned review we conclude that the beneficial effect of citicoline in the management of acute ischemic stroke may have diminished over time due to improved standards of care. The 2002 and 2010 reviews found significant benefits of citicoline, while the 2016 review supported some benefits. However, the 2020 review suggested minimal to no difference between citicoline and its controls. The latest review conducted in 2023 reached the conclusion that citicoline does not exhibit effectiveness in enhancing neurological function, promoting functional recovery, or improving independence and daily living activities in individuals with acute stroke. Meanwhile, choline alphoscerate may be beneficial in patients who have suffered an acute stroke. Therefore, we suggest further studies to focus on choline alphoscerate and stroke.

9.5 Homocysteine: Impact on Cardiovascular Health

Homocysteine (Hcy) an amino acid that is not obtained from the diet, can be recycled into methionine or converted into cystien, which is an essential amino acid, with the help of certain B vitamins. The normal range of homocysteine levels in male and female is typically 5–10 micromole/L. When the level of homocysteine in serum exceed 10 micromol/L, it is referred to as hyperhomocysteinemia [37, 38]. Several factors can contribute to elevated homocysteine levels, including deficiencies in vitamin B6, folate, and/or vitamin B12, which can be caused by inadequate intake or absorption, renal insufficiency, certain medications [39], lifestyle choices like smoking and alcohol consumption, or genetic factors [40].

More than four decades ago, it was observed that individuals with homocysteinuria had an increased chance of early vascular diseases, which resulted to the understanding that elevated homocysteine levels may contribute to arterial or venous thromboembolism [41]. Homocysteine, a naturally occurring amino acid contains sulfur and is strongly linked to vascular endothelial dysfunction and the proliferation of the extracellular matrix, both of which can lead to damage in blood vessels [42]. Recent studies have also indicated a potential link between hyperhomocyteinemia and vascular thrombotic events, including ischemic stroke [43].

9.6 Homocysteine and Stroke

Several meta-analyses and systematic reviews have consistently shown that individuals who have experienced an ischemic stroke tend to have significantly greater levels of homocysteine compared to healthy controls. Therefore, it is important to identify hyperhomocysteinemia and explore ways to reduce homocysteine levels in individuals with an increased risk of ischemic stroke [44–48]. A study found that elevated levels of total homocysteine in individuals suffering acute ischemic stroke are linked to higher long-term mortality rates [49]. A Cochrane review conducted in 2017 concluded that no significant difference were observed in the effects of homocysteine-lowering interventions, specifically through the use of supplements containing vitamin B6 (pyridoxine), B9 (folate), or B12 (cyanocobalamine), whether administered individually or in combination, compared to placebo, in terms of myocardial infarction (MI), overall mortality, or adverse events. However, concerning the occurrence of stroke, the review did find a slight favorable difference in the effect of homocysteine-lowering interventions compared to placebo, when using supplements of vitamins B6, B9, or B12. It was also noted that homocysteine-lowering interventions had not a significant impact on serious adverse events like cancer when compared to placebo or any other interventions [47].

A recent meta-analysis, which encompassed eight randomized controlled trials involving 8513 participants, examined the impact of supplementation of vitamin on homocysteine levels as well as the risk of stroke, cardiovascular diseases, and death due to vascular disorders. The findings of this meta-analysis provided considerable evidence supporting the positive effects of vitamin B supplementation, particularly in patients with stroke, in reducing homocystien levels without any reported adverse effects. The evidence indicates that the use of vitamin B supplements is effective in decreasing levels of homocysteine and diminishing the likelihood of stroke and vascular deaths. Nevertheless, the group that received vitamin supplementation did not exhibit a noteworthy reduction in the risk of cardiovascular events [50].

The results of the above study indicated that vitamin B supplementation decreased the risk of stroke by 13% in comparison to the control group who received placebo. Equally, another meta-analysis [51] conducted in 2016, which included twenty trails with 31,164 events between 77,816 participants, reported a 10% decrease in the risk of stroke in group which received folic acid compared to other group in patients with cardiovascular disease, coronary heart disease, or stroke ($p = 0.002$; $RR = 0.90$). Another meta-analysis [52] conducted in 2017, which included 11 randomized controlled trials with 2826 strokes among an overall population of 65,790, demonstrated a 10% decrease in the risk of stroke among individuals with CVD.

In conclusion, individuals who have experienced ischemic stroke generally exhibit elevated levels of homocysteine compared to control subjects. Additionally, the supplementation of vitamin B has demonstrated to be effective in lowering homocysteine levels and decreasing the risk of stroke.

9.7 Branched-Chain Amino Acids in Stoke

The proteins of human body are constructed using twenty different amino acids, with 9 of them being classified as "essential" because the human body cannot produce them from other substances. These include the 3 branched-chain amino acids (BCAAs): leucine, isoleucine, and valine. These BCAAs are termed for their unique branched structure instead of a straight chain. Foods that are particularly rich in BCAAs include eggs, fish, meat, and dairy products [53].

Branched-chain amino acids play a vital role as metabolic markers and are necessary for optimal cellular and overall growth and function [54]. However, imbalances or disorders in the metabolism of these branched-chain amino acids can have negative effects on human health, particularly in relation to cardiovascular disease (CVD) [55, 56]. The impact of these amino acids metabolism on stroke, among the various cardiovascular diseases, is not yet well understood.

Cahill et al. conducted groundbreaking studies more than 50 years ago, demonstrating that individuals suffering obesity and insulin resistance had higher levels of circulating branched-chain amino acids (BCAAs) [52]. With the advancement and application of metabonomics over the years, elevated levels of BCAAs and related metabolites have been recognized as metabolic markers for insulin resistance, diabetes mellitus type 2, and obesity in mankind [57]. Metabolic abnormalities have a significant role in the pathophysiology of cardiovascular disease (CVD), and evidence suggests a direct involvement of branched-chain amino acids in CVD [55, 56] However, the causal relationship between BCAAs and stroke, a cardiovascular disease associated with high mortality and disability rates, has not been thoroughly explored.

A study by Kimberly et al. found that lower levels of branched- chain amino acids were associated with the severity of cardio-metabolic stroke and worse neurological outcomes [58]. Nonetheless, another case-cohort study, after adjusting for potential confounding factors, observed that baseline concentrations of leucine and isoleucine, but not valine, were linked to occurrence of stroke, including both ischemic and hemorrhagic types [59].

In contrast, a study demonstrated a beneficial effect of supplementation of branched-chain amino acids on stroke-related sarcopenia. The study group, which received BCAAs supplementation along with intensive rehabilitation therapy, showed significant improvements in skeletal muscle mass index and functional status in comparison to the control group. Stroke often leads to intricate systemic metabolic alterations, resulting in muscle loss and malnutrition following the event [60].

Higher baseline levels of BCAAs were found to be significantly associated with incident stroke in the primary preventive PREDIMED study [59]. Additionally, BCAAs were identified as unique markers for the development of cardiovascular disease (CVD), including stroke, in a population-based matched case-control study conducted in the Malmö Diet and Cancer Cardiovascular Cohort [61]. However, in another prospective cohort study of American women, the correlation between total BCAAs and stroke was only marginally significant [62]. The conclusions drawn from these observational studies are uncertain and subject to debate, likely due to limitations such as small sample sizes and potential confounders.

A recent randomized controlled trial (RCT) study, using Mendelian randomization analysis, found that an increase in plasma isoleucine levels had a causal effect on the risk of cardio-embolic stroke but not on the risk of other subtypes of stroke. However, the study did not find any causal effect of plasma leucine or valine levels on the risk of any stroke subtype [63].

A recent systematic review and meta-analysis, which included a limited number of studies (2 prospective and 3 non-prospective studies), did not find an association between BCAAs circulating levels and stroke risk [4].

To summarize, based on the available published works up to the writing of this chapter, there is no significant associated between BCAAs and stroke, except for one randomized controlled trial (RCT) that showed a causal effect of increased plasma isoleucine levels on the risk of cardio-embolic stroke. It is important to note that there is a limited number of studies on BCAAs and stroke. Hence, additional study must prioritize investigating the associations between BCAAs and ischemic and hemorrhagic strokes.

9.8 Conclusion

Stroke is the second most common cause of death and a significant contributor to disability worldwide, resulting in substantial economic burdens. Consequently, enhancing therapeutic interventions and advancing post-stroke management are crucial global health priorities. According to the aims of this chapter and based on the available published works up to the writing of this chapter we concluded that citicoline does not exhibit effectiveness in enhancing neurological function, homocysteine is a risk factor for stroke. There is no significant associated between BCAAs and stroke.

9.9 Future Direction and Research Perspectives

We suggest further studies to focus on choline alphoscerate as some studies found it is beneficial effects in stroke. And using supplements of vitamins B6, B9, or B12, to lower homocysteine level, for decreasing risk of stroke.

References

1. Go, A. S., Mozaffarian, D., Roger, V. L., et al. (2013). Heart disease and stroke statistics–2013 update: A report from the American Heart Association. *Circulation, 127*, e6–e245.
2. Treede, I., Braun, A., Sparla, R., Kühnel, M., Giese, T., Jerrold, R., Anes, E., Kulaksiz, H., Füllekrug, J., Stremmel, W., et al. (2009). Anti-inflammatory effects of phosphatidylcholine Irina. *The Journal of Biological Chemistry, 282*, 27155–27164.
3. Sagaro, G. G., & Amenta, F. (2023). Choline-containing phospholipids in stroke treatment: A systematic review and meta-analysis. *Journal of Clinical Medicine, 12*(8), 2875.
4. Roger, V. L., Go, A. S., Lloyd-Jones, D. M., Adams, R. J., Berry, J. D., Brown, T. M., Carnethon, M. R., Dai, S., de Simone, G., Ford, E. S., et al. (2011). Heart disease and stroke statistics–2011 update: A report from the American Heart Association. *Circulation, 123*, e18–e209.
5. Collaborators, G. S. (2019). Global, regional, and national burden of stroke, 1990–2016: A systematic analysis for the Global Burden of Disease Study 2016. *Lancet Neurology, 18*, 439–458.

6. Kelly-Hayes, M. (2010). Influence of age and health behaviors on stroke risk: Lessons from longitudinal studies. *Journal of the American Geriatrics Society, 58*, S325–S328.
7. Boehme, A. K., Esenwa, C., & Elkind, M. S. (2017). Stroke risk factors, genetics, and prevention. *Circulation Research, 120*, 472–495.
8. Appelros, P., Stegmayr, B., & Terént, A. (2009). Sex differences in stroke epidemiology: A systematic review. *Stroke, 40*, 1082–1090.
9. Reeves, M. J., Bushnell, C. D., Howard, G., Gargano, J. W., Duncan, P. W., Lynch, G., Khatiwoda, A., & Lisabeth, L. (2008). Sex differences in stroke: Epidemiology, clinical presentation, medical care, and outcomes. *Lancet Neurology, 7*, 915–926.
10. Addo, J., Ayerbe, L., Mohan, K. M., Crichton, S., Sheldenkar, A., Chen, R., Wolfe, C. D., & McKevitt, C. (2012). Socioeconomic status and stroke: An updated review. *Stroke, 43*, 1186–1191.
11. Kuriakose, D., & Xiao, Z. (2020). Pathophysiology and treatment of stroke: Present status and future perspectives. *International Journal of Molecular Sciences, 21*(20), 7609.
12. Musuka, T. D., Wilton, S. B., Traboulsi, M., & Hill, M. D. (2015). Diagnosis and management of acute ischemic stroke: Speed is critical. *CMAJ, 187*, 887–893.
13. Broughton, B. R., Reutens, D. C., & Sobey, C. G. (2009). Apoptotic mechanisms after cerebral ischemia. *Stroke, 40*, e331–e339.
14. Woodruff, T. M., Thundyil, J., Tang, S. C., Sobey, C. G., Taylor, S. M., & Arumugam, T. V. (2011). Pathophysiology, treatment, and animal and cellular models of human ischemic stroke. *Molecular Neurodegeneration, 6*, 11.
15. Gelderblom, M., Leypoldt, F., Steinbach, K., Behrens, D., Choe, C. U., Siler, D. A., Arumugam, T. V., Orthey, E., Gerloff, C., Tolosa, E., et al. (2009). Temporal and spatial dynamics of cerebral immune cell accumulation in stroke. *Stroke, 40*, 1849–1857.
16. Suh, S. W., Shin, B. S., Ma, H., Van Hoecke, M., Brennan, A. M., Yenari, M. A., & Swanson, R. A. (2008). Glucose and NADPH oxidase drive neuronal superoxide formation in stroke. *Annals of Neurology, 64*, 654–663.
17. Qureshi, A. I., Ali, Z., Suri, M. F., Shuaib, A., Baker, G., Todd, K., Guterman, L. R., & Hopkins, L. N. (2003). Extracellular glutamate and other amino acids in experimental intracerebral hemorrhage: An in vivo microdialysis study. *Critical Care Medicine, 31*, 1482–1489.
18. Wang, J., Fields, J., Zhao, C., Langer, J., Thimmulappa, R. K., Kensler, T. W., Yamamoto, M., Biswal, S., & Doré, S. (2007). Role of Nrf2 in protection against intracerebral hemorrhage injury in mice. *Free Radical Biology & Medicine, 43*, 408–414.
19. Flaherty, M. L., Woo, D., Haverbusch, M., Sekar, P., Khoury, J., Sauerbeck, L., Moomaw, C. J., Schneider, A., Kissela, B., Kleindorfer, D., et al. (2005). Racial variations in location and risk of intracerebral hemorrhage. *Stroke, 36*, 934–937.
20. National Institute of Neurological Disorders and Stroke rt-PA Stroke Study Group. (1995). Tissue plasminogen activator for acute ischemic stroke. *The New England Journal of Medicine, 333*, 1581–1587.
21. Berglund, A., Svensson, L., Sjöstrand, C., et al. (2012). Higher prehospital priority level of stroke improves thrombolysis frequency and time to stroke unit: The Hyper Acute STroke Alarm (HASTA) study. *Stroke, 43*, 2666–2670.
22. Kleindorfer, D., de Los Rios La Rosa, F., et al. (2013). Temporal trends in acute stroke management. *Stroke, 44*, S129–S131. https://doi.org/10.1161/STROKEAHA.113.001457
23. Secades, J. J., Alvarez-Sabin, J., Castillo, J., Diez-Tejedor, E., Martinez-Vila, E., Rios, J., & Oudovenko, N. (2016). Citicoline for acute ischemic stroke: A systematic review and formal meta-analysis of randomized, double-blind, and placebo-controlled trials. *Journal of Stroke and Cerebrovascular Diseases, 25*(8), 1984–1996.
24. Tayebati, S. K., & Amenta, F. (2013). Choline-containing phospholipids: Relevance to brain functional pathways. *Clinical Chemistry and Laboratory Medicine, 51*, 513–521.
25. Roy, P., Tomassoni, D., Nittari, G., Traini, E., & Amenta, F. (2022). Effects of choline containing phospholipids on the neurovascular unit: A review. *Frontiers in Cellular Neuroscience, 16*, 1–18.

26. Bergamaschi, M. (1995). Deacylated phospholipids: Chemistry and therapeutic potential. In G. Cvec & F. Paltauf (Eds.), *Phospholipids: Characterization, metabolism, and novel biological applications* (pp. 228–247). AOCS Press.
27. Parnetti, L., Amenta, F., & Gallai, V. (2001). Choline alphoscerate in cognitive decline and in acute cerebrovascular disease: An analysis of published clinical data. *Mechanisms of Ageing and Development, 122*, 2041–2055.
28. Secades Ruiz, J. J. (2011). Citicolina: Revisión farmacológica y clínica, actualización 2010. *Revista de Neurologia, 52*, S1.
29. Álvarez-Sabín, J., & Román, G. C. (2013). The role of citicoline in neuroprotection and neurorepair in ischemic stroke. *Brain Sciences, 3*, 1395–1414.
30. Clark, W. M., Wechsler, L. R., Sabounjian, L. A., & Schwiderski, U. E. (2001). A phase III randomized efficacy trial of 2000 mg citicoline in acute ischemic stroke patients. *Neurology, 57*, 1595–1602.
31. Dávalos, A., Alvarez-Sabín, J., Castillo, J., Díez-Tejedor, E., Ferro, J., Martínez-Vila, E., Serena, J., Segura, T., Cruz, V. T., Masjuan, J., et al. (2012). Citicoline in the treatment of acute ischaemic stroke: An international, randomised, multicentre, placebo-controlled study (ICTUS trial). *Lancet, 380*, 349–357.
32. Saver, J. L., & Wilterdink, J. (2002). Choline precursors in acute and subacute human stroke: A meta-analysis. *Stroke, 33*, 353.
33. Secades, J. J. (2011). Citicoline: Pharmacological and clinical review, 2010 update. *Revista de Neurologia, 52*(Suppl 2), S1–S62.
34. Dávalos, A., Castillo, J., Alvarez-Sabín, J., et al. (2002). Oral citicoline in acute ischemic stroke: An individual patient data pooling analysis of clinical trials. *Stroke, 33*, 2850–2857.
35. Martí-Carvajal, A. J., Valli, C., Martí-Amarista, C. E., Solà, I., Martí-Fàbregas, J., & Cosp, X. B. (2020). Citicoline for treating people with acute ischemic stroke. *Cochrane Database of Systematic Reviews, 8*, CD013066.
36. Veeranki, S., Gandhapudi, S. K., & Tyagi, S. C. (2017). Interactions of hyperhomocysteinemia and T cell immunity in causation of hypertension. *Canadian Journal of Physiology and Pharmacology, 95*, 239–246. https://doi.org/10.1139/cjpp-2015-0568
37. Herrmann, W., & Obeid, R. (2011). Homocysteine: A biomarker in neurodegenerative diseases. *Clinical Chemistry and Laboratory Medicine, 49*, 435–441. https://doi.org/10.1515/CCLM.2011.084
38. McCully, K. S. (1996). Atherogenesis and the chemical pathology of homocysteine. *European Journal of Laboratory Medicine, 4*, 121–128.
39. Welch, G. N., & Loscalzo, J. (1998). Homocysteine and atherothrombosis. *The New England Journal of Medicine, 338*(15), 1042–1050.
40. Ansari, R., Mahta, A., Mallack, E., & Luo, J. J. (2014). Hyperhomocysteinemia and neurologic disorders: A review. *Journal of Clinical Neurology, 10*(04), 281–288.
41. Spence, J. D. (2007). Homocysteine-lowering therapy: A role in stroke prevention? *Lancet Neurology, 6*, 830–838. https://doi.org/10.1016/S1474-4422(07)70219-37
42. Spence, J. D. (2007). Homocysteine-lowering therapy: A role in stroke prevention? *The Lancet Neurology, 6*(9), 830–838.
43. Shi, Z., Guan, Y., Huo, Y. R., Liu, S., Zhang, M., Lu, H., ... & Ji, Y. (2015). Elevated total homocysteine levels in acute ischemic stroke are associated with long-term mortality. *Stroke, 46*(9), 2419–2425. https://doi.org/10.1161/STROKEAHA.115.009136
44. Pinzon, R. T., Wijaya, V. O., & Veronica, V. (2023). The role of homocysteine levels as a risk factor of ischemic stroke events: A systematic review and meta-analysis. *Frontiers in Neurology, 14*, 1144584.
45. Rabelo, N. N., Telles, J. P., Pipek, L. Z., Farias Vidigal Nascimento, R., Gusmão, R. C., Teixeira, M. J., & Figueiredo, E. G. (2022). Homocysteine is associated with higher risks of ischemic stroke: A systematic review and meta-analysis. *PLoS One, 17*(10), e0276087.
46. Marti-Carvajal, A. J., Sola, I., Lathyris, D., & Dayer, M. (2017). Homocysteine-lowering interventions for preventing cardiovascular events. *Cochrane Database of Systematic Reviews, 8*, CD006612.

47. Huang, S., Cai, J., & Tian, Y. (2021). The prognostic value of homocysteine in acute ischemic stroke patients: A systematic review and meta-analysis. *Frontiers in Systems Neuroscience, 14*, 600582.
48. Kataria, N., Yadav, P., Kumar, R., Kumar, N., Singh, M., Kant, R., & Kalyani, V. (2021). Effect of vitamin B6, B9, and B12 supplementation on homocysteine level and cardiovascular outcomes in stroke patients: A meta-analysis of randomized controlled trials. *Cureus, 13*(5), e14958.
49. Zhou, Z., Liang, Y., Qu, H., Zhao, M., Guo, F., Zhao, C., & Teng, W. (2018). Plasma homocysteine concentrations and risk of intracerebral hemorrhage: A systematic review and meta-analysis. *Scientific Reports, 8*(1), 2568.
50. Li, Y., Huang, T., Zheng, Y., Muka, T., Troup, J., & Hu, F. B. (2016). Folic acid supplementation and the risk of cardiovascular diseases: A meta-analysis of randomized controlled trials. *Journal of the American Heart Association, 5*, e003768. https://doi.org/10.1161/JAHA.116.003768
51. Tian, T., Yang, K. Q., Cui, J. G., Zhou, L. L., & Zhou, X. L. (2017). Folic acid supplementation for stroke prevention in patients with cardiovascular disease. *The American Journal of the Medical Sciences, 354*, 379–387. https://doi.org/10.1016/j.amjms.2017.05.020
52. White, P. J., & Newgard, C. B. (2019). Branched-chain amino acids in disease. *Science, 363*(6427), 582–583.
53. Nie, C., He, T., Zhang, W., Zhang, G., & Ma, X. (2018). Branched chain amino acids: Beyond nutrition metabolism. *International Journal of Molecular Sciences, 19*(4), 954.
54. Lynch, C. J., & Adams, S. H. (2014). Branched-chain amino acids in metabolic signalling and insulin resistance. *Nature Reviews Endocrinology, 10*(12), 723–736.
55. McGarrah, R. W., & White, P. J. (2023). Branched-chain amino acids in cardiovascular disease. *Nature Reviews Cardiology, 20*(2), 77–89.
56. Felig, P., Marliss, E., & Cahill, G. F., Jr. (1969). Plasma amino acid levels and insulin secretion in obesity. *New England Journal of Medicine, 281*(15), 811–816.
57. Kimberly, W. T., Wang, Y., Pham, L., Furie, K. L., & Gerszten, R. E. (2013). Metabolite profiling identifies a branched chain amino acid signature in acute cardioembolic stroke. *Stroke, 44*(5), 1389–1395.
58. Ruiz-Canela, M., Toledo, E., Clish, C. B., Hruby, A., Liang, L., Salas-Salvado, J., Razquin, C., Corella, D., Estruch, R., Ros, E., & Fito, M. (2016). Plasma branched-chain amino acids and incident cardiovascular disease in the PREDIMED trial. *Clinical Chemistry, 62*(4), 582–592.
59. Park, M. K., Lee, S. J., Choi, E., Lee, S., & Lee, J. (2022). The effect of branched chain amino acid supplementation on stroke-related sarcopenia. *Frontiers in Neurology, 13*, 744945.
60. Magnusson, M., Hedblad, B., Engström, G., Persson, M., Nilsson, P., & Melander, O. (2013). High levels of cystatin C predict the metabolic syndrome: The prospective Malmö Diet and Cancer Study. *Journal of Internal Medicine, 274*(2), 192–199.
61. Tobias, D. K., Lawler, P. R., Harada, P. H., Demler, O. V., Ridker, P. M., Manson, J. E., Cheng, S., & Mora, S. (2018). Circulating branched-chain amino acids and incident cardiovascular disease in a prospective cohort of US women. *Circulation: Genomic and Precision Medicine, 11*(4), e002157.
62. Zhang, Y., Duan, Y., Jiang, M., He, X., Xu, S., Guo, J., Li, M., Zhou, C., Wu, D., Liu, G., & Ji, X. (2023). Branched-chain amino acids and risk of stroke: A Mendelian randomization study. *Frontiers in Neuroscience, 17*, 1143718.
63. Wang, Y., Huang, K., Liu, F., Lu, X., Huang, J., & Gu, D. (2022). Association of circulating branched-chain amino acids with risk of cardiovascular disease: A systematic review and meta-analysis. *Atherosclerosis, 350*, 90–96.

The Ketogenic Diet and Stroke

10

Ebele Joan Ajagun, Abraham Olufemi Asuku, Toheeb O. Oyerinde, Hammed A. Sangodipe, Wasiu Sanusi, Lukman Akanni Jayeoba, and Taoheed Adedeji Abdulsalam

Abstract

The ketogenic diet is a dietary intervention that has been around for over a century, dating back to the nineteenth century. It was initially developed to treat refractory epilepsy, but has since evolved into a popular dietary plan with modifications that improve palatability and adherence for patients with chronic diseases or those seeking to manage their weight and allow for individualised patient goals. The keto diet has gained widespread use and acceptance, and it is now one of the most researched dietary plans online, recognising over the years the role of the diet in various disease conditions. Recent studies have demonstrated the clinical benefits of the diet in metabolic disorders associated with obesity, insulin resistance, high blood pressure, high cholesterol and triglycerides levels, and type 2 diabetes mellitus, which are all confounding risk factors for stroke. A stroke, also known as a brain attack, is a life-threatening emergency and the primary cause of long-term neurological disability in adults globally. The modifiable risk factors are crucial in predicting the risk and development of stroke; hence, early detection and implementation of strategies that limit these factors can significantly reduce stroke incidence, morbidity, mortality rates, and

E. J. Ajagun · A. O. Asuku (✉) · L. A. Jayeoba · T. A. Abdulsalam
Bioresources Development Centre (BIODEC), National Biotechnology Development Agency (NABDA), Ogbomoso, Nigeria

T. O. Oyerinde
Laboratory for Experimental and Translational Neurobiology, University of Medical Sciences, Ondo, Nigeria

H. A. Sangodipe
Department of Medicine, Ladoke Akintola University of Technology, Ogbomoso, Oyo State, Nigeria

W. Sanusi
Department of Biochemistry and Nutrition, Fountain University, Osogbo, Nigeria

W. Mohamed, M. B. Mat-Nor (eds.), *Nutrition and Stroke*, Nutritional Neurosciences, https://doi.org/10.1007/978-981-97-6818-9_10

quality of life, with several studies recently recognising the diet's ability to indirectly prevent and aid in stroke recovery by managing these modifiable risk factors. However, clinical trials are needed to understand the mechanism of action, determine the best time to initiate the diet and develop a step-by-step dietary intervention plan to implement the keto diet effectively.

Keywords

Ketogenic diet · Stroke · Mortality rate · Quality of life · Obesity

10.1 Introduction

The ketogenic diet, commonly called 'keto diet' coined in the 1920s by Russell Wilder (M.D.) was introduced as a nutritional treatment option for paediatric epilepsy by physicians [1]. The goal was to trick the body into believing it was fasting, thereby reaping the benefits of the body's fasting state-nutritional ketosis [1, 2]. Although initially developed as a treatment option for epilepsy, its use was replaced decades later by the discovery and development of modern-era antiepileptic agents [3]. Several decades later, the ketogenic diet has gained popularity as several researchers have reported its potential benefits in chronic disease conditions and weight management/control [2, 3].

The global prevalence of ketogenic diet use remains challenging due to limited data. However, online evidence supports its acceptance and use, with evidence of it being the most searched food-related topic globally [3, 4]. The diet has gained significant following/patronage in many countries, with the majority of persons using it for weight loss and as a fitness trend; others have reported using the diet for its potential health benefits in chronic diseases such as metabolic syndrome, neurodegenerative diseases and type 2 diabetes [2, 3].

Recently, research has shown the benefits of a ketogenic diet in stroke prevention and recovery. The exact method by which a Ketogenic diet provides benefits has yet to be fully understood, but the diet is said to have a neuroprotective effect that can alleviate excitotoxicity, relieve oxidative stress, and prevent apoptosis and improve early motor and behavioural outcomes by providing preconditioning [5–7].

10.2 Ketogenic Diet

The ketogenic diet (KD) is a diet that comprises of high-fat, moderate-protein, and low-carbohydrate [4]; while other low-carbohydrate diets such as the Paleo, South Beach, and Dukan diets exist, the Ketogenic diet differs significantly due to its very high-fat (70% to 80% fat) and moderate protein contents [8]. A ketogenic diet has been reported to cause weight loss, enhance mental clarity, and improve body energy levels, achieving all these by changing the body's primary fuel source from carbohydrates to fat, resulting in the production of ketones that the body uses for energy instead of glucose [8, 9]. The Ketogenic diet aims to enhance an individual's metabolic health by reducing, controlling, or maintaining the percentage of body fat [8, 10]. Additionally, the ketogenic diet is beneficial in chronic disease conditions such as cancer, diabetes, polycystic ovary syndrome, and Alzheimer's disease [8].

10.2.1 Types of Ketogenic Diet

The first Ketogenic diet had a macronutrient composition ranging from 55–75% fat, 15–35% protein, and 0–10% carbs. By varying the ratios of fat, protein, and carbohydrates, various modifications to the Ketogenic diets have been developed, and include the classic long-chain triglycerides (LCT) ketogenic diet, medium-chain triglycerides (MCT) ketogenic diet, modified Atkins diet (MAD), and low glycemic index treatment (LGIT) [8, 11]. These modifications are made to provide flexibility, achieve individualised goals, and increase patient compliance, with recent studies showing these variations in Ketogenic diets have effectively treated various health conditions, making it an attractive option for those seeking dietary interventions [12–14].

10.2.1.1 Classic Long-Chain Triglycerides (LCT) Ketogenic Diet

The classical long-chain triglycerides (LCT) ketogenic diet was developed in the 1920s to manage refractory epilepsy. It comprises a macronutrient ratio of 4:1 or 3:1 of fat (in grams) to protein plus carbohydrates (in grams). Food-based LCT fat contributes to 90% of the daily calorie intake, while protein (6%) and carbohydrates (4%) make up the remaining 10% [13]. The LCT ketogenic diet promotes weight loss and reduces cardiovascular risks, including cholesterol levels and blood pressure [15]. However, its palatability is a significant challenge, hindering adherence by patients due to its low intake of carbohydrates; therefore, a significant drawback to the diet [14].

10.2.1.2 Medium-Chain Triglycerides (MCT) Ketogenic Diet

This ketogenic diet utilising medium-chain triglycerides (MCT) is a variation of the standard ketogenic diet, where dietary MCT oils are used as fat supplements instead of long-chain triglycerides in the dietary plan [16]. The primary advantage of the MCT ketogenic diet is that MCT fats/oils readily produce ketone bodies compared to LCT fat, allowing for flexibility in diet ratios and also improving diet palatability, therefore, making it easier for patients to comply with the diet plans [13, 16, 17]. However, it is frequently associated with gastrointestinal adverse effects when used for prolonged periods [13, 17].

10.2.1.3 Modified Atkins Diet (MAD)

The Modified Atkins Diet (MAD), a more flexible variation of the ketogenic diet, consists of high fat, moderate protein, and low carbohydrate portions and induce ketosis by replicating the metabolic state of fasting while yet providing adequate nutrient intake [18, 19]. The macronutrient ratio used in MAD ranges from 1:1 to 2:1 and does not involve any restriction of fluid, carbohydrate, or protein [8, 20–27].

Although initially designed for weight loss, clinical evidence shows that the Atkins diet's modification (MAD) is effective in managing intractable epilepsy in children [8, 21, 23, 24, 26, 27].

10.2.1.4 Low Glycemic Index (LGI) Diet

The low glycemic index (LGI) diet is a type of ketogenic diet that emphasises high fat (60%), moderate protein (30%), and low carbohydrates, with complex carbs preferred to simple sugars due to their low glycemic index [28]. The goal of this diet is not to cause ketosis but rather to reduce postprandial glycemic index. The Ketogenic diet's protective effect is due to stable blood glucose levels allowing for increased carbohydrate consumption with the diet [16, 29, 30]. Although this diet contains a higher amount of carbohydrates, it is as effective as other ketogenic diets in managing various diseases and promoting weight loss in obese patients [28, 31, 32].

10.3 Mechanism of Action

The human body requires various sources such as carbohydrates, fats, amino acids, lactate, and ketone bodies to generate the required energy for survival. However, the intake of carbohydrates decreases when one starts a Ketogenic diet, leading to a drop in glucose levels in the blood. As a result, the body's endogenic glucose metabolism becomes inadequate to provide the necessary energy as Adenosine triphosphate (ATP) required for daily activities [14, 33].

This glucose restriction forces the body to mimic the state of fasting, leading it to shift its primary source of fuel/energy towards fat (ketones), which is the ultimate goal of a ketogenic diet. Over time, the body exhausts the stored glucose (glycogen) for energy through glycogenolysis. On exhaustion of the body's glycogen stores, it turns to stored body fat and fatty acids for energy, transitioning from gluconeogenesis to ketogenesis [8].

Ketogenesis is a metabolic pathway that converts fatty acids and certain amino acids into ketone bodies and ketoacids, an alternative energy source for the body [8, 34]. These include acetoacetate, 3-beta-hydroxybutyrate, and acetone, produced in the liver by metabolising lipids and fats. All major organs in the body, such as the heart, brain, and skeletal muscles, can utilise energy derived from ketogenesis except for the liver, as it lacks the necessary enzyme, β-keto-acyl-CoA transferase [8, 35, 36].

Ketogenesis is a natural process that occurs in every healthy individual, and increases in situations such as low blood glucose levels (due to fasting or starvation), depletion of carbohydrate stores, or insufficient insulin (as in Type 1 diabetes). However, excessive ketone bodies synthesis and accumulation may lead to a fatal condition called ketoacidosis [37–40]. Blood insulin levels regulate the Ketogenic process, while glucagon, catecholamine, thyroid hormones, and cortisol increase the rate of ketogenesis by promoting the metabolism of free fatty acids [41].

10.4 The Ketogenic Pathway

The major ketogenic pathway takes place within the mitochondria of the liver cells and produces three byproducts—acetoacetate (ACA), 3-beta-hydroxybutyrate (BHB), and acetone [14, 42–44]. Fatty acids are moved by a protein called carnitine palmitoyltransferase (CPT-1) into the mitochondria. Here, they undergo β-oxidation to produce acetyl coenzyme A (acetyl-CoA). Two molecules of acetyl-CoA catalysed via thiolase enzyme form acetoacetyl-CoA. HMG-CoA synthesis catalysed by the HMG-CoA synthase forms 3-hydroxy-3-methylglutaryl-CoA (HMG-CoA) [44–46]. HMG-CoA undergoes conversion into two compounds, acetoacetate and acetyl-CoA, through HMG-CoA lyase. Specifically, HMG-CoA lyase is responsible for producing acetoacetate. Finally, acetoacetate undergoes decarboxylation to form acetone and reduction to form 3-beta-hydroxybutyrate, which is used by the body as energy [14, 44, 47].

In extra-hepatic tissues, 3-β-hydroxybutyrate (BHB) is converted to acetoacetate by β-hydroxybutyrate dehydrogenase, which is then converted to acetyl CoA by β-keto-acyl-CoA transferase. Acetyl-CoA is then used directly in the Krebs cycle (the tricarboxylic acid cycle or TCA cycle) to produce ATP energy [47]. Without glucose, on entering the bloodstream, the organs in our body, such as the muscles and heart, use ketone bodies as energy source. Furthermore, the ketone bodies can pass through the blood-brain barrier and serve as an alternative energy source [14, 48].

10.5 Adverse Effects of the Ketogenic Diet

Although the Ketogenic diet aims to improve overall well-being, it can cause specific adverse effects at the onset of the diet. These time-resolving effects include nausea, vomiting, dizziness, constipation, sluggishness, hypoglycemia, halitosis, abdominal pain, muscle pain, and fever [8, 49]. These effects are due to carbohydrate withdrawal and are called the "keto-flu." Symptoms of the keto-flu, such as irritability, insomnia, fatigue, headache, and mood swings, typically occur within days of starting the Ketogenic diet and resolve on their own [8, 14, 50]. Ketonemia, ketonuria, weight loss, and bad breath are signs of ketosis [8, 11, 51, 52].

Although some patients experience self-limiting adverse effects, others may experience severe hypoglycemia, dehydration, and electrolyte imbalances that may require hospital visits [53–55]. It is, therefore, essential to limit these adverse effects by maintaining proper fluid and electrolyte intake. Extended use of the Ketogenic diet, although rare, may result in several harmful effects, such as hypercalciuria, hepatic steatosis, hypocitraturia, kidney stones, hypoproteinemia, and deficiencies of essential minerals and vitamins. Although these complications are rare, they can be life-threatening and require immediate discontinuation of the diet [11, 52].

Diabetic patients on insulin and or oral blood glucose-lowering drugs may experience severe hypoglycemia if on a ketogenic diet; hence, medication must be adjusted accordingly before the initiation of the diet. Also, it is essential to educate patients on early signs of severe hypoglycemia for early detection and intervention [8]. Patients with disorders of metabolism, pancreatitis, liver failure, as well as pyruvate kinase deficiency and primary carnitine deficiency, should avoid a ketogenic diet [8]. The Ketogenic diet is known to cause a false positive for alcohol breath levels as the liver converts acetone to isopropanol through the alcohol dehydrogenase enzyme [8].

10.6 Health Benefits of the Ketogenic Diet

Although the ketogenic diet has been known for several decades, research into its health advantages is still ongoing. The ketogenic diet has several short- and long-term advantages, including its influence on metabolic problems caused by obesity, insulin resistance, high blood pressure, high cholesterol and triglycerides. Recent studies have shown it can also help manage type 2 diabetes mellitus [11, 46, 56–58].

The ketogenic diet is effective in children and adults with non-responsive epilepsy, as an adjunct to conventional chemotherapy for some forms of cancer, to relieve or ameliorate neurodegenerative diseases, and to provide neuroprotection in chronic disease conditions [8, 14, 47, 57]. Although the mechanism by which the Ketogenic diet exerts these effects is unclear, several hypotheses have been postulated, which include the ketone hypothesis, metabolic hypothesis (increased brain activity, increased number of mitochondria), GABA shunt/amino acid hypothesis, and gut microbiome hypothesis, among others [58–65]. However, despite the benefits of the Ketogenic diet, there are potential pitfalls such as an increased risk of kidney stones, osteoporosis, and gout [47, 57, 58]. Hence, consuming a diverse range of approved meats, fish, vegetables, fruits, nuts, and seeds is essential. These foods are rich in fibre, vitamin B, and essential minerals. Examples include iron, magnesium, and zinc, found in whole grains and other foods restricted in ketogenic diets [47, 66].

10.7 Stroke

A stroke, also referred to as a cerebrovascular accident (CVA), is a serious medical emergency that can be fatal. It occurs when the brain is damaged due to a disruption in blood flow [67]. Strokes are commonly known as brain attacks and are the primary cause of long-term neurological impairments in people [68]. The different types of strokes are based on aetiology and include ischemic, haemorrhagic, transient ischemic (mini-stroke), brain stem and cryptogenic stroke [69]. Haemorrhagic stroke is a severe form of stroke and accounts for approximately 20% of all new cases annually [70]; globally, ischemic stroke is responsible for more than 80% of all cases of stroke [71]. Motor deficits are common after a stroke irrespective of the underlying aetiology and are the sequelae after a stroke occurs, causing paralysis and difficulty in walking, speaking, and understanding essential functions [69, 72]. Recovery varies across patients, with spontaneous recovery occurring in some patients within the first three months. However, recovery tends to be slower and less common in patients with more severe motor deficits [72, 73].

Stroke is the main cause of long-term disability in adults globally and the second leading cause of death, accounting for 11.6% of deaths with a predicted lifetime risk of ≥18.3% in adults aged 25 years and above and a projected increase in prevalence by 3.4 million cases between the years 2012 and 2030 with a more significant impact in developing nations [73–75].

10.8 Contributing Factors to Stroke

Although identifying the risk factors for stroke is difficult due to its many varieties, several risk factors have been implicated based on its aetiology [71, 76]; and are classified as modifiable (e.g., diet and comorbid conditions- diabetes, hypertension, hyperlipidemia, atrial fibrillation (AF)) and non-modifiable risk factors (e.g. age, race) [71]. These risks are either short-term triggers (e.g. infectious events, sepsis, and stress), intermediate-term triggers (e.g. hyperlipidemia and hypertension) or long-term triggers (e.g. sex and ethnicity/race) [71, 76–78].

Stroke is a complex and multifactorial disease initially believed to be associated with the ageing process. However, recent research suggests that ischemic stroke is becoming more common among persons aged 20–54 years, while haemorrhagic stroke is becoming more prevalent in individuals aged 45 and above [79, 80]. The relationship between sex/gender and stroke is highly dependent on the age of an individual, with younger women being more likely to experience increased risks due to hormonal factors such as the use of hormonal contraceptives and pregnancy-related states [81]. Additionally, race and ethnicity are essential predictive risk factors for stroke, with Black, Hispanic, and Latino Americans having a higher risk of stroke than those of White ethnicity due to their genetic makeup [80, 82].

While non-modifiable risk factors play an important role in predicting the risk and development of stroke, modifiable risk factors are also crucial. Early detection of these factors and implementing strategies aimed at limiting them can significantly lower the incidence of stroke, improve morbidity, reduce mortality rates, and improve the quality of life of patients [83–85]. Several researches have reported the modifiable risk factors as responsible for most stroke cases [83, 86].

Hypertension is considered the most significant modifiable risk factor for stroke due to the firm, direct, linear, and continuous relationship between elevated blood pressure and stroke risk, and present in over 65% of stroke cases [85, 87, 88]. In addition, diabetes mellitus is another important independent predictive risk factor [85, 87]. Diabetes mellitus, as well as pre-diabetic conditions, are independent risk factors for stroke, as it is one of the micro-vascular complications that lead to poorer post-stroke outcomes in chronic cases of impaired glucose tolerance [86, 89]. Patients with diabetes have twice the risk of stroke, which accounts for 20% of diabetes-related deaths; hence, diabetic patients with stroke are usually younger, most likely to be black and have one or two additional stroke risk factors [89, 90].

Obesity is a disorder characterised by an excessive and abnormal accumulation of fat within the body, leading to an increased risk of various diseases, including dyslipidemia, type 2 diabetes, hypertension, and cardiovascular diseases [91]. There is significant research linking obesity to an increased risk and incidence of stroke, ranking obesity within the top ten risk factors for stroke [92, 93].

Available evidence now suggests that obesity is associated with 1 in 5 stroke cases. Obesity accounts for a 64% increase in stroke risk, while being overweight increases stroke risk by over 22% [93, 94]. The exact mechanism by which obesity increases the incidence of stroke is unclear. Obesity, on the other hand, is linked to other risk factors, such as diabetes, hypertension, and dyslipidemia, all of which are known to increase the risk of stroke [95, 96].

The Framingham Stroke Risk Profile is a tool that helps predict an individual's stroke risk within the next ten years [71, 87]. It factors in several predictors, modifiable and non-modifiable risk factors such as age, diabetes, smoking, cardiovascular diseases, systolic blood pressure, antihypertensive therapy, and other related health conditions [71, 87, 97]. Patients at risk of stroke are also at risk of other cardiovascular events; hence the profile also includes cardiac-related events [97–99]. Also, the recently concluded American College of Cardiology (ACC) and the American Heart Association (AHA) Pooled Cohort 10-year atherosclerotic cardiovascular disease study factors in the role in stroke, making the Atherosclerotic Cardiovascular (ASCVD) risk estimator invaluable within black communities [97, 99]. The profile is beneficial for predicting the risk of stroke in black communities, where race/ethnicity is a contributing factor.

10.9 Ketogenic Diet in Stroke Prevention and Recovery

Several decades of research have shown that the Ketogenic diet has a positive impact on various medical conditions, including diabetes, hypertension, and obesity, which can increase the risk of stroke [8, 47]. A ketogenic diet may reduce stroke risk by managing cardiovascular health markers, including blood pressure, oxidative stress, inflammation and lipid profiles [47, 56]; which are also known markers for stroke risk and include obesity, elevated blood pressure, lipid profile, inflammation, and diabetes [71, 89].

The exact way in which the Ketogenic diet helps to improve hypertension is yet to be determined. However, studies have shown that dietary modifications such as the DASH and low-carb diets can significantly reduce blood pressure in hypertensive patients. By reducing body fat and improving lipid profiles, including lowering LDL cholesterol, triglyceride levels, and total cholesterol, the Ketogenic diet indirectly improves cardiovascular risk factors and blood pressure control [100–102].

The Ketogenic diet has the added advantage of providing antioxidant effects. An imbalance between the production of reactive oxygen species and the body's antioxidant defences leads to oxidative stress. BHB and ACA, the major ketone bodies, have scavenging abilities, which help reduce oxidative stress and protect against damage caused by stroke [103, 104].

A stroke occurs when there is not enough oxygen and nutrients reaching the brain, leading to hypoxia and cell and neuron death. This results in permanent brain damage [105]. However, a Ketogenic diet provides an abundance of ketone bodies (BHB and ACA), which are the preferred energy source of the brain during hypoxia, thus preventing neuron death and permanent brain damage caused by lack of carbohydrate and oxygen deprivation [106, 107].

The ketogenic diet can have a positive impact on cardiovascular health markers providing a quick supply of ketone bodies to the brain immediately after a stroke. Thus, it may help in reducing the risk of stroke (prevention) and prevent brain damage (permanent neuronal death) leading to a faster recovery [6, 108].

10.10 Research Studies on the Effect of Ketogenic Diet on Stroke

Recent studies have shown that the ketogenic diet may have various beneficial effects on stroke prevention, recovery, and rehabilitation [8, 108]. Ketogenic and keto-like diets have been studied for their clinical benefits in chronic health conditions, such as refractory epilepsy, metabolic disorders, and weight loss [5, 46]. These positive findings have renewed interest in other health conditions, including neurodegenerative disorders, Alzheimer's disease, and stroke, among others [5, 57]. The major ketone bodies produced during the process of ketogenesis, BHB and ACA, have been found to have a direct effect on inflammatory proteins, reactive oxygen species (ROS), and transcription factors, as well as having neuroprotective effects [5, 14, 42, 109]. Although the precise mechanism by which the ketogenic diet, carbohydrate derivation and ketone bodies prevent stroke is unknown, it improves mitochondrial function, decreases inflammation, and increases neurotrophins [110, 111]. Brain injury after a stroke involves glutamate-mediated excitotoxicity, intracellular calcium overload, mitochondrial dysfunction, and the formation of reactive oxygen species (ROS), leading to neurological dysfunction [112]. Hence, treatment modalities aimed at any of these pathophysiological mechanisms may significantly improve treatment outcomes and prognosis [113].

Several animal studies have shown the benefits of exogenously administered BHB and a ketogenic diet before injury using various animal models, although the studies conducted so far have proved inconclusive [5, 108, 114, 115]. In one study, researchers used the middle cerebral artery occlusion (MCAO) rat model to demonstrate how exogenous administration of β-hydroxybutyrate (BHB) significantly reduced damage caused by stroke, including reduced infarct area, swelling, lipid peroxidation, and neurological issues, hence providing a cerebroprotective effect [116].

In a different study, a ketogenic diet showed preconditioning properties, allowing for early recovery of motor-behavioral outcomes in ischemic stroke [117]. Similarly, another study reported that initiating a ketogenic diet before injury prevented neurodegeneration caused by cerebral ischemia following cardiac arrest in an animal study [7]. These studies provide promising insights into the potential benefits of

BHB and a ketogenic diet in preventing and treating neurological injuries, reducing brain damage and improving functional recovery.

The ketogenic diet may reduce stroke risk by improving cardiovascular health markers. These markers include lowering blood pressure, reducing inflammation, and minimising oxidative stress. These effects may help in preventing the development of risk factors for stroke [14, 111]. For instance, one study found that a ketogenic diet can regulate the inflammatory pathway by stabilising Hypoxia-inducible factor alpha (HIF1α), which modulates the expression of inflammatory cytokines [111].

10.11 Implementing a Ketogenic Diet After Stroke

The Ketogenic diet has numerous benefits for men with chronic disease conditions, making it an exciting new dietary option for managing these conditions [8, 9, 11]. Each patient's Ketogenic diet is personalised based on unique needs, age, comorbidity, and disease conditions [8, 11, 118]. However, certain disease conditions preclude using the Ketogenic diet; hence, it might not be suitable for all patients [8]. Therefore, when considering a Ketogenic diet after a stroke, several factors must be considered, including the patient's existing disease conditions, metabolic requirements, and physical activity [8, 118]. Before embarking on a Ketogenic diet, it is crucial to consult healthcare professionals, particularly the medical team responsible for the patient's care, who are familiar with their medical history and can guide personalised patient-centred objectives [118]. The International Ketogenic Diet Study Group recommends the following steps for initiating a Ketogenic diet: patient selection, timing for the initiation of diet, pre-diet evaluation, diet initiation, monitoring and adjustments, and follow-ups [118].

10.11.1 Patient Selection

Although there is no standard guide for the use of the ketogenic diet in treating stroke and other chronic diseases, a practical guide created by The International Ketogenic Diet Study Group is available for treating epilepsy in children and may serve as a valuable reference for initiating a ketogenic diet in other disease conditions [118].

It is possible to initiate a ketogenic diet in patients of any age and gender as long as there is no contraindication. However, before starting the diet, all patients should undergo screening for primary carnitine deficiency, pyruvate carboxylase deficiency, and other conditions that can prevent the formation of ketone bodies [8, 13, 118]. It is essential to consult with a healthcare provider before starting a ketogenic diet, as it may not be safe for everyone.

10.11.2 Timing for the Initiation of Diet

As for the timing of the diet, it can be considered as soon as possible after screening for comorbidities and conducting basic routine tests. The optimal timing for starting the diet is still undetermined [118].

10.11.3 Pre-Diet Evaluation

Before starting a diet, it is essential to conduct a pre-diet evaluation to determine the patient's nutritional requirements and overall health status. The pre-diet evaluation involves a series of tests such as complete blood count, lipid profile, height, weight, liver and kidney function tests, and others. These tests provide a comprehensive understanding of the patient's health status. They are used during the monitoring phase to determine if the diet is effective or harmful for the patient over time. Several studies have recommended conducting a pre-diet evaluation to ensure the diet is safe and beneficial for the patient [13, 118, 119].

10.11.4 Diet Initiation

After completing all the necessary preliminary tests, the next step is to create and implement a diet plan. The patient's metabolic requirement is determined by calculating their caloric and fluid intake needs, setting diet goals, determining the type of diet required, establishing any necessary supplement intake (vitamins and minerals), and finally initiating the diet plan [118].

10.11.5 Monitoring and Adjustments

After starting a ketogenic diet, it is essential to monitor and make necessary adjustments regularly. Monitoring serum glucose, blood BHB levels, and urine ketones is crucial to determine if the patient has achieved the desired ketosis [120, 121]. Additionally, it is essential to reexamine all tests carried out at the beginning of the diet to compare current values with baseline results and evaluate the diet's effectiveness [118].

Monitoring patients for potential life-threatening adverse effects is also essential. In such cases, it may be necessary to discontinue the diet and consider alternative treatments to achieve individual goals [8]. If the ketogenic diet fails to achieve ketosis, changing the diet type might be helpful. This can involve modifying the ratios of fat-to-carbohydrate and protein or transitioning to a less strict diet with better palatability if compliance is challenging [118].

10.11.6 Follow-Up

The Follow- up process in patients after the acute phase of treatment is crucial. If dietary adjustments have failed to achieve the desired individualised patient goals and harm the intended patient, discontinuing therapy may be necessary. Additionally, it is essential to educate patients and their caregivers on the diet, signs of ketosis and toxicity, and the need for strict adherence to the dietary plans designed by the dieticians [13, 118, 119].

10.12 Conclusion

Stroke is a severe medical condition that can be life-threatening, especially in developing countries where there are several risk factors. The Ketogenic diet is known to offer many benefits in managing chronic diseases like retractable epilepsy, type 2 diabetes mellitus, obesity, stroke prevention, and recovery. Therefore, dietary intervention has become a critical part of stroke management. However, more research is required to determine the exact mechanism of action of the diet in stroke prevention and recovery. It is also essential to describe in detail the steps for initiating a Ketogenic diet for stroke prevention and recovery and imperative to study the role of the Ketogenic diet in enhancing the effectiveness of the existing therapeutic strategies for stroke prevention and recovery. Furthermore, it is necessary to develop guidelines to help clinicians in the step-by-step initiation of a Ketogenic diet for stroke prevention and management.

References

1. Wheless, J. W. (2008). History of the ketogenic diet. *Epilepsia, 49*(suppl. 8), 3–5. https://doi.org/10.1111/j.1528-1167.2008.01821.x
2. Dowis, K., & Banga, S. (2021). The potential health benefits of ketogenic diet: A narrative review. *Nutrients, 13*, 1654. https://doi.org/10.3390/nu13051654
3. Algamdi, K. A., Alsaud, S. H. A., Alghubaini, G. B., Alsaggaf, M. M., Altalhi, I. A., Asali, J. S., et al. (2020). The prevalence of ketogenic diet among adults with chronic disease in Saudi Arabia. *International Journal of Innovative Research in Medical Sciences, 6*(1), 82–85. https://doi.org/10.23958/ijirms/vol6-i01/1061
4. Kossoff, E. H., & McGrogan, J. R. (2005). Worldwide use of the ketogenic diet. *Epilepsia, 46*(2), 280–289. https://doi.org/10.1111/j.0013-9580.2005.42702.x
5. Gough, S. M., Casella, A., Ortega, K. J., & Hackam, A. S. (2021). Neuroprotection by the ketogenic diet: Evidence and controversies. *Frontiers in Nutrition, 8*, 782657. https://doi.org/10.3389/fnut.2021.782657
6. Shaafi, S., Mahmoudi, J., Pashapour, A., Farhoudi, M., Sadigh-Eteghad, S., & Akbari, H. (2014). Ketogenic diet provides neuroprotective effects against ischemic stroke neuronal damages. *Advanced Pharmaceutical Bulletin, 4*(2), 479–481. https://doi.org/10.5681/apb.2014.071
7. Shaafi, S., Sharifi-Bonab, M., Ghaemian, N., Mokhtarkhani, M., & Abbari, H. (2019). Early motor-behavioural outcome of ischemic stroke with ketogenic diet precondition-

ing: Interventional animal study. *Journal of Stroke and Cardiovascular Diseases, 28*(4), 1032–1039. https://doi.org/10.1016/j.jstrokecerebrovasdis.2018.12.024

8. Masood, W., Annamaraju, P., Khan Suheb, M. Z., et al. (2023). Ketogenic diet. [Updated 2023 Jun 16]. In *StatPearls [internet]*. StatPearls Publishing. Available from: https://www.ncbi.nlm.nih.gov/books/NBK499830/
9. O'Neill, B., & Raggi, P. (2020). The ketogenic diet: Pros and cons. *Atherosclerosis, 292*, 119–126.
10. Weber, D. D., Aminzadeh-Gohari, S., Tulipan, J., Catalano, L., Feichtinger, R. G., & Kofler, B. (2020). Ketogenic diet in the treatment of cancer—Where do we stand? *Molecular Metabolism, 33*, 102–121. https://doi.org/10.1016/j.molmet.2019.06.026
11. Kim, J. M. (2017). Ketogenic diet: Old treatment, new beginning. *Clinical Neurophysiology Practice, 2*, 161–162.
12. Blackford, R. (2020). Not your parents' ketogenic diet—Flexibility in 2020. *Epilepsy Research, 162*, 106307.
13. Kossoff, E. H., Zupec-Kania, B. A., Auvin, S., Ballaban-Gil, K. R., Christina Bergqvist, A. G., Blackford, R., Buchhalter, J. R., Caraballo, R. H., Cross, J. H., Dahlin, M. G., Donner, E. J., Guzel, O., Jehle, R. S., Klepper, J., Kang, C., Lambrechts, D. A., Christiana Liu, Y. M., Nathan, J. K., Nordli, D. R., & Wirrell, E. C. (2018). Optimal clinical management of children receiving dietary therapies for epilepsy: Updated recommendations of the International Ketogenic Diet Study Group. *Epilepsia Open, 3*(2), 175–192. https://doi.org/10.1002/epi4.12225
14. Zhu, H., Bi, D., Zhang, Y., Kong, C., Du, J., Wu, X., Wei, Q., & Qin, H. (2022). Ketogenic diet for human diseases: The underlying mechanisms and potential for clinical implementations. *Signal Transduction and Targeted Therapy, 7*(1), 1–21. https://doi.org/10.1038/s41392-021-00831-w
15. Batch, J. T., Lamsal, S. P., Adkins, M., Sultan, S., & Ramirez, M. N. (2020). Advantages and disadvantages of the ketogenic diet: A review article. *Cureus, 12*(8), e9639. https://doi.org/10.7759/cureus.9639
16. Lin, T. Y., Liu, H. W., & Hung, T. M. (2021). The ketogenic effect of medium chain triglycerides. *Frontiers in Nutrition, 8*, 747284. https://doi.org/10.3389/fnut.2021.747284
17. Liu, Y. M., & Wang, H. S. (2013). Medium-chain triglyceride ketogenic diet, an effective treatment for drug-resistant epilepsy and a comparison with other ketogenic diets. *Biomedical Journal, 36*, 9–15.
18. Calcaterra, V., Verduci, E., Pascuzzi, M. C., Magenes, V. C., Fiore, G., Di Profio, E., Tenuta, E., Bosetti, A., Todisco, C. F., D'Auria, E., & Zuccotti, G. (2021). Metabolic derangement in pediatric patient with obesity: The role of ketogenic diet as therapeutic tool. *Nutrients, 13*(8), 2805. https://doi.org/10.3390/nu13082805
19. Kossoff, E. H., McGrogan, J. R., Bluml, R. M., Pillas, D. J., Rubenstein, J. E., & Vining, E. P. (2006). A modified Atkins diet is effective for the treatment of intractable pediatric epilepsy. *Epilepsia, 47*(2), 421–424. https://doi.org/10.1111/j.1528-1167.2006.00438.x
20. Kang, C., Lee, H. S., You, S. J., Kang, D. C., Ko, S., & Kim, H. D. (2007). Use of a modified Atkins diet in intractable childhood epilepsy. *Epilepsia, 48*(1), 182–186. https://doi.org/10.1111/j.1528-1167.2006.00910.x
21. Kim, J. A., Yoon, R., Lee, E. J., Lee, J. S., Kim, J. T., Kim, H. D., & Kang, C. (2016). Efficacy of the classic ketogenic and the modified Atkins diets in refractory childhood epilepsy. *Epilepsia, 57*(1), 51–58. https://doi.org/10.1111/epi.13256
22. Kossoff, E. H., Turner, Z., Bluml, R. M., Pyzik, P. L., & Vining, E. P. (2007). A randomized, crossover comparison of daily carbohydrate limits using the modified Atkins diet. *Epilepsy & Behavior, 10*(3), 432–436. https://doi.org/10.1016/j.yebeh.2007.01.012
23. Mehta, R., Goel, S., Sharma, S., Jain, P., Mukherjee, S. B., & Aneja, S. (2016). Efficacy and tolerability of the modified Atkins diet in young children with refractory epilepsy: Indian experience. *Annals of Indian Academy of Neurology, 19*(4), 523–527. https://doi.org/10.4103/0972-2327.194463

24. Sharma, S., Goel, S., Jain, P., Agarwala, A., & Aneja, S. (2016). Evaluation of a simplified modified Atkins diet for use by parents with low levels of literacy in children with refractory epilepsy: A randomized controlled trial. *Epilepsy Research, 127*, 152–159. https://doi.org/10.1016/j.eplepsyres.2016.09.002
25. Sharma, S., Sankhyan, N., Gulati, S., & Agarwala, A. (2012). Use of the modified Atkins diet in infantile spasms refractory to first-line treatment. *Seizure, 21*, 45–48.
26. Sharma, S., Sankhyan, N., Gulati, S., & Agarwala, A. (2013). Use of the modified Atkins diet for treatment of refractory childhood epilepsy: A randomized controlled trial. *Epilepsia, 54*, 481–486.
27. Sondhi, V., Agarwala, A., Pandey, R. M., Charkrabarty, B., Jauhari, P., Lodha, R., et al. (2020). Efficacy of ketogenic diet, modified Atkins diet, and low glycemic index therapy diet among children with drug-resistant epilepsy: A randomized clinical trial. *JAMA Pediatric, 174*, 944–951.
28. Tereshko, Y., Bello, S. D., Lorenzo, C. D., Pez, S., Pittins, A., Sartor, R., et al. (2023). 2:1 Ketogenic diet and low-glycemic index diet for the treatment of chronic and episodic migraine: A single center real-life retrospective study. *Journal of Headache and Pain, 24*, 95. https://doi.org/10.1186/s10194-023-01635-9
29. Hussain, T. A., Mathew, T. C., Dashti, A. A., Asfar, S., Al-Zaid, N., & Dashti, H. M. (2012). Effect of low-calorie versus low-carbohydrate ketogenic diet in type 2 diabetes. *Nutrition, 28*(10), 1016–1021. https://doi.org/10.1016/j.nut.2012.01.016
30. Vranceanu, M., Pictering, C., Filip, L., Pralea, I. E., Sundaram, S., Al-Saleh, A., Popa, D. S., & Grimaldi, K. A. (2020). A comparison of a ketogenic diet with a low GI/nutrigenetic diet over 6 months for weight loss and a 18-month follow-up. *BMC Nutrition, 6*, 53.
31. Evcili, G., Utku, U., Oün, N. M., & Ozdemir, G. (2018). Early and long period follow-up results of low glycemic index diet for migraine prophylaxis. *Agriculture, 30*, 8–11. https://doi.org/10.5505/agri.2017.62443
32. Finsterer, J., & Frank, M. (2019). Low-glycemic-index diet relieving migraine but inducing muscle cramps. *Journal of Neuroscience Rural Practicals, 10*, 552–554.
33. Yuen, A. W., & Sander, J. W. (2014). Rationale for using intermittent calorie restriction as a dietary treatment for drug resistant epilepsy. *Epilepsy Behaviour, 33*, 110–114.
34. Laffel, L. (1999). Ketone bodies: A review of physiology, pathophysiology and application of monitoring to diabetes. *Diabetes Metabolism Research Review, 15*(6), 412–426. https://doi.org/10.1002/(sici)1520-7560(199911/12)
35. Dilliraj, L. N., Schinuma, G., Lara, D., Strazzobasco, G., Clement, J., Giovannini, P., et al. (2022). The evolution of ketosis: Potential impact on clinical conditions. *Nutrients, 14*(17), 3613. https://doi.org/10.3390/nu14173613
36. Evans, M., McClure, T. S., Koutnik, A. P., & Egan, B. (2022). Exogenous ketone supplements in athletic contexts: Past, present, and future. *Sports Medicine, 52*(Suppl 1), 25–67. https://doi.org/10.1007/s40279-022-01756-2
37. Cardoso, L., Vicente, N., Rodrigues, D., Gomes, L., & Carritho, F. (2017). Controversies in the management of hyperglycemic emergencies in adults with diabetes. *Metabolism, 68*, 43–54.
38. Nakagawa, Y., & Shimano, H. (2018). CREBH regulates systemic glucose and lipid metabolism. *International Journal of Molecular Sciences, 19*(5), 1396. https://doi.org/10.3390/ijms19051396
39. Qui, H., Novikor, A., & Vallon, V. (2017). Ketosis and diabetic ketoacidosis in response to SGLT2 inhibitors: Basic mechanism and therapeutic perspective. *Diabetes/Metabolism Research and Reviews, 33*(5), e2886. https://doi.org/10.1002/dmrr.2886
40. Youngson, N. A., Morris, M. J., & Ballard, J. W. O. (2017). The mechanism mediating the antiepileptic effects of the ketogenic diet and potential opportunities for improvement with metabolism-altering drugs. *Seizure, 52*, 15–19.
41. Grabacka, M., Pierzchalska, M., Dean, M., & Reiss, K. (2016). Regulation of ketone body metabolism and the role of PPARα. *International Journal of Molecular Sciences, 17*(12), 2093. https://doi.org/10.3390/ijms17122093

42. Dhilton, K. K., & Gupta, S. (2023). Biochemistry, ketogenesis. In *StatPearls [Internet]*. StatPearls Publishing. www.ncbi.nlm.gov/books/NBK493179
43. Newman, J. C., & Verdin, E. (2014). Ketone bodies as signaling metabolites. *Trends Endocrinology Metabolism, 25*, 42–52.
44. Veyrat-Durebex, C., Reynier, P., Procaccio, V., Hergesheimer, R., Corcia, P., Andres, C. R., & Blasco, H. (2018). How can a ketogenic diet improve motor function? *Frontiers in Molecular Neuroscience, 11*, 315811. https://doi.org/10.3389/fnmol.2018.00015
45. Ghimire, P., & Dhamoon, A. S. (2022). Ketoacidosis. In *StatPearls [Internet]*. StatPearls Publishing.
46. Westman, E. C., Tondt, J., Maguine, E., & Yancy, W. S. (2018). Implementing a low carbohydrate ketogenic diet to manage type 2 diabetes mellitus. *Expert Review Endocrinology & metabolism, 13*(5), 263–273.
47. Maguire, E., & Youngson, N. A. (2022). Review on the benefits and drawbacks of ketogenic diet in humans. *IDOSR Journal of Experimental Sciences, 7*(1), 19–23.
48. Elamin, M., Ruskin, D. N., Masino, S. A., & Sacchetti, P. (2017). Ketone-based metabolic therapy: Is increased NAD (+) a primary mechanism? *Frontier in Molecular Neuroscience, 10*, 377.
49. Shalabi, H., Alotaibi, A., Alqahtani, A., Alattas, H., & Alghamdi, Z. (2021). Ketogenic diets: Side effects, attitude, and quality of life. *Cureus, 13*(12), e20390. https://doi.org/10.7759/cureus.20390
50. Crosby, L., Davis, B., Joshi, S., Jardine, M., Paul, J., Neola, M., & Barnard, N. D. (2021). Ketogenic diets and chronic diseases: Weighing the benefits against the risks. *Frontiers Nutrition, 8*, 702802. https://doi.org/10.3389/fnut.2021.702802)
51. Achanta, L. B., & Rae, C. D. (2017). β-Hydroxybutyrate in the brain: One molecule, multiple mechanisms. *Neurochemical Research, 42*, 35–49.
52. Luong, T. V., Abild, C. B., Bangshaab, M., Gormsen, L. C., & Sondergaard, E. (2022). Ketogenic diet and cardiac substrate metabolism. *Nutrients, 14*(7), 1322. https://doi.org/10.3390/nu14071322
53. Armeno, M., Araujo, C., Stomontesano, B., & Carabello, R. H. (2018). Update on the adverse effects during therapy with a ketogenic diet in paediatric refractory epilepsy. *Review Neurology, 66*(6), 193–200.
54. Guzel, O., Uysal, U., & Arlsan, N. (2019). Efficacy and tolerability of olive oil based ketogenic diet in children with drug-resistant epilepsy: A single center experience from Turkey. *European Journal of Paediatric Neurology, 23*(1), 143–151.
55. Wlodarek, D. (2019). Role of ketogenic diets in neurodegenerative diseases (Alzheimer's disease and Parkinson's disease). *Nutrients, 11*(1), 169. https://doi.org/10.3390/nu11010169
56. Paoli, A. (2014). Ketogenic diet for obesity: Friend or foe? *International Journal of Environmental Research Public Health, 11*(2), 2092–2107.
57. Alharbi, A., & Al-Sowayan, N. (2020). The effect of ketogenic-diet on health. *Food and Nutrition Sciences, 11*, 301–313. https://doi.org/10.4236/fns.2020.114022
58. Nuwaylati, D., Eldakhakhny, B., Bima, A., Sakr, H., & Elsamanoudy, A. (2022). Low-carbohydrate high-fat diet: A SWOC analysis. *Metabolites, 12*(11), 1126. https://doi.org/10.3390/metabo12111126
59. Appleton, D. B., & DeVivo, D. C. (1979). An animal model for the ketogenic diet. *Epilepsia, 15*, 211–227.
60. Bough, K., Wetheringlon, J., Hassel, B., Pare, J. F., Gawryluk, J. W., Greene, J. G., et al. (2006). Mitochondrial biogenesis in the anticonvulsant mechanism of the ketogenic diet. *Annals of Neurology, 60*, 223–235.
61. Greene, A. E., Todorova, M. T., & Seyfriend, T. N. (2003). Perspectives on the metabolic management of epilepsy through dietary reduction of glucose and elevation of ketone bodies. *Journal of Neurochemistry, 86*, 529–537.
62. Nemani, K., Ghomi, H. R., McCormick, B., & Fan, X. (2015). Schizophrenia and the gut-brain axis. *Progress in Neuro-Psychopharmacology and Biological Psychiatry, 56*, 155–160.

63. Nordi, D. R., & DeVivo, D. C. (2004). Effects of the ketogenic diet on cerebral energy metabolism. In C. E. Stafstrome & J. M. Rho (Eds.), *Epilepsy and the ketogenic diet* (1st ed., pp. 179–1184). Humana Press.
64. Seyfriend, T. N., Greene, A. E., & Todorova, M. T. (2004). Caloric restriction and epilepsy: Historical perspective, relationship to the ketogenic diet and analysis in epileptic EL mice. In C. E. Stafstrome & J. M. Rho (Eds.), *Epilepsy and the ketogenic diet* (1st ed., pp. 247–264). Humana Press.
65. Vuong, H. E., & Hsiao, E. Y. (2017). Emerging roles for the gut microbiome in autism spectrum disorder. *Biological Psychiatry, 18*, 411–423.
66. Andrewski, E., Cheng, K., & Vanderpool, C. (2022). Nutritional deficiencies in vegetarian, gluten-free and ketogenic diets. *Paediatrics in Review, 43*(2), 61–70.
67. Amado, B., Melo, L., Pinto, R., Lobo, A., Barros, P., & Gomes, J. R. (2022). Ischemic stroke, lessons from the past towards effective preclinical models. *Biomédica, 10*(10), 2561. https://doi.org/10.3390/biomedicals10102561
68. French, B., Thomas, L. H., Coupe, J., McMahon, N. E., Connell, L., Harrison, J., Sutton, C. J., Tishkovskaya, S., & Watkins, C. L. (2016). Repetitive task training for improving functional ability after stroke. *Cochrane Database of Systematic Review, 11*(11), CD006073. https://doi.org/10.1002/14651858.CD006073.pub3
69. Chung, C.-P. (2017). Types of stroke and their differential diagnosis. In L. R. Caplan, J. Biller, M. C. Leary, E. H. Lo, A. J. Thomas, M. Yenari, & J. H. Zhang (Eds.), *Primer on cerebrovascular diseases* (2nd ed., Ch. 77, pp. 372–376). Academic. https://doi.org/10.1016/B978-0-12-803058-5.00077-1
70. Chen, S., Zeng, L., & Hu, Z. (2014). Progressing haemorrhagic stroke: Categories, causes, mechanisms and managements. *Journal of Neurology, 261*, 2061–2078. https://doi.org/10.1007/s00415-014-7291-1
71. Boehme, A. K., Esenwa, C., Elkind, V., & M. S. (2017). Stroke risk factors, genetics, and prevention. *Circulation Research, 120*(3), 472. https://doi.org/10.1161/CIRCRESAHA.116.308398
72. Bai, Z., Fong, K. N. K., Zhang, J. J., et al. (2020). Immediate and long-term effects of BCI-based rehabilitation of the upper extremity after stroke: A systematic review and meta-analysis. *Journal of NeuroEngineering Rehabilitation, 17*, 57. https://doi.org/10.1186/s12984-020-00686-2
73. Feigin, V. L., Lawes, C. M., Bennett, D. A., Barker-Collo, S. L., & Parag, V. (2009). Worldwide stroke incidence and early case fatality reported in 56 population-based studies: A systematic review. *Lancet Neurology, 8*, 355–369. https://doi.org/10.1016/S1474-4422(09)70025-0
74. Mozzafarian, D., Benjamin, E. J., Go, A. S., et al. (2016). On behalf of the American Heart Association Statistics Committee and Stroke Statistics Subcommittee. Heart disease and stroke statistics–2016 update: A report from the American Heart Association. *Circulation, 133*, e38–e360.
75. Pu, L., Wang, L., Zhang, R., Zhao, T., Jiang, Y., & Han, L. (2023). Projected global trends in ischemic stroke incidence, deaths and disability- adjusted life years from 2020 to 2023. *Stroke, 54*, 1330–1339. https://doi.org/10.1161/STROKEAHA.122.040073
76. Powers, B. J., Danus, S., Grubber, J. M., Olsen, M. K., Oddone, E. Z., & Bosworth, H. B. (2011). The effectiveness of personalized coronary heart disease and stroke risk communication. *American Heart Journal, 161*, 673–680. https://doi.org/10.1016/j.ahj.2010.12.021
77. Lau, H., Lew, J., Borschmann, K., Thijs, V., & Ekinci, E. I. (2019). Prevalence of diabetes and its effects on stroke outcomes: A meta-analysis and literature review. *Journal of Diabetes Investigation, 10*(3), 780–792. https://doi.org/10.1111/jdi.12932
78. Zhao, D., Liu, J., Wang, W., Zeng, Z., Cheng, J., Liu, J., Sun, J., & Wu, Z. (2008). Epidemiological transition of stroke in China: Twenty-one-year observational study from the Sino-MONICA-Beijing Project. *Stroke, 39*, 1668–1674. https://doi.org/10.1161/STROKEAHA.107.502807
79. Bejot, Y., Daubail, B., Jacquin, A., Durier, J., Osseby, G. V., Rouaud, O., & Giroud, M. (2014). Trends in the incidence of ischaemic stroke in young adults between 1985 and 2011: The Dijon Stroke Registry. *Journal of Neurology, Neurosurgery and Psychiatry, 85*, 509–513.

80. Cui, Q., & Naikoo, N. A. (2019). Modifiable and non-modifiable risk factors in ischemic stroke: A meta-analysis. *African Health Sciences, 19*(2), 2121–2129. https://doi.org/10.4314/ahs.v19i2.36
81. Rexrode, K. M., Madsen, T. E., Yu, A. Y. X., Carcel, C., Lichtman, J. H., & Miller, E. C. (2022). The impact of sex and gender on stroke. *Circulation Research, 130*, 512–528. https://doi.org/10.1161/CIRCRESAHA.121.319915
82. Song, Y., Long, Y., Long, L., Zhang, N., & Liu, Y. (2014). Polymorphism Ala54Thr of fatty acid-binding protein 2 gene is not associated with stroke risk in Han population of Hunan China. *Medical Science Monitor, 20*, 1751–1717.
83. Kaddumukasa, M., Kayima, J., Nakibuuka, J., Blixen, C., Welter, E., Katabira, E., et al. (2017). Modifiable lifestyle risk factors for stroke among a high risk hypertensive population in Greater Kampala, Uganda; A cross-sectional study. *BMC Research Notes, 10*, 675. https://doi.org/10.1186/s13104-017-3009-7
84. Maredza, M., Bertram, M. Y., Gomez-Olive, X. F., & Tollman, S. M. (2016). Burden of stroke attributable to selected lifestyle risk factors in rural South Africa. *BMC Public Health, 16*, 143.
85. Yao, H., Zhang, J., Wang, Y., Wang, Q., Zhao, F., & Zhang, P. (2021). Stroke risk and its association with quality of life: A cross-sectional study among Chinese urban adults. *Health Quality of Life Outcomes, 19*, 236. https://doi.org/10.1186/s12955-021-01868-9
86. O'Donnell, M. J., Chin, S. L., Rangarajan, S., Xavier, D., Liu, L., Zhang, H., Rao-Melacini, P., Zhang, X., Pais, P., Agapay, S., et al. (2016). Global and regional effects of potentially modifiable risk factors associated with acute stroke in 32 countries (INTERSTROKE): A case–control study. *Lancet, 388*(10046), 761–775.
87. Dufouil, C., Beiser, A., McLure, L. A., Wolf, P. A., Tzourio, C., Howard, V. J., et al. (2017). Revised Framingham stroke risk profile to reflect temporal trends. *Circulation, 135*, 1145–1159. https://doi.org/10.1161/CIRCULATIONAHA.115.021275
88. Wajngarten, M., & Silva, G. S. (2019). Hypertension and stroke: Update on treatment. *European Cardiology Review, 14*(2), 111–115. https://doi.org/10.15420/ecr.2019.11.1
89. Chen, R., Ovbiagele, B., & Feng, W. (2016). Diabetes and stroke: Epidemiology, pathophysiology, pharmaceuticals and outcomes. *The American Journal of the Medical Sciences, 351*(4), 380. https://doi.org/10.1016/j.amjms.2016.01.011
90. Banerjee, C., Moon, Y. P., Paik, M. C., Rundek, T., Mora-McLaughlin, C., Vieira, J. R., Sacco, R. L., & Elkind, M. S. V. (2012). Duration of diabetes and risk of ischemic stroke: The Northern Manhattan Study. *Stroke, 43*, 1212–1217. https://doi.org/10.1161/STROKEAHA.111.641381
91. Purnell, J. Q. (2023). Definitions, classification, and epidemiology of obesity. [Updated 4 May 2023]. In K. R. Feingold, B. Anawalt, M. R. Blackman, et al. (Eds.), *Endotext [Internet]*. MDText.com. Available from: https://www.ncbi.nlm.nih.gov/books/NBK279167/
92. Katsiki, N., Ntaios, G., & Vemmos, K. (2011). Stroke, obesity and gender: A review of the literature. *Maturitas, 69*(3), 239–243. https://doi.org/10.1016/j.maturitas.2011.04.010
93. World Stroke Organisation (WSO). (2023). Understanding weight and stroke. Stroke risk factors—Weight. www.world-stroke.org/assets/downloads/WSo_DontBeTheOne-PI-Leaflets_-_Weight.pdf. Accessed 5 Nov 2023.
94. Cong, X., Liu, S., Wang, W., Ma, J., & Li, J. (2022). Combined consideration of body mass index and waist circumference identifies obesity patterns associated with risk of stroke in a Chinese prospective cohort study. *BMC Public Health, 22*, 347. https://doi.org/10.1186/s12889-022-12756-2
95. Kernan, W. N., Inzucchi, S. E., Sawan, C., Macko, R. F., & Furie, K. L. (2013). Obesity: A stubbornly obvious target for stroke prevention. *Stroke, 44*, 278–286. https://doi.org/10.1161/STROKEAHA.111.639922
96. Onwuegbuzie, G. A., Alabi, P., & Abdulai, F. (2021). Abdominal obesity as a risk factor for stroke in Abuja, Nigeria. *International Journal of Medical Research & Health Sciences, 10*(8), 170–177.

97. Azevedo, T. A., Moreira, M. L. V., & Nucera, A. P. C. S. (2018). Cardiovascular risk estimation by the ASCVD risk estimator application in a university hospital. *International Journal of Cardiovascular Sciences, 31*(5), 492–498. https://doi.org/10.5935/2359-4802.20180052
98. Damen, J. A., Pajouheshnia, R., Heus, P., Moons, K. G. M., Reitsma, J. B., Scholten, R. J. P. M., et al. (2019). Performance of the Framingham risk models and pooled cohort equations for predicting 10-year risk of cardiovascular disease: A systematic review and meta-analysis. *BMC Medicine, 17*, 109. https://doi.org/10.1186/s12916-019-1340-7
99. Osoro, I., Kumar, R., & Sharma, A. (2023). Ten-year risk assessment for cardiovascular diseases using ASCVD risk estimator plus: Outcomes from hypertension and diabetes patients. *Diabetology & Metabolic Syndrome, 15*, 216. https://doi.org/10.1186/s13098-023-01170-2
100. Dynka, D., Kowalaze, K., Charuta, A., & Paziewska, A. (2023). The ketogenic diet and cardiovascular diseases. *Nutrients, 15*(15), 3368. https://doi.org/10.3390/nu15153368
101. Raimondo, D. D., Buscemi, S., Musiari, G., Rizzo, G., Pirera, E., Corleo, D., Pinto, A., & Tuttolomondo, A. (2021). Ketogenic diet, physical activity, and hypertension—A narrative review. *Nutrients, 13*(8), 2567. https://doi.org/10.3390/nu13082567
102. Santos, F. L., Esteves, S. S., da Costa, P. A., Yancy, W. S., Jr., & Nunes, J. P. (2012). Systematic review and meta-analysis of clinical trials of the effects of low carbohydrate diets on cardiovascular risk factors. *Obesity Review, 13*, 1048–1066. https://doi.org/10.1111/j.1467-789X.2012.01021.x
103. Greco, T., Glenn, T. C., Hovda, D. A., & Prins, M. L. (2015). Ketogenic diet decreases oxidative stress and improves mitochondrial respiratory complex activity. *Journal of Cerebral Blood Flow & Metabolism, 36*(9), 1603–1613. https://doi.org/10.1177/0271678X15610584
104. Haces, M. L., Hernández-Fonseca, K., Medina-Campos, O. N., Montiel, T., Pedraza-Chaverri, J., & Massieu, L. (2008). Antioxidant capacity contributes to protection of ketone bodies against oxidative damage induced during hypoglycemic conditions. *Experimental Neurology, 211*(1), 85–96. https://doi.org/10.1016/j.expneurol.2007.12.029
105. Guo, M., Wang, X., Zhao, Y., Yang, Q., Ding, H., Dong, Q., Chen, X., & Cui, M. (2018). Ketogenic diet improves brain ischemic tolerance and inhibits NLRP3 inflammasome activation by preventing Drp1-mediated mitochondrial fission and endoplasmic reticulum stress. *Frontiers in Molecular Neuroscience, 11*, 335605. https://doi.org/10.3389/fnmol.2018.00086
106. Camberos-Luna, L., Geronimo-Olvera, C., Montiel, T., Rincon-Heredia, R., & Massieu, L. (2016). The ketone body, β-hydroxybutyrate stimulates the autophagic flux and prevents neuronal death induced by glucose deprivation in cortical cultured neurons. *Neurochemistry Research, 41*, 600–609. https://doi.org/10.1007/s11064-015-1700-4
107. Kirsch, J. R., & D'Alecy, L. G. (1984). Hypoxia induced preferential ketone utilization by rat brain slices. *Stroke, 15*(2), 319–323. https://doi.org/10.1161/01.str.15.2.319
108. Makievskaya, C. I., Popkov, V. A., Andrianova, N. V., Liao, X., Zorov, D. B., & Plotnikov, E. Y. (2023). Ketogenic diet and ketone bodies against ischemic injury: Targets, mechanisms, and therapeutic potential. *International Journal of Molecular Sciences, 24*(3), 2576. https://doi.org/10.3390/ijms24032576
109. Dhamija, R., Eckert, S., & Wirrell, E. (2013). Ketogenic diet. *Canadian Journal of Neurological Science, 40*, 158–167. https://doi.org/10.1017/S0317167100013676
110. Gibson, C. L., Murphy, A. N., & Murphy, S. P. (2012). Stroke outcome in the ketogenic state–A systematic review of the animal data. *Journal of Neurochemistry, 123*(0 2), 52–57. https://doi.org/10.1111/j.1471-4159.2012.07943.x
111. Sethuraman, A., Rao, P., Pranay, A., Xu, K., LaManna, J., & Puchowicz, M. A. (2022). Chronic ketosis modulates HIF1α-mediated inflammatory response in rat brain. *Advances in Experimental Medicine and Biology, 1395*, 75–79. https://doi.org/10.1007/978-3-031-14190-4_13
112. Gasior, M., Rogawski, M. A., & Hartman, A. L. (2006). Neuroprotective and disease-modifying effects of the ketogenic diet. *Behavioural Pharmacology, 17*(5–6), 431. https://doi.org/10.1097/00008877-200609000-00009

113. Pinto, A., Bonucci, A., Maggi, E., Corsi, M., & Businaro, R. (2018). Anti-oxidant and anti-inflammatory activity of ketogenic diet: New perspectives for neuroprotection in Alzheimer's disease. *Antioxidants, 7*(5), 63. https://doi.org/10.3390/antiox7050063
114. Wood, T. R., Stubbs, B. J., & Juul, S. E. (2019). Exogenous ketones bodies as promising neuroprotective agents for developmental brain injury. *Developmental Neuroscience, 40*(5–6), 451–462. https://doi.org/10.1159/000499563
115. Yarar-Fisher, C., Li, J., Womack, E. D., Alharbi, A., Seira, O., Kolehmainen, K. L., Plunet, W. T., Alaeiilkhchi, N., & Tetzlaff, W. (2021). Ketogenic regimens for acute neurotraumatic events. *Current Opinion in Biotechnology, 70*, 68–74. https://doi.org/10.1016/j.copbio.2020.12.009
116. Suzuki, M., Suzuki, M., Kitamura, Y., Mori, S., Sato, K., Dohi, S., Sato, T., Matsuura, A., & Hiraide, A. (2002). β-Hydroxybutyrate, a cerebral function improving agent, protects rat brain against ischemic damage caused by permanent and transient focal cerebral ischemia. *The Japanese Journal of Pharmacology, 89*(1), 36–43.
117. Tai, K. K., Nguyen, N., Pham, L., & Truong, D. D. (2008). Ketogenic diet prevents cardiac arrest-induced cerebral ischemic neurodegeneration. *Journal of Neural Transmission, 115*, 1011–1017. https://doi.org/10.1007/s00702-008-0050-7
118. Tong, X., Cai, Q., Cao, D., Yu, L., Sun, D., Yang, G., et al. (2022). Chinese expert recommendation on ketogenic diet therapy for super-refectory status epilepticus. *Acta Epileptologica, 4*, 8. https://doi.org/10.1186/s42494-021-00078-6
119. Farias-Moeller, R., Bartolini, L., Pasupuleti, A., Brittany Cines, R. D., Kao, A., & Carpenter, J. L. (2017). A practical approach to ketogenic diet in the paediatric intensive care unit for super-refractory status epilepticus: A system review. *Neurocritical Care, 26*(2), 267–272.
120. Nguyen, K. T., Xu, N. Y., Zhang, J. Y., Shang, T., Basu, A., Bergenstal, R. M., Castorino, K., Chen, K. Y., Kerr, D., Koliwad, S. K., Laffel, L. M., Mathioudakis, N., Midyett, L. K., Miller, J. D., Nichols, J. H., Pasquel, F. J., Prahalad, P., Prausnitz, M. R., Seley, J. J., & Klonoff, D. C. (2021). Continuous ketone monitoring consensus report 2021. *Journal of Diabetes Science and Technology, 16*(3), 689–715. https://doi.org/10.1177/19322968211042656
121. Zhang, J. Y., Shang, T., Koliwad, S. K., & Klonoff, D. C. (2021). Continuous ketone monitoring: A new paradigm for physiologic monitoring. *Journal of Diabetes Science and Technology, 15*(4), 775–780. https://doi.org/10.1177/19322968211009860

Macronutrients, Micronutrients, Trace Elements, Magnesium and Stroke

11

Huda Aljumayi, Nazeha A. Khalil, and Manal Almatrafi

Abstract

Stroke is one of the major challenges serious health condition for the global healthcare system affecting individual's health and well-being. However, proper nutrition with healthy balanced diet is playing an effective role in either preventing and/or recovering healthy conditions due to their nutrients levels. E.g. macronutrients, micronutrients and trace elements (magnesium, selenium, and zinc … etc.) consumptions played a key role in contribution to cardiovascular health and reduce the risk of stroke. However, all previous studies conclusions of prospective observational in association to dietary factors remained unclear. Therefore, the current chapter is illustrating the dietary and nutritional factors affecting the stroke risk from different food groups with special refer to selenium as trace elements associated with stroke prevention. Food groups and required nutrients (macro/micro-nutrients) in addition to the role of different nutrients and diets in stroke prevention and management have been covered. DASH diet as well has been explained with its benefits and importance to the stroke prevention and/or recovery. Indeed, eating fresh fruits and vegetables in addition to the low saturated fats, trans-fat and cholesterol free diets with high dietary fiber levels reduced the blood pressure and stroke incidence as it has been recorded and obtained from different previous studies therefore, DASH diet could be recommended an effective strategy for stroke prevention.

H. Aljumayi · M. Almatrafi
Department of Food Science and Nutrition, College of Sciences, Taif University, Taif, Saudi Arabia
e-mail: huda.a@tu.edu.sa; msmatrafi@tu.edu.sa

N. A. Khalil (✉)
Nutrition and Food Sciences Department, Faculty of Home Economics, Menoufia University, Shibin El-Kom, Egypt
e-mail: nazih.khalil@hec.menofia.edu.eg

W. Mohamed, M. B. Mat-Nor (eds.), *Nutrition and Stroke*, Nutritional Neurosciences, https://doi.org/10.1007/978-981-97-6818-9_11

Keywords

Food groups · Fruits and vegetables · Antioxidant · Fibers and DASH diet

Abbreviations

WHO	World Health Organization
AHA	American Heart Association
CHOs	Carbohydrates
T2D	Type 2 diabetes
SCFAs	Short chain fatty acids
GL	Glycemic load
GI	Glycemic index
DAP	Dietary animal proteins
DVP	Dietary vegetable proteins
MUFA	Monounsaturated fatty acids
PUFA	Polyunsaturated fatty acids
LDL-c	Low density lipoproteins cholesterol
DASH	Dietary approaches to stop hypertension

11.1 Introduction

Stroke has been recognized as one of the major challenges for the global healthcare system and also has been stated by the World Health Organization (WHO) as the second biggest death cause worldwide [1] About 5.5 million causes of deaths people in 2016 were from stroke[2, 3]. Additionally, the incidence of stroke in American adults has been estimated to reach about 4% by the American Heart Association (AHA) within 2030 affecting huge economic expenses by the stroke-related medical up to $183 billion [4]. Different certain association between stroke prevention or recovery conditions and dietary factors has been recognized in different ages, genders, countries and cultures because of unhealthy diet consumptions. The healthy diet well known as a diet that helps to improve or maintain overall human health by providing the adequate required essential nutrients to the body such as fluid, macronutrients, micronutrients, and calories [5]. However, the conclusions of previous studies within the nutritional factors remained controversial and unclear. Indeed, the appropriate nutrition has a vital role in the occurrence of stroke that is not well focused; it is playing as well an effective role in either preventing and/or recovering post-stroke conditions. Accordingly, the current work herein synthesis the available evidence of the associations between dietary factors and stroke. Food groups, nutrients as macro and micro ones have been highlighted with stroke occurrence in different epidemiological evidence. Fruits and vegetables were in high priority foods for stroke preventions and/or recovery in addition to the omga-3 fatty foods in contrast to meats, coffee and tea consumptions. Also, both fruits and vegetables show to have different important minerals and vitamins that all very vital for the human health and stroke incidence; have high anti-oxidant levels. Their fore, the current chapter exploring the relationship of various modifiable factors and stroke risk especially the nutritional or dietary factors.

11.2 Nutrition and Healthy Diet

Nutrition is well known by the provision of different nutrients to the human body depending on the age, gender, activities and the health status. While the healthy diet known by the sufficient and adequate supplementations of all nutrients. It also can be maximized by obtaining the right amount of all different nutrients including all macro and micro ones that could be obtained from fruits, vegetables, and whole grains in contrast to the processed food and sweetened beverages consumptions [6, 7]. It is could be met from different plant and animal-based foods. However, the low consumed levels of fruits and vegetables are attributed to the unhealthy diet that is associated within huge numbers of chronic diseases such as diabetes, obesity, cardiovascular diseases and cancer. The WHO estimates diet low in fruits and vegetables attributable to about 2.7 million deaths annually in addition to about 19% of gastrointestinal cancer and 11% of strokes [8, 9].

11.3 Lifestyle Choices and Stroke

Regarding the healthy diet consumption in order to maintain and prevent the risk of different diseases especially the stroke, lifestyle choices are very important factor to reduce the risk of stroke in about 80%. Dietary foods mainly by choosing the healthy meal and snack can help stroke prevention. For instance, eating plenty of fresh fruits and vegetables with foods low in saturated fats, trans fat, and cholesterol in contrast to foods high in fiber can help. Such dietary condition is mainly has been found and attributed within the Mediterranean diet that is mainly recommended for stroke prevention [10]. Also, limiting salt (sodium) consumption in the diet should be able to lower the blood pressure as it has been shown in recent studies that high sodium intake is associated with stroke increased risk factor. Additionally and beside healthy diet consumptions daily exercise are required for optimizing healthy lifestyle that in total will be able to lower different disease risk factors such as obesity, diabetes, heart disease, hypertension and stoke [5, 11]. The healthy dietary choices should be achieved when any of such diseases is occurred containing all the required macro and micronutrients that will be discussed as following focusing on the trace elements and stroke condition.

11.4 Macronutrients

The macronutrients are providing energy with various bodily functions support so they are essential nutrients. They are well known by the macronutrients as they are required in large quantities for all the human health. They include energy, water, carbohydrates, fibers, proteins, and fats that each plays a vital unique role within our bodies for maintaining longevity and health [12]. However, macronutrients most often simply refer to carbohydrates, proteins, and fats. Have all macronutrients are required daily in a balanced ratio of the consumed diets depending on the main individual goals such as weigh management and overall health status or condition [13]. Therefore, obtaining macronutrients balanced intake is crucial in balanced dietary pattern for health and well-being optimization due to their effects [14, 15].

11.4.1 Carbohydrates

Dietary carbohydrates (CHOs) composed of carbon, hydrogen, and oxygen molecules that mainly are the body's primary source of energy. It also classified in two main groups (simple or complex) depending on their chemical structure and especially the quickly of their digested and absorbed time by the body [13, 16]. The simple ones can be found in some foods and drinks as added sugar that called 'free sugars' in addition to sugars found naturally in foods such as honey, syrups, unsweetened fruit and vegetable juices, smoothies and purées. The simple sugar form known by its quick absorption while the other complex carbohydrates can be found in whole grains and vegetables and that are digested more slowly resulting in a steady release of energy by the time [17]. Variety of foods has both simple and/or complex macronutrient form such as grains, fruits, vegetables, and dairy products. Having much sugary foods or diets means having many calories causing body weight gain increasing the consequences risk of different disorders such as stroke, heart disease and type 2 diabetes (T2D). Indeed increasing the food consumptions often contain added sugar can increase the stroke risk by about 10%. So it is important to limit the dietary sugar intake at no more than 30 g of added sugar daily (about seven teaspoons of sugar) [18, 19].

Fibers also are counted as a type of complex carbohydrate because it has high molecule weigh and that cannot be digested by the body even with long time. It only can be fermented by the colonic microbiota resulting in metabolic end products called short chain fatty acids (SCFAs) that give it the prebiotic properties [20]. The fiber daily requirements for adults are about 30 g. Additionally foods contain fibers such as whole meal foods (e.g. whole meal pasta and bread) and plant-based foods shown to reduce the stroke risk by up to 10% over time either in insoluble and/or soluble fiber [17]. So it is important to know and evaluate the carbohydrates levels and sources that needed for the human health conditions in order to regulate the blood sugar levels and different human disorders.

11.4.2 Proteins

The second macronutrient required daily is protein which is essential for building and repairing mainly the body tissues especially muscles, organs in addition to the body skin. Proteins are mainly consists of amino acids; the main building blocks/units counted twenty different amino acids [13]. The amino acids divided to nine essential amino acids that are can't be produced by the body and must be obtained from the diet. Therefore, it is important that the body got different dietary protein sources ensuring that the body received all the essential needed amino acids. Having enough protein amounts daily is especially vital for many cases such as active individuals, athletes and body-building goals. Such dietary protein sources mainly are animal products such as meat, poultry, fish, and dairy in addition to plant-based sources such as legumes, nuts, and seeds. It is recommended to consume at least 90 g protein daily in about one portion and size of hand palm. There are different dietary protein sources such as meat, fish, eggs, pulses and beans in addition to dairy products, nuts, and meat alternatives [17]. However, to include protein in the daily diets it should be considered to keep saturated fat intake at low levels by choosing lean cuts of meat. Also, fish consumption is important at least once a week such as makerel, salmon or trout. Otherwise not having meat or fish the plant-based diets such as beans and pulses are a good alternative. They are good sources for soluble fibre as well resulting in reducing the blood cholesterol levels additionally to have vitamins and minerals.

11.4.3 Fats

The final macronutrient needed for the human body is fat. It is another important macronutrient that plays an important role in the body because of their high valuable energy levels. Also, fats are composed essentially of fatty acids required for different many bodily functions such as hormone production additionally to cell membrane structure and nutrient absorption [13]. The essential fatty acids is required from consumed foods as the human body can't make itself, however, high fat consumptions can increase different human health problems such as stroke. Fats divided into different kinds including saturated, unsaturated and trans fats that each has different dietary sources and importance to the human health [21]. For instance, the unsaturated fats found in different foods such as fish, avocados and plant-based foods mainly seeds, nuts, oils (olive oil) that is mainly tend to be in oils not solid formats [22]. Having such type of fats it important even in small and/or moderate amounts as they can help to prevent blocked arteries and blood clots that in order preventing heart problems such as strokes [17]. Additionally, the second fat types are the saturated fats that are usually in solid formats and that is normally known as the bad ones because they can cause high cholesterol blood levels causing blocked to the arteries and heart disease when consumed in excess. Such saturated fats are found in different foods, mainly meat and dairy products including fatty red meat, processed meats (e.g. sausages and meat pies) in addition to butter, cream, cheese and different oils (e.g. palm oil and coconut oil) [21, 23]. Finally, the trans fats should be in limited consumed levels as they can increase the heart disease levels. They normally could be found in processed and fried foods [17].

To conclude up, all the macronutrients are needed in high levels for building blocks of a healthy diet supporting their crucial role for overall health and well-being such as the essential fatty acids and amino acids. Each macronutrient of carbohydrates, proteins, and fats is needed in basic daily needed for optimal health promotion with low health problem levels. Therefore, a healthy diet is the good dietary source including a variety of nutrient-rich foods presented from all food groups that meet the body's macronutrient requirements as it will be described within the following sections.

11.5 Micronutrients and Trace Elements

The dietary micronutrients are very well known to be needed in very small amounts by the body such as vitamins and minerals. However, they have vital impact on the body's health and any deficiency can cause different life-threatening conditions. They could be used in different quantities to regulate the body cell's physiological functions as essential dietary elements called vitamins and minerals (trace elements). The trace elements are varying and important for the stroke condition so it is important to draw attention on their importance and dietary sources. There are many different trace elements needed for the human body's functions to promote optimal health. Such trace elements rich foods can support the overall human health ensuring to meet the body's nutritional needs. Therefore, essential trace elements should be included in daily human foods or meals incorporating ways such as magnesium, zinc, copper, selenium, iron and manganese [24]. They all can support the overall health with stroke low levels contributing potentially with different important physiological processes in the heart health and antioxidant defense in addition to the energy production that all in total can affect the stroke risk [17]. However, other trace element such as sodium that is the main contain in the daily used salt can increase the stroke risk. Sodium is important to keep the body fluids at the right level and any increases will increase the fluid levels that the body stores that in turn raising the blood pressure (hypertension). Such unhealthy condition is the biggest stroke risk factor. The walls of the arteries become harden and narrow increasing forming blood clots risk. Then the clot can move and travel to the brain causing a stroke (ischaemic stroke). It also can increase the risk of the blood vessel bursting and bleeding into the brain (called a hemorrhagic stroke). Too much salt or salty food intake can elevate the stroke by 25%; salt intake should be daily in no more than 6 g (about a teaspoon). Different changes needed to reduce the amount of salt in the diet such as taking foods with the lowest salt content. Also, salt addition to the food during cooking or the eating table is not recommended. However, using herbs, spices, garlic or lemon juice to add flavor is good to be used instead of salt [25]. Additionally, potassium is another element that could be increased to help blood vessels relaxation with excessive activation of platelets inhibition. So, high-potassium diet consumption is in advance for delaying the development of vascular damage that causes by restraining the production of reactive oxygen species [17]. The following section is going to explain the importance, needed and effects of different trace elements within the human health conditions especially in the stroke conditions.

11.5.1 Trace Elements Importance and Rich Foods

There are many different trace elements presented within our foods that in total are very important for the human body functions. Maintaining a balanced healthy diet should include different foods rich in many trace elements such as green leafy vegetables that are essential for promoting cardiovascular well-being and reducing stroke occurrence [17, 26]. Each element is required for specific role in the human's muscles and cells. The following table is presenting some different elements that are highly needed for different functions in addition to their dietary sources (Table 11.1).

11.5.1.1 Magnesium

Table 11.1 illustrating that Magnesium is requires for muscle and nerve function's support in addition to blood sugar levels regulation and bone health contributions. Also, it could be found in many dietary food sources such as Spinach, Almonds, Pumpkin Seeds, Avocado, Cashews, Quinoa, Dark Chocolate, Black Beans, Salmon and Bananas.

11.5.1.2 Zinc

The second presented element in the same Table 11.1 is Zinc that is known as one of the most effective nutrients playing a major role in the responses of immune system functions, wound healing and DNA synthesis. Recent studies recommended daily zinc allowance for the American subjects to be 12 mg\day for women and 15 mg\day for men, with the highest limitation of 21 mg\day. Other data from different researches presented that the more consumption of rich-zinc foods is the low levels of zinc absorptions, which is considered as universal behavior within the gut [27]. Such requirements are in high demand for human health with different ages and conditions in contrast to low zinc levels that have been associated with many other symptoms such as hypogeusia, infections, and delayed wounds healing in addition to the unbalanced sense of taste [28]. Zinc deficiency has been also associated with mainly poor dietary-protein sources, any abnormal mal-absorption and diarrhea. Therefore, providing the zinc-deficiency patients with different zinc-sources either from zinc-rich foods, and\or zinc supplementations can be helpful for treating many related symptoms in addition to protecting the immunity system; inhibiting the replication of different viruses in the body cells. Zinc rich-foods as shown in Table 11.1. The main zinc rich foods are known in pumpkin seeds and dairy products.

Table 11.1 Different trace elements needed for the body's functions with their importance and dietary sources

Element	Importance/functions	Dietary sources/rich foods
Magnesium	• Supports muscle and nerve function • Regulates blood sugar levels • Contributes to bone health	• Green leafy vegetables; Spinach • Nuts; cashews and almonds • Seeds and whole grains; Pumpkin seeds • Avocado • Quinoa • Dark chocolate • Black beans • Salmon • Bananas
Zinc	• Important for immune function • Wound healing • DNA synthesis	• Whole grains and seeds; Pumpkin seeds • Shellfish • legumes • Nuts
Iron	• Essential for oxygen transport in the blood • Energy production	• Red meat • Spinach • Pumpkin Seeds • Poultry • Fish • Lentils • Fortified cereals
Selenium	• Acts as an antioxidant • Supports thyroid function • Immune response	• Whole grains • Nuts • Seafood • Organ meats
Manganese	• Supports bone health • Antioxidant activity	• Whole grains • Nuts • Seeds • Legumes • Green leafy vegetables
Copper	• Involved in energy production • Iron metabolism • Antioxidant defense	• Whole grains and seeds • Nuts and cashews • Organ meats • Seafood

11.5.1.3 Iron

No one can deny that how iron is an essential trace element required for oxygen transport in the blood in addition to energy production. It is well known by ages and ages ago also to be found in main dietary sources especially spinach, red meat and pumpkin seeds as demonstrated again in Table 11.1.

11.5.1.4 Selenium

Again, selenium is another important trace element that is required to the human body that playing an important role in the immune response and acts as an antioxidant with thyroid function support (Table 11.1). The perfect dietary selenium sources could be recommended to be in whole grains, nuts, seafood, and organ meats.

11.5.1.5 Manganese

Manganese is another important trace element supporting the bone health in addition to its antioxidant activities. It has different dietary sources such as whole grains, nuts, seeds, legumes, and green leafy vegetables.

11.5.1.6 Copper

The final trace element that has been covered within the work herein is copper. It is important for the human body health as it acts as antioxidant within iron metabolism in addition to be involved in energy production. The main dietary source that could be recommended for copper is whole grains and seeds in addition to cashews, organ meats, seafood and nuts (Table 11.1).

All previous macronutrients and micronutrients (trace elements) are mainly presented within the dietary foods especially the green leafy vegetables needed for promoting cardiovascular well-being that all could be achieved from different food groups in healthy lifestyle and healthy manner [17, 26].

11.6 The Food Groups

The food groups used up to for dietary composition containing different essential nutrients needed for growth, development, and overall well-being therefore food could be called the fuel that powers our bodies. Indeed, foods are essential for nutrients supplementations to fuel the human bodies that needed for human survival with maintaining optimal health. The healthy balanced diet should be consists of various food groups that each providing unique vital sets of nutrients in promoting overall health and well-being [29]. Knowing and understanding the incorporating of the food groups within different meals, this will meet the nutritional needs in order to promoting overall health and longevity [11]. Therefore, providing the human body with the essential balanced and nutritious diet is important and that will be described in the following sections that are illumination the different food groups with their required portion/units in maintaining good health (cups as unit) in addition to their dietary sources as shown in Table 11.2. The main five food groups in any healthy diet are fruits, vegetables, grains, protein, and dairy products that each group provides unique nutrients (either macronutrients and/or micronutrients) and benefits needed for overall human health and well-being.

It could be seen from the previous table (Table 11.2) that all the five groups is easy access in daily usage between all the human communities. They could be achieved from different foods as each group has many dietary sources. For instance, fruits could be obtained from apples, oranges, berries and bananas. Such group is rich in different vitamins, minerals and fibers that in total should be consumed daily as recommended by the dietary guidelines by 1.5–2 cups of the foods. The second group is the vegetables that should be consumed at 2–3 cups daily and could supply

Table 11.2 Summarizing the food groups, their dietary guidelines, and sources

Food groups	Dietary guidelines	Dietary sources
Fruits	Consume 1.5–2 cups per day, rich in vitamins, minerals, and fiber	Apples, oranges, berries, bananas
Vegetables	Consume 2–3 cups per day, rich in vitamins, minerals, and fiber	Spinach, broccoli, carrots, bell peppers
Grains	Make half your grains whole, consume 3–5 ounces per day	Brown rice, whole wheat bread, oats
Protein	Consume 5–6.5 ounces per day, lean protein sources	Chicken, fish, beans, nuts
Dairy	Consume 3 cups per day, choose low-fat or fat-free options	Milk, yogurt, cheese, fortified soy milk

the body with different vitamins, minerals and fibers that are much more similar to the fruits group so some references and guidelines collect both groups in one called fruits and vegetables group. Such vegetable group could be achieved by Spinach, Broccoli, Carrots and peppers consumptions [11].

Grains is the third following food groups that again supply the body with different macronutrients especially the carbohydrates and that could be obtained by rice, Paste and bread consumptions. Additionally, whole grains should count up to the half of the healthy diet and that could be achieved from brown rice, whole wheat bread, oats consumptions. The total required daily amounts are counted from 3–5 ounces per day [30, 31]. The final required two food groups are protein and dairy that both needed to be included in the healthy daily diets. Proteins need to be consumed from 5 up to 6.5 ounces daily and that could be found in the different dietary protein sources such as lean, chicken, fish, beans and nuts. Dairy group is required to be consumed up to 3 cups/day and the main foods are low-fat or fat-free options of milk, yogurt, cheese and fortified soy milk.

All the food groups are needed and required daily in the healthy daily diets. However, the most important ones for the heart health in order to reduce the stroke risk are fruits and vegetables in addition to fibers so the following sections are explaining much more data about both groups such as their importance for the human body.

11.6.1 Fruits and Vegetables

The groups of fruits and vegetables are mainly rich in all of vitamins, minerals, and antioxidants that all very vital to protect the human bodies from different diseases in addition to immune system support. Both fruits and vegetables are also rich in dietary fiber that is in need and important for digestive health additionally to help the stomach to be full for long time. They also rich sources for the dietary antioxidants especially different vitamins such as V. A, V.C, and V.E with the beta-carotene that in total may prevent the arteries damage. Having wide range of the essential nutrients especially variety of fruits and vegetables (fibers, vitamins and minerals) could be achieved by different colored foods consumption [11, 32]. Different dietary sources needed at least in one portion that about 80 g of either fruits and/or vegetables such as one apple or two plums or a handful of berries or three tablespoons of different fruit salad. It also could be achieved by about 30 g or one tablespoon of dried fruit consumption. A one daily portion also equals a glass of fruit juice or smoothie in about 150 ml or three tablespoons of raw or cooked vegetables or even one bowl dessert of fruit salad [17].

11.6.2 Grains and Whole-Grains

Different dietary sources known as grains include rice, oats, and whole wheat bread that in total are major carbohydrate sources in addition to B vitamins and iron. The whole grains are the grain rich outer skin such as brown rice, quinoa, and whole wheat that have much more nutrients. For instance vitamins, minerals and fibers (soluble and insoluble) in dietary rich foods with B-vitamins and folic acid such as cereals, brown rice, brown pasta and whole-wheat couscous [17]. Such whole grains are important within the human diets in order to regulate the blood glucose levels in addition to improve the digestion with reducing the chronic diseases (e.g. diabetes and heart disease) with especial focus on the stroke.

11.7 Balanced Meal Plan: As Sample

Regarding the following meal plan, as the following table (Table 11.3) shows the dietary sample plan that includes all macronutrients and micronutrients [29, 33].

Consuming a balanced diet is an essential issue for patients recovering from stroke that can promote the overall health with recommended serving sizes for every food group including all the macronutrients and micronutrients. It is an example for

Table 11.3 Sample for balanced meal shows their foods and macronutrients

The meals	Foods	The macronutrients
Breakfast	Scrambled eggs, spinach and tomatoes	Protein and fats
	Brown toast (whole grain) and avocado spread	Carbohydrates and fats
	Fresh fruit salad	Carbohydrates and fiber
Mid-morning snack	Greek yogurt, mixed berries and a sprinkle of nuts	Protein, carbohydrates and fats
Lunch	Grilled chicken breast, quinoa and roasted vegetables	Protein, carbohydrates and fiber
	Side salad, cucumber and dressing	Fiber and fats
Afternoon snack	Hummus and baby carrots with whole grain crackers	Protein, carbohydrates and fats
Dinner	Baked salmon, brown rice and steamed broccoli	Protein, carbohydrates and fats
	Kale and garlic with olive oil	Fiber and fats
Evening snack	Apple slices and almond butter	Carbohydrates and fats

balanced healthy meal provides a balance daily macronutrients including different many colorful fruits and vegetables, lean proteins with whole grains, and healthy fats that in total supporting the overall health and well-being providing the necessary daily energy and nutrients for different activities [29, 33]. Table 11.3 incorporates different of nutrient-rich foods from all food groups however; the portion sizes needed are based on the individual energy needs.

11.8 Macro/Micro-Nutrients and Stroke

The different macronutrients including carbohydrates, proteins, and fats are playing an important role within the overall health especially affecting the stroke risk factor. Firstly, carbohydrates as known by the body's primary source of energy and is classified in simple (sugars) or complex (e.g. whole grains, fruits, and vegetables) that affect the quality of carbohydrates in the diet. It has influenced extremely the postprandial glycemic response. In another words the glycemic load (GL) that has been calculated as the product of glycemic index (GI) of any food item with its available carbohydrate content levels; widely has been used to assess the glycemic effect of CHO rich foods within the human diets. The possible associations between CHO intake/GI/GL and stroke risk were evaluated in two meta-analyses that both illustrated that high GL has been associated with stroke increased risk factor and also similar data were reported on 225,000 participants free from diabetes with high CHO intake [32, 34]. The mechanism behind the high GL and increased stroke risk may be due to the higher oxidative stress with sub-clinical systemic inflammation with oxidized lipoproteins production that consist in the vascular injury induced by chronic blood glucose increase. Therefore, having and consuming high refined carbohydrate diets and added sugars with processed foods could be contribute to obesity, diabetes, and high blood pressure that in total are risk factors for stroke. However and on the other hand, having complex carbohydrates in dietary rich fiber, vitamins, and minerals shown to support the heart health in contrast to stroke risk reduction by blood sugar regulating levels in addition to promoting satiety. Fibers include whole grains in the daily diet shows previously to prevent many diseases such as T2D, heart disease and body weight control as the risk factors for stroke. Fibers in the diet can be consumed to help stroke risk reduction as it helps to keep blood sugar levels stable with body weight management in addition to help with constipation and immune system stance. It could be work by reducing the gastric emptying process with satiety promotion in addition to reducing the food absorption and thus body overweight and blood lipid reduction levels [17].

The main dietary fibers' sources are fruits and vegetables so a daily portion of fruits and vegetables is shown in the diet to reduce the stroke risk by 10% and any extra eaten portion is important to reduce the stroke risk more. So five portions consumption have been recommended from variety of fruits and vegetables daily to get the required fiber levels in addition to their antioxidant, minerals and vitamins levels [17, 35].

Regarding the proteins consumptions as the second macronutrients, it is essential in the body for tissues, enzymes, and hormones building and repairing. Red meat can be a good source of protein and nutrients. However, consuming red meat more than four portions weekly can increase stroke risk. Lean sources of protein consuming as poultry, fish, legumes, and nuts can be used to help muscle mass maintenance with blood pressure regulation and overall cardiovascular health support. However, red and processed meats rich diets such as sausages have been associated with stroke increased risk by around 17%because of their higher saturated fat content levels causing cholesterol increased levels. The possible biologically reasonable reason could be the high circulating of LDL-C and triglycerides levels causing atherosclerotic plaques with interrupt blood flow to the brain that in total leading to stroke occurrence [17]. Also, red meat is the main heme iron source that is a redox active substance promoting oxygen free radicals production leading to LDL-C peroxidation and subsequent vascular inflammation and damage [36]. Additionally, usually the processed meat contains sodium and nitrite preservatives in big levels [37] that can reduce arterial compliance, vascular stiffness then subsequently cause high blood pressure and stroke. Also, different human studies were measuring the stroke risk of dietary proteins by their types; dietary animal proteins (DAP) *VS.* dietary vegetable proteins (DVP) has found that DAP consumption was associated with stroke risk in addition to no evidence with the total protein levels and stroke risk, whereas DVP intake was associated in significantly lower rate. Additionally, the blood pressure-lowering levels show to be a possible explanation for dietary vegetable protein intake association and stroke risk [17].

Finally, the last macronutrient is the fats in healthy monounsaturated fatty acids (MUFA) and polyunsaturated fatty acids (PUFA) that could be easily found in olive oil, avocados and nuts in addition to the fatty fish. The dietary sources of MUFA shows to help lowering the blood pressure and reducing the low-density lipoprotein cholesterol with the triglyceride levels that lower the cardiovascular risk [38]. Indeed, MUFA dietary sources shown to primary prevent the stroke especially in high intake levels. PUFA has both the omega-3 fatty acids and omega-6 fatty acids known as the essential fatty acids that has been linked to stroke risk and cardiovascular disease reduction levels [19]. They show to help the artery walls keep healthy in addition to regulating the blood clotting and lowering the blood pressure. Both omega-3 and omega-6 fatty acids found in fishy and plant oils (oils obtained from fish or plants). E.g. oils collected from salmon, nuts and seeds such as walnuts with flax seeds and soya products. They all are playing a protective role against stroke; their consumptions are improving the body cholesterol levels, supporting brain health and reducing the inflammation levels because of their anti-inflammatory and blood-thinning properties. On the other hand, too much saturated fats consumptions from animal products in addition to the trans fats obtained from processed foods can increase stroke risk; raising low density lipoproteins (LDL-c) cholesterol in addition to plaque buildup promotion in the arteries.

To conclude up, different studies measured and shows the possible protective evidence for the consumption of fruits, vegetables [11], fish, and chocolate in association to the risk of stroke [39]. Then the healthy balanced maintenance diet including variety of macronutrients rich foods is the key to reduce the stroke risk additionally to overall healthy cardiovascular promoting [35]. For instance, consuming whole grains, lean proteins and healthy fats with colorful fruits and vegetables can support and help optimal brain function, blood flow levels and heart health; personalized required nutritional recommendations is a vital help for individual risk factors vs. their healthy promotions.

11.9 Trace Elements and Stroke

Different trace elements are essential minerals needed and required in small amounts to the body different physiological functions. One of the key trace elements is Magnesium that plays a crucial role in maintaining overall health. Magnesium shows as one of the most important elements in stroke prevention by the crucial role in maintaining the cardiovascular health and stroke risk reduction with regulating the blood pressure levels. it also show to reduce the inflammation with improving the blood vessel function that in total are important factors in stroke prevention [40]. Moreover, zinc is an essential trace element illustrated to support the immune system function due to its antioxidant activities and properties. The adequate zinc intake demonstrated to reduce the oxidative stress and inflammation levels that are both associated to stroke risk. So the recommended daily intake should be 8–11 mg for adults with different dietary sources as mentioned in Table 11.1. Another important element that can help in stroke prevention is selenium that is an important antioxidant helping to protect the cells from damage [35]. Also, it may contribute to inflammation reducing with blood vessel health improvement that all in total factors lowering stroke risk and it has been recommended daily to about 55 mcg for adults. Additionally, iron is an essential for transporting oxygen with blood and energy production. Therefore, iron can help to support healthy blood flow with reducing the risk of conditions that can lead to stroke, such as anemia. So iron has been recommended to be consumed daily from 8 to 18 mg for adults. Again, manganese is another element that can help stroke prevention with helping to supports bone health with different metabolism and antioxidant activities. It is again can contribute to the overall health of cardiovascular due to promoting proper blood clotting and inflammation reduction. It is daily intake recommendation reached 1.8–2.3 mg for the adults obtained from any dietary sources. Ensuring adequate trace elements supplementations is very important in their serving sizes that are designed to provide stroke patients with the nutrients needed for recovery and overall well-being.

11.10 Magnesium and Stroke

As it has been mentioned before, magnesium is so important for stroke prevention especially when the total dietary magnesium intake was associated with the stroke risk inversely specifically ischemic stroke. Magnesium is a key mineral involved in more than 300 enzymatic reactions within the body. It has a vital role in muscle function and energy production with nerve transmission. Different previous and present scientific studies and clinical research and evidence demonstrated the association between dietary magnesium intake/supplementation and stroke incidence vs. stroke prevention. For instance, published study with atherosclerosis participants examined the association between dietary magnesium intake and serum magnesium levels in association to the risk of stroke [41, 42]. While another comprehensive review discusses the role of magnesium in cardiovascular health with its impact on stroke risk including findings from epidemiological, clinical, and laboratory trials [43]. Also, clinical Trial measured the impact of magnesium supplementation on blood pressure as a key risk factor for stroke and shows that lowering blood pressure through magnesium intake may indirectly reduce the risk of stroke [44, 45]. All such referencing studies support with their scientific evidence the relationship between magnesium levels and stroke incidence/prevention. All reported the relationship between magnesium levels and the risk of stroke; low magnesium levels increased risk of stroke incidence in addition to highlighted the potential role of magnesium in stroke prevention or risk reduction. Understanding such findings can contribute to developing strategies for stroke prevention with the importance of adequate magnesium consumption in order to maintain cardiovascular health. So the following key points are important to be considered and recommended for dietary magnesium intake in association with stroke prevention.

11.10.1 Adequate Magnesium Intake

Maintaining an adequate intake of magnesium from diet shows an important role for overall health, including cardiovascular health. It is involved in many biochemical processes in the body and plays a role in maintaining normal heart function and over all cardio-protective effects.

11.10.2 Magnesium Dietary Sources

There are different foods rich in magnesium such as leafy green vegetables, nuts, seeds and whole grains in addition to legumes and fish. So adequate incorporating such magnesium-rich foods into the consumed diet can help ensure an adequate intake of this essential mineral.

11.10.3 Recommended Daily Allowance (RDA)

The recommended daily allowance (RDA) for magnesium is different from person to other depending on the age and gender. E.g. the RDA for adult males is about 400–420 mg daily where RDA for adult females is less at around 310–320 mg daily. It is very important to have such recommended daily intake levels included within a balanced diet.

11.10.4 Consultation with Healthcare Provider

To consult with a healthcare provider or a registered dietitian is very important and recommended due to the differences between individuals such as underlying health conditions or medications that could influence the optimal amounts of magnesium requirements.

11.10.5 Supplementation Consideration

In some different cases, peoples may require supplemented magnesium intake especially if they diagnosed by magnesium deficiencies or specific health conditions then affect the absorption or utilization of magnesium. Therefore, magnesium supplementation should be under supervisor of healthcare provider for the appropriate dosage and form determinations.

11.10.6 Lifestyle Factors

The lifestyle factors are very important especially the dietary choices; affecting dietary magnesium intake in addition to other factors such as regular physical activity, managing stress and body weight for maintaining a healthy weight in addition to avoiding smoking that in total can contribute to reduce stroke risk. So the dietary choices and/or meal planning is the key point for stroke management and preventions.

11.11 Stroke-Prevention and DASH Diet

The dietary requirements for stroke prevention are very important to be followed and that should have variety in fruits and vegetables that both rich foods in vitamins and minerals in addition to the high fiber levels in contrast to low sugar and calorie intake levels [35]. Therefore, the best diet is mostly the plant-based diet in small amounts of meat and fish that called and described as Mediterranean Diet. Also, the plant-based diet is known as DASH (Dietary Approaches to Stop Hypertension) diet that helps stroke risk reduction with lowering or stopping the blood pressure (Hypertension; [46, 47]). The Mediterranean diet widely is recognized with its beneficial impacts on cardiovascular health especially stroke prevention; reducing the risk of stroke [48, 49].

11.11.1 The DASH Dietary Patterns

The DASH diets is a traditional Mediterranean diet that is characterized by increasing levels of plant-based foods such as olive oil and nuts in addition to fruits, vegetables, and cereals. It also recognized by moderate consumed levels of fish and poultry in addition to low consumption amounts of dairy products or low-fat dairy products with red and processed meats that are all low in saturated and total fats in addition to sweets low amounts. Therefore, DASH diet is high in especial nutrients especially the dietary fiber, vitamins, minerals, phenolics, and unsaturated fatty acids in contrast to low sodium and saturated and Trans fatty acids levels. Again, all such nutrients as a high adherence in the Mediterranean-style diet pattern was associated with about 30% stroke risk reduction. Indeed, previous Spanish study between adults at high risk of cardiovascular disease randomly assigned to Mediterranean diet consumptions and shown that blood pressure was reduced in addition to total cholesterol and fasting blood glucose levels. Again, another studies have demonstrated an inverse association between DASH dietary consumed pattern stroke risk [50, 51]. Therefore, incorporating magnesium-rich foods within the balanced diet in recommended daily intake levels in addition to adopting a healthy lifestyle can all contribute to supporting cardiovascular health and potentially reducing stroke risk.

11.12 Conclusion

To conclude up, dietary management with especial required nutrient needs is needed for stroke management and/or prevention that recommended in the way of a healthy diet. Such healthy balanced diet should include all macronutrients, micronutrients and trace elements in adequate amounts from different food groups suggest lifestyle modifications to meet the RDA individually. Magnesium show good associated relationship to the stroke risk in different studies so provide dietary recommendation is very affective in order to reduce stroke risk through food sources. The best to be followed is the Mediterranean diet or that called DASH diet. It has been shown with many benefits and importance to the stroke prevention and/or recovery. Therefore, having fresh fruits and vegetables with low saturated fats, trans-fat and cholesterol free diets and high dietary fiber helped to reduce the blood pressure and stroke incidence and that could be recommended an effective strategy for stroke prevention. However, it could be recommended to discuss and consult healthcare providers for personalized dietary recommendations individually.

References

1. World Health Organization. (2022). The Top 10 Causes of Death. https://www.who.int/news-room/fact-sheets/detail/the-top-10-causes-of-death
2. Feigin, V. L., Norrving, B., & Mensah, G. A. (2017). Global burden of stroke. *Circulation Research, 120*(3), 439–448.
3. Feigin, V. L., Stark, B. A., Johnson, C. O., Roth, G. A., Bisignano, C., Abady, G. G., Abbasifard, M., Abbasi-Kangevari, M., Abd-Allah, F., Abedi, V., et al. (2021). Global, regional, and national burden of stroke and its risk factors, 1990–2019: A systematic analysis for the global burden of disease study 2019. *The Lancet Neurology, 20*, 795–820.

4. Pavlidou, E., Fasoulas, A., Mantzorou, M., & Giaginis, C. (2022). Clinical evidence on the potential beneficial effects of probiotics and prebiotics in cardiovascular disease. *International Journal of Molecular Sciences, 23*(24), 15898. https://doi.org/10.3390/ijms232415898
5. Healthy Food Display: Image Details. (2021). *NCI Visuals Online*. National Cancer Institute.
6. Parlasca, M. C., & Qaim, M. (2022). Meat consumption and sustainability. *Annual Review of Resource Economics, 14*, 17–41. https://doi.org/10.1146/annurev-resource-111820-032340. ISSN 1941-1340
7. Wang, X., Ouyang, Y., Liu, J., et al. (2014). Fruit and vegetable consumption and mortality from all causes, cardiovascular disease, and cancer: Systematic review and dose-response meta-analysis of prospective cohort studies. *BMJ (Clinical Research Ed.), 349*, g4490.
8. Ridder, D., Kroese, F., Evers, C., Adriaanse, M., & Gillebaart, M. (2017). Healthy diet: Health impact, prevalence, correlates, and interventions. *Psychology & Health, 32*(8), 907–941. https://doi.org/10.1080/08870446.2017.1316849. PMID 28447854
9. Walker, C., & Reamy, B. V. (2009). Diets for cardiovascular disease prevention: What is the evidence? *American Family Physician, 79*(7), 571–577. PMID 19378874
10. Vasto, S., Barera, A., Rizzo, C., Di Carlo, M., Caruso, C., & Panotopoulos, G. (2014). Mediterranean diet and longevity: An example of nutraceuticals? *Current Vascular Pharmacology, 12*(5), 735–738. https://doi.org/10.2174/1570161111666131219111818. PMID 24350926
11. Bechthold, A., Boeing, H., Schwedhelm, C., Hoffmann, G., Knüppel, S., Iqbal, K., et al. (2019). Food groups and risk of coronary heart disease, stroke and heart failure: A systematic review and dose-response meta-analysis of prospective studies. *Critical Reviews in Food Science and Nutrition, 59*(7), 1071–1090.
12. Ekmekcioglu, C. (2020). Nutrition and longevity-From mechanisms to uncertainties. *Critical Reviews in Food Science and Nutrition, 60*(18), 30633082. https://doi.org/10.1080/10408398.2019.1676698. PMID 31631676. S2CID 204815279
13. Venn, B. J. (2020). Macronutrients and human health for the 21st century. *Nutrients, 12*(8), 2363. https://doi.org/10.3390/nu12082363. PMID: 32784664; PMCID: PMC7468865.
14. Aversa, R., Petrescu, R. V., Apicella, A., & Petrescu, F. I. (2016). The basic elements of life's. *American Journal of Engineering and Applied Sciences, 9*(4), 1189–1197.
15. Niazi, P., & Monib, A. (2023). Function of macronutrients in plant growth and human. *International Journal of Scientific Development and Research (IJSDR; IJSDR2304205), 8*(4), 1265. ISSN: 2455-2631. https://doi.org/10.1729/Journal.33883
16. Mohammadifard, N., Mansourian, M., Firouzi, S., Taheri, M., & Haghighatdoost, F. (2021). Longitudinal association of dietary carbohydrate and the risk cardiovascular disease: A dose-response meta-analysis. *Critical Reviews in Food Science and Nutrition, 62*, 1–16.
17. Guo, N., Zhu, Y., Tian, D., et al. (2022). Role of diet in stroke incidence: An umbrella review of meta-analyses of prospective observational studies. *BMC Medicine, 20*, 194. https://doi.org/10.1186/s12916-022-02381-6
18. Seidelmann, S. B., Claggett, B., Cheng, S., Henglin, M., Shah, A., Steffen, L. M., Folsom, A. R., Rimm, E. B., Willett, W. C., & Solomon, S. D. (2018). Dietary carbohydrate intake and mortality: A prospective cohort study and meta-analysis. *The Lancet Public Health, 3*, e419–e428. https://doi.org/10.1016/S2468-2667(18)30135-X
19. Zhao, B., Zeng, L., Zhao, J., Wu, Q., Dong, Y., Zou, F., et al. (2020). Association of magnesium intake with type 2 diabetes and total stroke: An updated systematic review and meta-analysis. *BMJ Open, 10*(3), e032240.
20. Khalil, N. A., Walton, G. E., Gibson, G. R., Tuohy, K. M., & Andrews, S. C. (2014). In vitro batch cultures of gut microbiota from healthy and ulcerative colitis (UC) subjects suggest that sulphate-reducing bacteria levels are raised in UC and by a protein-rich diet. *International Journal of Food Sciences and Nutrition, 65*(1), 79–88. https://doi.org/10.3109/09637486.2013.825700
21. Sacks, F. M., Lichtenstein, A. H., Wu, J. H. Y., Appel, L. J., Creager, M. A., Kris-Etherton, P. M., Miller, M., Rimm, E. B., Rudel, L. L., Robinson, J. G., Stone, N. J., & Van Horn, L. V. (2017).

Dietary fats and cardiovascular disease: A presidential advisory from the American heart association (PDF). *Circulation, 136*(3), e1–e23. https://doi.org/10.1161/CIR.0000000000000510
22. Kang, Z. Q., Yang, Y., & Xiao, B. (2020). Dietary saturated fat intake and risk of stroke: Systematic review and dose-response meta-analysis of prospective cohort studies. *Nutrition, Metabolism and Cardiovascular Diseases, 30*(2), 179–189.
23. Jakobsen, M. U., Trolle, E., Outzen, M., Mejborn, H., Grønberg, M. G., Lyndgaard, C. B., et al. (2021). Intake of dairy products and associations with major atherosclerotic cardiovascular diseases: A systematic review and meta-analysis of cohort studies. *Scientific Reports, 11*(1), 1303.
24. Islam, M., Akash, S., Jony, M. H., Nowrin, F. T., Rahman, M., Rauf, A., & Thiruvengadam, M. (2023). Exploring the potential function of trace elements in human health: A therapeutic perspective. *Molecular and Cellular Biochemistry, 478*, 1–31.
25. Zhu, Y., Zhang, J., Li, Z., Liu, Y., Fan, X., Zhang, Y., et al. (2018). Association of sodium intake and major cardiovascular outcomes: A dose-response meta-analysis of prospective cohort studies. *BMC Cardiovascular Disorders, 18*(1), 192.
26. Ojagbemi, A., Okekunle, A. P., Olowoyo, P., Akpa, O. M., Akinyemi, R., Ovbiagele, B., et al. (2021). Dietary intakes of green leafy vegetables and incidence of cardiovascular diseases. *Cardiovascular Journal of Africa, 32*(4), 215–223.
27. Wessels, I., Rolles, B., & Rink, L. (2020). The potential impact of zinc supplementation on COVID-19 pathogenesis. *Frontiers in Immunology, 11*, 1712.
28. Ozturk, H., Niazi, P., Mansoor, M., Monib, A. W., Alikhail, M., & Azizi, A. (2023). The function of zinc in animal, plant, and human nutrition. *Journal for Research in Applied Sciences and Biotechnology, 2*(2), 35–43. https://doi.org/10.55544/jrasb.2.2.6
29. Anderson, J. V., Palombo, R. D., & Earl, R. (1998). Position of the American Dietetic Association: The role of nutrition in health promotion and disease prevention programs. *Journal of the American Dietetic Association, 98*(2), 205–208. https://doi.org/10.1016/s0002-8223(98)00052-2
30. Aune, D., Giovannucci, E., Boffetta, P., Fadnes, L. T., Keum, N., Norat, T., et al. (2017). Fruit and vegetable intake and the risk of cardiovascular disease, total cancer and all-cause mortality-a systematic review and dose-response meta-analysis of prospective studies. *International Journal of Epidemiology, 46*, 1029–1056. https://academic.oup.com/ije/article/3039477
31. Chen, G. C., Lv, D. B., Pang, Z., Dong, J. Y., & Liu, Q. F. (2013). Dietary fiber intake and stroke risk: A meta-analysis of prospective cohort studies. *European Journal of Clinical Nutrition, 67*(1), 96–100.
32. Cheng, P., Wang, L., Ning, S., Liu, Z., Lin, H., Chen, S., et al. (2018). Vitamin E intake and risk of stroke: A meta-analysis. *British Journal of Nutrition, 120*(10), 1181–1188.
33. Grajek, M., Krupa-Kotara, K., Białek-Dratwa, A., Sobczyk, K., Grot, M., Kowalski, O., & Staśkiewicz, W. (2022). Nutrition and mental health: A review of current knowledge about the impact of diet on mental health. Frontiers in Nutrition, 22(9), 943998. https://doi.org/10.3389/fnut.2022.943998. PMCID: PMC9441951.
34. Wehrli, F., Taneri, P. E., Bano, A., Bally, L., Blekkenhorst, L. C., Bussler, W., et al. (2021). Oat intake and risk of type 2 diabetes, cardiovascular disease and all-cause mortality: A systematic review and meta-analysis. *Nutrients, 13*(8), 2560.
35. Larsson, S. C. (2017). Dietary approaches for stroke prevention. *Stroke, 48*(10), 2905–2911. https://doi.org/10.1161/STROKEAHA.117.017383. Epub 2017 Aug 17. PMID: 28818862.
36. Hori, A., Mizoue, T., Kasai, H., Kawai, K., Matsushita, Y., Nanri, A., et al. (2010). Body iron store as a predictor of oxidative DNA damage in healthy men and women. *Cancer Science, 101*(2), 517–522.
37. Strazzullo, P., D'Elia, L., Kandala, N. B., & Cappuccio, F. P. (2009). Salt intake, stroke, and cardiovascular disease: Meta-analysis of prospective studies. *BMJ, 339*, b4567.
38. Mazidi, M., Katsiki, N., Mikhailidis, D. P., & Banach, M. (2019). Dietary choline is positively related to overall and cause-specific mortality: Results from individuals of the national health and nutrition examination survey and pooling prospective data. *British Journal of Nutrition, 122*(11), 1262–1270.

39. Morze, J., Schwedhelm, C., Bencic, A., Hoffmann, G., Boeing, H., Przybylowicz, K., et al. (2020). Chocolate and risk of chronic disease: A systematic review and dose-response meta-analysis. *European Journal of Nutrition, 59*(1), 389–397.
40. Smyth, A., Dehghan, M., O'Donnell, M., Anderson, C., Teo, K., Gao, P., Sleight, P., Dagenais, G., Probstfield, J. L., Mente, A., Yusuf, S., & ONTARGET and TRANSCEND Investigators. (2015). Healthy eating and reduced risk of cognitive decline: A cohort from 40 countries. *Neurology, 84*(22), 2258–2265. https://doi.org/10.1212/WNL.0000000000001638. Epub 2015 May 6. PMCID: PMC4456656.
41. Adebamowo, S. N., Spiegelman, D., Flint, A. J., et al. (2014). Association between serum magnesium and risk of stroke: A meta-analysis. *Stroke, 45*(1), 288–293. https://pubmed.ncbi.nlm.nih.gov/24302599/
42. Folsom, A. R., Hong, C., Myung, S. K., et al. (1999). Serum and dietary magnesium and the risk of stroke: The atherosclerosis risk in communities study. *Stroke, 30*(8), 1772–1778. https://pubmed.ncbi.nlm.nih.gov/10436112/
43. Del Gobbo, L. C., Imamura, F., Wu, J. H., et al. (2013). Magnesium and cardiovascular disease: A review of epidemiological, clinical, and laboratory evidence. *Nutrients, 5*(10), 5172–5182. https://www.ncbi.nlm.nih.gov/pmc/articles/PMC3820047/
44. Jee, S. H., Miller, E. R., 3rd, Guallar, E., et al. (2002). Effect of magnesium supplementation on blood pressure: A meta-analysis. *European Journal of Clinical Nutrition, 56*(3), 240–245. https://pubmed.ncbi.nlm.nih.gov/11960293/
45. Larsson, S. C., Orsini, N., & Wolk, A. (2012). Dietary magnesium intake and risk of stroke: A meta-analysis of prospective studies. *The American Journal of Clinical Nutrition, 95*(2), 362–366.
46. Appel, L. J., Moore, T. J., Obarzanek, E., Vollmer, W. M., Svetkey, L. P., Sacks, F. M., et al. (1997). A clinical trial of the effects of dietary patterns on blood pressure. DASH collaborative research group. *New England Journal of Medicine, 336*, 1117–1124. https://doi.org/10.1056/NEJM199704173361601
47. Bloomfield, H. E., Kane, R., Koeller, E., Greer, N., MacDonald, R., & Wilt, T. (2015). Benefits and harms of the Mediterranean diet compared to other diets (PDF). VA Evidence-Based Synthesis Program Reports. PMID 27559560.
48. Katsiki, N., Kolovou, G., & Mikhailidis, D. P. (2018). The role of the mediterranean diet in the prevention of stroke. *International Journal of Stroke*.
49. Martínez-González, M. A., Gea, A., & Ruiz-Canela, M. (2019 Mar). The Mediterranean diet and cardiovascular health. *Circulation Research, 124*(5), 779–798. https://doi.org/10.1161/CIRCRESAHA.118.313348
50. Paterson, K. E., Myint, P. K., Jennings, A., Bain, L. K. M., Lentjes, M. A. H., Khaw, K. T., & Welch, A. A. (2018). Mediterranean diet reduces risk of incident stroke in a population with varying cardiovascular disease risk profiles. *Stroke, 49*(10), 2415–2420. https://doi.org/10.1161/STROKEAHA.117.020258. PMCID: PMC6159687.
51. Sureda, A., Bibiloni, M. D. M., Julibert, A., et al. (2019). Mediterranean diet and risk of stroke: A systematic review and meta-analysis. *Nutrients*.

Nutrigenomics and Stroke

12

Abraham Olufemi Asuku, Toheeb O. Oyerinde, Ebele Joan Ajagun, Hammed A. Sangodipe, Wasiu Sanusi, Taoheed Adedeji Abdulsalam, and Lukman Akanni Jayeoba

Abstract

Stroke, a significant public health concern causing considerable morbidity and mortality worldwide, stands as a primary instigator of functional disability. Its multifaceted and enigmatic nature emerges from an intricate interplay of specific risk factors, including genetic elements. Delving into the intricate landscape of stroke causes, this review illuminates both monogenic and polygenic origins, meticulously elucidating the nexus between genes and diet. Nutrigenomics, an emerging field studying how nutrients interact with genes, holds promise in stroke prevention and management. This review aims to synthesize the current state of knowledge on nutrigenomics in stroke, encompassing the genetic foundations, nutrient impact on risk and outcomes, and the potential for personalized prevention and treatment. Within this exploration, a keen focus is directed towards unravelling the dynamics of CADASIL and MELAS. Investigating the latest research on nutrient-gene interactions in stroke, including dietary patterns and individual nutrients, this paper underscores the importance of a multidisciplinary approach to stroke care, involving healthcare professionals, patients, and caregivers. The findings suggest that nutrigenomics could revolutionize stroke

A. O. Asuku (✉) · E. J. Ajagun · T. A. Abdulsalam · L. A. Jayeoba
Bioresources Development Centre (BIODEC), National Biotechnology Development Agency (NABDA), Ogbomoso, Nigeria

T. O. Oyerinde
Laboratory for Experimental and Translational Neurobiology, University of Medical Sciences, Ondo, Nigeria

H. A. Sangodipe
Department of Medicine, Ladoke Akintola University of Technology, Ogbomoso, Oyo State, Nigeria

W. Sanusi
Department of Biochemistry and Nutrition, Fountain University, Osogbo, Nigeria

W. Mohamed, M. B. Mat-Nor (eds.), *Nutrition and Stroke*, Nutritional Neurosciences, https://doi.org/10.1007/978-981-97-6818-9_12

management, enhancing the quality of life for survivors and families. However, a call for further research echoes throughout, urging advancements in nutrigenomics to translate findings into clinical practice. Ultimately, this review serves as a valuable resource for healthcare professionals, researchers, and policymakers, contributing significantly to the global mission of alleviating the societal burden imposed by stroke.

CADASIL: Cerebral autosomal dominant arteriopathy with subcortical infarcts and leukoencephalopathy

MELAS: Mitochondrial encephalopathy, lactic acidosis, and stroke-like episodes

Keywords

Stroke · Nutrigenomics · Nutrients · Genes · Diet

12.1 Introduction to Stroke

The assessment conducted by the Global Burden of Disease, Injuries, and Risk Factors Study [1] has illuminated a significant shift away from infectious diseases, maternity and nutritional causes towards noncommunicable diseases, specifically highlighting the prominence of conditions such as stroke. This transition can be attributed to factors such as the global population's increase in size and age, coupled with a notable reduction in global mortality rates over the past few decades [1]. It is worth noting that the predominant causes of death have a vascular origin [1]. According to the World Health Organization (WHO), cardiovascular diseases are defined as disorders stemming from disorders affecting organs of the cardiovascular system; arteries, veins and the heart. These encompass conditions like coronary heart disease, cerebrovascular disease (stroke), raised blood pressure (hypertension), peripheral artery disease, rheumatic heart disease, congenital heart disease, and heart failure. Remarkably, CVDs have taken the forefront in terms of both morbidity and mortality during the twenty-first century, with forecasts suggesting that this status quo will persist in the coming decades [2].

Stroke, being a type of cardiovascular disease, ranks as the second most common reason for mortality and a significant contributor to disability on a global scale. Its incidence is on the rise, largely due to the aging population, and there is a troubling trend of stroke affecting younger individuals in low- and middle-income nations [3]. Of the 795,000 new stroke cases, 26% result in persistent disability in fundamental daily activities, based on the Framingham cohort study, and 50% lead to reduced mobility owing to hemiparesis [4].

Ischemic stroke is the most prevalent stroke subtype, accounting for an estimated 80% of cases [5]. A less common variant is hemorrhagic stroke, which arises from subarachnoid hemorrhage (SAH) and/or intracerebral hemorrhage (ICH).

In a general sense, stroke is a complex, multifactorial disease influenced by a combination of specific risk factors and genetic elements [6]. Traditional risk factors, including hypertension, cigarette smoking, diabetes, atrial fibrillation, and obesity, explain a substantial portion of the stroke risk, although they cover only around 60% of cases [7]. Notably, nearly half of the risk associated with ischemic stroke remains unexplained by these conventional risk factors [8], sparking speculation that genetic predisposition plays a role [9]. Various studies on genetically related individuals and animal models have hinted at a genetic background contributing to the risk of development of stroke [10].

It is essential to recognize that both genetic factors and the environment are integral in shaping the disease phenotype and its manifestation in an individual [11]. Undeniably, among environmental factors, the diet holds the highest quantitative importance, as it is the most modifiable aspect based on individual needs [12]. While it is evident that nutritional factors as well as genetic factors play distinct roles in defining the individual's health, the intricate interactions between genetics, dietary factors, and downstream networks remain largely uncharted [13]. Nutrigenomics, a field exploring the nutritional influence on gene expression, holds promise in offering more personalized diets for improved treatment and prevention of diseases, including stroke, given the burgeoning discoveries concerning the human genome's role in nutrition [14].

12.1.1 Overview of Nutrigenomics

Nutrition involves the acquisition and utilization of nutrients by living organisms for energy, maintenance, and growth purposes [15]. Nutrition encompasses the study of how the body absorbs, processes, metabolizes, and utilizes nutrients [16]. These nutrients are essential substances that provide the necessary energy and vital components for biological systems to function optimally [17]. In nutrition, an organism's diet constitutes the totality of substances it ingests or consumes [18]. Improving an organism's mental and physical well-being is attainable through a balanced diet, incorporating essential elements like vitamins, minerals, and crucial amino acids found in proteins, and necessary fatty acids from high-fat foods. These components are crucial for overall health improvement [19]. Proteins, fats, and carbohydrates each play an important role in an organism's longevity, health, and overall quality of life [20]. The investigation of inherited molecules passed from one generation to another, known as genes, defines the field of genetics [21]. These genetic components intricately shape and determine numerous characteristics and traits within living organisms. Genes orchestrate the formation of proteins in our bodies, simultaneously defining the unique distinctions among various entities [22]. Due to the recognized significance of nutrition and genetics in the onset and prevention of diverse ailments like cancer, diabetes, neurodegenerative diseases, hypertension, cardiovascular disease, and osteoporosis, there's an increasing focus on the interplay between the genome and nutrition, a field known as nutrigenomics or nutritional genomics [23]. The burgeoning field of nutrigenomics enables the understanding of how specific foods communicate with genes. Understanding the language of one's genes and gaining control over the signals and instructions they transmit to the body and metabolism can lead to substantial changes in the interaction between food and the body. This, in turn, may result in weight reduction and enhanced health [24]. Nutrigenomics, often termed nutritional genomics, encompasses the broad association between genes, nutrition, and diet [25]. The genesis of contemporary nutrigenomics traces back to the initiation of the Human Genome Project in the 1990s. This groundbreaking endeavor resulted in the mapping of human DNA sequencing, marking the advent of the "era of big science" and laying the foundation for nutrigenomics as recognized today [26]. In recent times, nutritional science has increasingly delved into exploring the impact of nutrients on DNA stability, repair mechanisms, and various processes related to gene expression [27]. Various food components possess the capability to influence genetic and epigenetic processes, potentially impacting an individual's health [28]. The establishment of nutrigenomics stemmed from extensive discoveries highlighting the role of the human genome in nutrition. This field now presents personalized diet plans, potentially enhancing health promotion and therapy through tailored nutritional strategies [29, 30]. For a considerable time, it's been evident that individuals react differently to various foods. An instance of this diversity lies in individuals with lactose intolerance, where some experience digestive discomfort after consuming specific dairy products, while others can consume dairy without encountering any issues. Individuals with lactose intolerance face difficulties in digesting the natural sugars present in milk products due to inadequate amount of enzyme lactase [31]. Nutrigenomics aims to offer personalized dietary guidance to individuals according to their unique genetic composition [32]. Several key factors contribute to nutritional genomics. Foremost is the recognition that certain individuals bear a significantly heightened risk of developing specific diseases owing to their dietary habits [33]. Moreover, common dietary nutrients can alter the expression or structure of genes in humans, either directly or indirectly impacting the genome [34]. The impact of nutrition on the balance between healthy and unhealthy conditions might vary based on an individual's genetic configuration. Most genes influenced by diet likely play a role in the initiation, advancement, and severity of chronic ailments. Consequently, the concept of "individualized nutrition" emerges as a potential tool in the prevention, management, or treatment of chronic diseases such as stroke

12.2 Nutrient and Stroke

Stroke affects a significant majority of survivors and is the leading cause of functional disability. Within 3 months of having a stroke, over 20% of people need institutional care, and 15–30% of these people have a permanent disability [35]. This tragic occurrence has far-reaching effects on the affected person, as well as on their family and support system. According to utility assessments, more than half of at-risk people believe that a severe stroke is a fate worse than death [36]. Primary prevention is especially crucial because approximately 70% of strokes occur for the first time [35].

Nutrition has a far bigger impact on stroke risk than is typically acknowledged in the medical community. With dietary choices taking center stage, adopting a healthy lifestyle can result in an amazing 80% reduction in predisposition to stroke development [37]. Notably, according to the US Health Professionals Study and the Nurses Health Study, leading a healthy lifestyle entails quitting smoking, consuming alcohol in moderation, maintaining a healthy body mass index (BMI < 25), getting daily exercise of 30 min, and consumption of a diet that is within the upper 40% of recommended dietary guidelines. Unhealthy habits accounted for 50% of the stroke risk among a sizable cohort of 43,685 men and 71,243 women in the United States, whereas those who adhered to all five healthy lifestyle criteria had an 80% reduction in their risk of stroke [38]. Additionally, a considerable advancement has been made in our understanding of the function of nutrition in preventing stroke, including the impact of intestinal bacteria, the advantages of Mediterranean diet, the benefits of B vitamins and the adverse effects of egg yolk [37].

Observational studies provide information on a variety of dietary influences on stroke risk. Increased fruit and vegetable consumption has been observed to have the tendency to lower the risk of stroke, according to a body of prospective studies [39–42]. The evidence from randomized control trials does not consistently support the efficacy of single antioxidant supplementation in stroke or cardiovascular events risk reduction, despite the fact that antioxidants like β-carotene and vitamins C and E have been observed to have a converse relationship with stroke risk in observational studies [43]. In contrast, individuals in a separate randomized controlled trial who received both vitamin E and C had fewer stroke events than a placebo group for both treatments (RR 0.69; 95% CI 0.49–0.98) [44]. The B vitamins, which include vitamins B6, B12 and folate, have positive impacts on plasma homocysteine levels, antioxidant defences, and endothelial function [45]. With its ability to cause reduction in blood pressure, improve endothelial function, have anti-inflammatory effects, and improve insulin sensitivity, vitamin D is emerging as another possible defender against cardiovascular disease [46]. Compared to higher concentrations, low 25-hydroxyvitamin D levels are directly related with a higher risk of stroke [47].

Consumption of sodium in higher level has been linked to a higher chance of stroke development [48–50]; and there exist a relationship between higher potassium intake and a lower risk of stroke [51, 52]. Blood pressure plays a part in how sodium and potassium affect stroke risk. However, there is a biologically reasonable explanation that suggests their effects on stroke risk also function without the blood pressure pathway, which is confirmed by findings from animal research [53].

Dietary cholesterol, which is mostly present in animal products including eggs, liver, and other meats, has been linked to a heightened stroke risk. High red meat and processed meat intake has been linked to an elevated risk of stroke in Western populations [54]. Notably, a prospective observational study of Swedish women found that the highest quintile of cholesterol intake (median, 168 mg/day) was linked with a statistically significant 29% higher risk of ischemic stroke than the lowest quintile (median, 302 mg/day), with this association remaining independent of red meat consumption [55]. On the other hand, a small Japanese cohort study found a inverse relationship between cholesterol intake and mortality from ischemic stroke [55]. Other prospective studies have not shown a link between cholesterol consumption and the risk developing of ischemic stroke, total hemorrhagic stroke, or intraparenchymal hemorrhage [56–59].

12.3 Genes and Stroke

While conventional risk factors for stroke are well-understood, roughly 50% of the risk remains elusive, with genetic predisposition thought to play a role in this unexplained portion [60]. However, differentiating between risk stemming from genetic mutations and risk resulting from shared familial exposure remains a complex task [61]. Research indicates that individuals, especially those at a younger age, who have a family history of stroke, are at an increased risk of experiencing a stroke themselves [62–64]. This phenomenon is further substantiated by twin studies and animal research, revealing the presence of genes that autonomously impact the risk of stroke [65, 66]. Stroke encompasses a wide array of risk factors, including those that can be modified, those that cannot be modified and genetic risk factors. In this chapter, our emphasis is on genetic risk factors, with a particular focus on how nutrients impact these genetic elements. Researchers have made significant advancements in pinpointing particular genes associated with stroke disorders. However, addressing this specific aspect remains a formidable challenge as identifying specific genes associated with the stroke disorder has been hindered by the heterogeneity of stroke, the numerous conventional risk factors that can lead to strokes, and the variations within different populations and research studies [61, 67].

The application of technical advancements, such as novel testing techniques like whole-exome sequencing and whole-genome sequencing, has expedited the discovery of new genes linked to various monogenic diseases, including stroke [68, 69]. In this chapter, our aim is to provide a concise summary of extensively researched genetic stroke risk factors and the influence of diet on their onset by categorizing them based on various potential mechanisms. We will explore various genes that influences stroke disorder, including single-gene disorders that are the primary culprits behind genetic stroke, the genetic disorders that manifest as multisystem issues and incorporate stroke as one of their symptoms, the common genetic variants are linked to stroke risk. Following that, we will examine the role of genetics in contributing to traditional stroke risk factors like diabetes, hypertension, and atrial fibrillation.

Single Gene Disorders Associated with Stroke: Stroke may represent one of the most severe manifestations in a monogenic disorder and it can result from a multitude of single gene disorders. However, it is important to note that the majority of these conditions are exceptionally uncommon [9, 70]. Accurate identification of single-gene disorders linked to stroke not only facilitates effective treatment but also enables the provision of dietary recommendations, collectively enhancing the overall well-being of patients. The gene's precise function in humans remains unclear, but it is known to encode a membrane receptor that potentially enhances vascular smooth muscle cell structural stability. Patients exhibit a progressive accumulation of the Notch3 receptor, leading to granular osmiophilic deposits in vascular smooth muscle cells, ultimately causing arterial thickening and fibrosis, resulting in cerebral infarction.

12.3.1 CADASIL

Cerebral autosomal dominant arteriopathy with subcortical infarcts and leukoencephalopathy (CADASIL) is the most common single gene disorder cause of stroke with an approximate prevalence of around 0.002% in the UK population [71]. Additionally, research has approximated the CADASIL's prevalence at approximately 0.004–0.002% in Scotland [72], and also, an analysis of over 60,000 exome sequences from various global regions indicated that genetic mutations with the potential to induce CADASIL were present in about 0.34% of exomes [73]. CADASIL is as a result of mutation in NOTCH3 gene and its clinical features are confined to central nervous system. The gene's precise function in humans is not well understood, but it is known to encode a membrane receptor (A type I transmembrane protein, comprising 2321 amino acids, and plays a role in the NOTCH signaling pathway) that potentially enhances vascular smooth muscle cell structural stability [74, 75]. Patients with CADASIL exhibit a progressive accumulation of the NOTCH3 receptor, leading to granular osmiophilic deposits in intracranial and extracranial vascular smooth muscle cells, ultimately causing arterial thickening and fibrosis, resulting in cerebral infarction [76].

CADASIL typically manifests around the age of 45, and patients often have a life expectancy ranging from 60 to 65 years [77]. Extensive research consistently indicates that brain ischemia is the predominant symptom in this condition, affecting well over half of the patients [78, 79]. The condition can present discreetly, with some patients experiencing migraine with aura in 20–40% of cases, while others may exhibit classic signs of vascular impairment in the central nervous system, including transient ischemic attack, repeated episodes of ischemic strokes and cognitive dysfunction, which are observed in 60–85% of cases [77, 78]. Moreover, CADASIL patients may encounter ocular symptoms, albeit less precisely defined, along with neuropsychiatric symptoms such as depression, anxiety, and apathy [76, 80]. CADASIL diagnosis often involves neuroimaging techniques, which reveal ischemic changes in the white matter, basal ganglia, external capsules, and anterior temporal pole [81, 82]. However, it's worth noting that these changes are not exclusive to CADASIL, and they may not be evident during the early stages of the disease. Therefore, the sequencing of the NOTCH3 gene plays a crucial role in specifying the diagnosis. Additional diagnostic methods include molecular genetics testing and histological and immunohistochemical findings, such as vascular degeneration with the accumulation of proteinaceous non-amyloidotic material in the vessel's middle layer [70]. Furthermore, skin biopsy stands as an effective method for diagnosing CADASIL, offering a high level of sensitivity and specificity [83].

12.3.2 CARASIL

Cerebral autosomal recessive arteriopathy with subcortical infarcts and leukoencephalopathy (CARASIL) is one of the single-gene disorders that are the primary culprits behind genetic stroke and can also be called Maeda syndrome. CARASIL is a monogenic disorder that is typically presents with its characteristic symptoms between the ages of 20 and 45 years and caused by an autosomal recessive mutation in the HTRA1 gene, been the only gene known to be associated with CARASIL, located on chromosome 10q [84, 85]. Although it was originally documented in Japanese and Chinese communities, instances have also been noted in European populations [84, 86]. CARASIL impacts the small blood vessels of the brain and is marked by chronic ischemia and widespread white matter abnormalities. Its primary pathological characteristics involve, significant atherosclerosis without the build-up of granular osmiophilic material recurring strokes, the gradual decline of cognitive and motor functions, seizures, and psychiatric disruptions [84, 87, 88].

12.3.3 Familial Cerebral Amyloid Angiopathy

Cerebral Amyloid Angiopathy (CAA) encompasses a range of medical conditions linked to the buildup of congophilic material, specifically amyloid fibrils, in the walls of small blood vessels within the cerebral cortex and leptomeninges [89, 90]. The predominant clinical presentation of CAA that is widely acknowledged is the occurrence of spontaneous intracerebral hemorrhages, primarily observed in the elderly population, and it has been associated with cognitive decline [89, 91]. CAA exists in both hereditary and sporadic forms, displaying similar clinical characteristics. However, the hereditary form tends to manifest at a younger age [92]. In most instances, hereditary CAA follows an autosomal dominant inheritance pattern, and the primary cause is the accumulation of amyloid beta (Ab), which is derived from the amyloid precursor protein (APP). This protein is encoded by a gene located on chromosome 21 [93]. The sporadic form of CAA is usually associated with Alzheimer disease. Published research has provided confirmation that there are shared risk factors in the development of both CAA and AD, indicating a certain degree of overlap between the two conditions [94]. The diagnosis of CAA typically relies on clinical data and specific neuroimaging indicators, particularly the observation of cerebral microbleeds on T2*-weighted MRI scans [95]. Current diagnostic criteria based on neuroimaging are specific, but frequently lack the required sensitivity. However, there is potential to enhance diagnostic sensitivity through the in vivo detection of white matter perivascular spaces using MRI [96]. While it is possible to also utilize pre-mortem brain biopsies to diagnose CAA, they are rarely accessible.

Diet-Gene and Disease Onset Interaction; Compelling evidence supports the notion that the prolonged consumption of a high-fat Western diet is associated with a heightened occurrence of metabolic conditions like obesity, diabetes, hyperinsulinemia, and cardiovascular diseases (Stroke as an example).

Other rare single gene disorders that can lead to stroke include arterial tortuosity syndrome, resulting from mutations in the SLC2A10 gene, responsible for encoding the glucose transporter GLUT10 [97]. Autosomal-dominant retinal vasculopathy with cerebral leukodystrophy is linked to mutations in TREX1, which codes for a DNA-specific exonuclease [98], as well as mutations in COL4A1, responsible for encoding the type IV collagen alpha 1 chain. COL4A1 mutations are known to cause autosomal-dominant porencephaly and infantile hemiparesis [99].

12.3.4 Multifactorial Genetic Disorders with Stroke as a Significant Manifestation

Fabry disease results from a mutation in the alpha-galactosidase A (GLA) gene, located on the long arm of the X chromosome, responsible for encoding the alpha-galactosidase A enzyme [100]. This leads to the impaired breakdown of globotriaosylceramide (Gb3), causing the buildup of Gb3 within lysosomes found in vascular endothelial cells and smooth muscle cells across various organ systems [101]. Fabry disease ranks as the second most prevalent lysosomal storage disorder. This disorder often presents at a young age, with cerebrovascular complications affecting both major and minor blood vessels, resulting in neuropathies and acroparesthesias [6]. Its clinical spectrum is extensive, encompassing chronic abdominal pain, angiokeratoma, renal dysfunction, cardiac arrhythmias, hypertrophy, and various other symptoms [101]. Stroke mechanisms include cardioembolism due to cardiac involvement, large artery thromboembolism from the elongation and tortuosity of major blood vessels, and small vessel occlusive disease resulting from glycolipid accumulation in endothelial and vascular smooth muscle cells [102]. MRI is the preferred imaging technique for identifying cerebrovascular issues in Fabry disease, offering detailed assessment of brain damage in patients with specific neurological problems. Advanced imaging methods like diffusion tensor imaging, functional MRI, MRI spectroscopy, and Positron emission tomography are being used to enhance our understanding of the Fabry disease's cerebrovascular mechanisms [103].

Sickle-cell disease comprises a set of genetic disorders caused by mutations in the HBB gene (hemoglobin subunit β), inherited in an autosomal recessive manner [104]. It leads to the formation of polymerized red blood cells under low oxygen conditions, which, in turn, causes blockages in small blood vessels and contributes to sickle-related arterial conditions [61]. The likelihood of experiencing a stroke varies across age groups, with children having a 5–10% risk, while young adults face an increased risk: 11% by the age of 20, 15% by age 30, and 25% by age 45. When considering the type of stroke, around 75% of cases are ischemic strokes, with the remaining 25% being hemorrhagic strokes [105].

MELAS, Mitochondrial encephalopathy, lactic acidosis, and stroke-like episodes, arises from a mutation in mitochondrial DNA (mtDNA) that is inherited exclusively from the mother [106]. This mutation results in respiratory chain dysfunction and a decrease in energy production [107]. The clinical diagnostic guidelines for MELAS syndrome have been established, defining this syndrome based on three consistent criteria: The occurrences of stroke-like episodes before the age of 40, the encephalopathy marked by seizures and/or dementia, and then, the presence of mitochondrial myopathy, which is characterized by the presence of lactic acidosis and/or ragged-red fibers (RRFs). Further, to confirm the diagnosis, it is necessary to fulfill at least two additional criteria from the following: The normal early psychomotor development, the recurrent headaches, or the recurrent episodes of vomiting [108]. However, the new diagnostic criteria have been established for MELAS syndrome more recently. A definitive diagnosis requires meeting at least two "category A" criteria, which include headaches with vomiting, seizures, hemiplegia, cortical blindness, and acute focal lesions in neuroimaging. Additionally, at least two "category B" criteria should be satisfied, such as elevated lactate levels in plasma or cerebrospinal fluid (CSF), mitochondrial abnormalities in muscle biopsy, or a MELAS-related gene mutation [109]. Symptoms typically begin to manifest between the ages of 2 and 20. Stroke-like episodes represent a prominent hallmark of MELAS syndrome, affecting a significant majority of those affected, ranging from 84% to 99% [109]. Stroke-like episodes in MELAS syndrome do not adhere to a specific vascular distribution, and they occur without any identified embolic source or stenotic lesions as observed in angiography. Typically, these episodes involve asymmetric cortical areas, with the parietal, occipital, and temporal lobes being the most frequently affected regions [110]. Two theories have been proposed in a published research paper by Koga et al. [110] to elucidate the pathogenesis of stroke-like episodes in MELAS syndrome. The first is the mitochondrial cytopathy theory, which posits that the mutation in leucine transfer RNA results in decreased protein synthesis, leading to oxidative phosphorylation failure and, ultimately, adenosine triphosphate (ATP) depletion and energy deficiency. The second is the mitochondrial angiopathy theory, which suggests that the accumulation of abnormal mitochondria in the endothelium and smooth muscle disrupts cerebral small vessel function. MRI, MR spectroscopy (MRS), and other diagnostic instruments and methods such as Mitochondrial DNA sequencing, histological studies and genetic testing are valuable in assessing MELAS syndrome. MRI typically reveals hyperintense cortical multifocal laminar lesions that are not confined to a specific vascular territory, while MR spectroscopy shows decreased N-acetyl aspartate signal and the accumulation of lactate due to anaerobic metabolism. These diagnostic tools, along with others, aid in the evaluation of MELAS syndrome [110–112].

Other multifactorial genetic disorders with stroke as a significant manifestation include Ehlers-Danlos syndrome type 4 arises from mutations occurring in the COL3A1 gene, which encodes Type III procollagen, located on chromosome 2q31 [113], Marfan syndrome; an inherited connective tissue disorder that follows an autosomal dominant pattern and primarily attributed to mutations found in the FBN1 gene; a gene that encode fibrillin-1, a glycoprotein serving as the primary component of the extracellular matrix's microfibrils [114], and mutations occurring in the ACTA2 gene result in alterations to the smooth muscle cell-specific isoform of α-actin protein [115].

12.4 Nutrigenomics and Stroke

While no concrete evidence exists to directly link diet to the modulation of monogenic causes of stroke, such as CADASIL and CARASIL, exploring the interplay between nutrition and these specific genetic stroke predispositions remains an intriguing avenue in nutrigenomic research. Despite the lack of conclusive findings pinpointing dietary influence on these monogenic stroke cases, investigating potential connections may shed light on nuanced aspects of nutrigenomics in stroke pathology. This exploration could offer insights into underlying mechanisms or interactive effects between genetic predispositions and dietary factors, contributing to a comprehensive understanding of stroke development within the context of nutrigenomics. Certainly, the emerging body of evidence suggests a significant role played by gut microbiota in both the onset and progression of CADASIL. This revelation challenges the conventional understanding that genetics solely dictate the trajectory of monogenic stroke conditions. The intricate interplay between gut microbiota and CADASIL development signifies a potentially crucial link between the gut-brain axis and stroke pathology. Understanding how the gut microbiome modifies the course of CADASIL could open avenues for novel therapeutic interventions and dietary modifications tailored to mitigate the impact of this monogenic stroke condition.

Emerging evidence indicates that nutrients can influence the prevalence of various gut microbiota. For instance, research suggests that whole grains and dietary fibers have been shown to modify the composition of intestinal microbiota and promote the growth of beneficial bacteria like Bifidobacteria and Lactobacilli, which have demonstrated capabilities in reducing inflammation and oxidative stress. This reduction potentially lowers the risk of stroke by diminishing the abundance of harmful bacteria such as Clostridia and Enterobacteria, which are known to cause inflammation and oxidative stress, factors associated with stroke development [116]. On the other hand, high-fat and high-sugar diets have demonstrated the capacity to encourage the proliferation of harmful gut bacteria, such as Firmicutes and Proteobacteria, while concurrently decreasing the presence of beneficial bacteria [117]. This can lead to an imbalance in gut microbiota, known as dysbiosis, which has been linked to a range of health problems, including obesity, diabetes, stroke and inflammatory bowel disease [118, 119]. In the context of CADASIL, the investigation into gut microbiota revealed notable taxonomic variances between CADASIL patients and control subjects. This indicates a potential link between the makeup of gut microbiota and the onset as well as progression of CADASIL [120]. While the study did not directly investigate the impact of specific nutrients on gut microbiota in CADASIL patients, it is possible that certain nutrients could indirectly affect the development of CADASIL by altering the composition of gut microbiota. In essence, these discoveries imply that sustaining a nourishing diet conducive to fostering beneficial gut bacteria might serve as a significant factor in lessening the likelihood of neurological conditions, including CADASIL. Additional exploration is imperative to comprehensively grasp the intricate interplay among diet, gut microbiota, and neurological disorders. Nonetheless, these outcomes pave a promising path for forthcoming research endeavors [15]

MELAS is an uncommon and severe genetic condition impacting the mitochondria, the cell's energy-generating organelles [107]. Individuals affected by MELAS encounter a spectrum of symptoms, such as seizures, stroke-like episodes, cognitive decline and muscle weakness [121]. Though a definitive cure for MELAS remains elusive, however, recent research highlights a potential breakthrough in symptom management through the adoption of a modified ketogenic diet. The ketogenic diet, a regimen high in fats and low in carbohydrates, has been utilized for many years as a therapeutic approach to managing epilepsy in children [122]. This diet functions by initiating a state called ketosis, where the body utilizes fat as its primary fuel source instead of glucose. This metabolic shift has demonstrated the ability to decrease both the frequency and intensity of seizures in certain individuals with epilepsy [123]. In recent times, scientists have started investigating the potential advantages of implementing the ketogenic diet for various neurological conditions beyond epilepsy, including its possible relevance in managing disorders like MELAS. An essential characteristic of MELAS is mitochondrial dysfunction, often resulting in a deficiency within complex I of the mitochondrial respiratory chain [124]. Such a deficiency can hinder the cells' capacity to generate energy, causing an accumulation of lactic acid alongside other potentially harmful byproducts. The ketogenic diet potentially offers a pathway to circumvent this deficiency by providing an alternative energy production route via complex II within the respiratory chain [125]. In a recent case study by Steriade et al. [126], a 22-year-old woman diagnosed with MELAS encountered recurring seizures and stroke-like lesions. Treatment involved a modified ketogenic diet coupled with magnesium supplementation, resulting in freedom from seizures and a reduction in the frequency of stroke-like episodes. The authors introduced a metabolic model suggesting that the diet specifically ketogenic diet might enhance the performance of respiratory chain complexes, potentially circumventing the deficiencies in complex I and redirecting energy production to complex II. Though this case study stands as a singular example, it presents encouraging evidence indicating the potential efficacy of a modified ketogenic diet as an additional therapy for individuals with MELAS. Nonetheless, further research is essential to comprehensively grasp the mechanisms driving the potential advantages of the ketogenic diet for MELAS patients. Additionally, investigating any potential risks or side effects linked to this dietary intervention remains crucial.

12.4.1 Limitation

One limitation of this research article is that the current state of knowledge on nutrigenomics and stroke is still in its early stages, and there is a lack of large-scale, long-term studies on the topic. The majority of studies in this field are observational or small-scale intervention studies, which limit the generalizability of the findings. More also, Research on the genetic causes of stroke predominantly emphasizes therapeutic approaches, often neglecting the captivating influence of dietary factors. Additionally, the complexity of the gene-nutrient interactions involved in stroke also makes it challenging to identify specific dietary recommendations for stroke prevention and management.

12.4.2 Future Perspective

Despite these limitations, the field of nutrigenomics and stroke holds great promise for the future. Future research should focus on large-scale, long-term studies to establish the causal relationship between nutrients and stroke risk and outcomes. Advances in technology, such as genome sequencing and metabolomics, will enable a more comprehensive understanding of the gene-nutrient interactions involved in stroke. The development of personalized nutrition interventions based on individual genetic profiles will allow for more effective stroke prevention and management. Additionally, the integration of nutrigenomics into clinical practice will require the collaboration of healthcare professionals, researchers, and policymakers to ensure that the benefits of this approach are accessible to all stroke patients. Overall, the future of nutrigenomics and stroke research is promising, and has the potential to significantly improve the health outcomes of stroke patients.

References

1. GBD 2015 Mortality and Causes of Death Collaborators. (2016). Global, regional, and national life expectancy, all-cause mortality, and cause-specific mortality for 249 causes of death, 1980–2015: A systematic analysis for the global burden of disease study 2015. *Lancet (London, England), 388*(10053), 1459–1544. https://doi.org/10.1016/S0140-6736(16)31012-1
2. Writing Group Members, Mozaffarian, D., Benjamin, E. J., Go, A. S., Arnett, D. K., Blaha, M. J., Cushman, M., Das, S. R., de Ferranti, S., Després, J. P., Fullerton, H. J., Howard, V. J., Huffman, M. D., Isasi, C. R., Jiménez, M. C., Judd, S. E., Kissela, B. M., Lichtman, J. H., Lisabeth, L. D., … Stroke Statistics Subcommittee. (2016). Heart disease and stroke statistics-2016 update: A report from the American Heart Association. *Circulation, 133*(4), e38–e360. https://doi.org/10.1161/CIR.0000000000000350
3. Katan, M., & Luft, A. (2018). Global burden of stroke. *Seminars in Neurology, 38*(2), 208–211. https://doi.org/10.1055/s-0038-1649503
4. Kelly-Hayes, M., Beiser, A., Kase, C. S., Scaramucci, A., D'Agostino, R. B., & Wolf, P. A. (2003). The influence of gender and age on disability following ischemic stroke: The Framingham study. *Journal of Stroke and Cerebrovascular Diseases: the Official Journal of National Stroke Association, 12*(3), 119–126. https://doi.org/10.1016/S1052-3057(03)00042-9
5. Rosamond, W., Flegal, K., Furie, K., Go, A., Greenlund, K., Haase, N., Hailpern, S. M., Ho, M., Howard, V., Kissela, B., Kittner, S., Lloyd-Jones, D., McDermott, M., Meigs, J., Moy, C., Nichol, G., O'Donnell, C., Roger, V., Sorlie, P., Steinberger, J., et al. (2008). Heart disease and stroke statistics—2008 update: A report from the American heart association statistics committee and stroke statistics subcommittee. *Circulation, 117*(4), e25–e146. https://doi.org/10.1161/CIRCULATIONAHA.107.187998

6. Dichgans, M. (2007). Genetics of ischaemic stroke. *The Lancet Neurology, 6*(2), 149–161. https://doi.org/10.1016/S1474-4422(07)70028-5
7. Rosengren, A., Hawken, S., Ounpuu, S., Sliwa, K., Zubaid, M., Almahmeed, W. A., Blackett, K. N., Sitthi-amorn, C., Sato, H., Yusuf, S., & INTERHEART investigators. (2004). Association of psychosocial risk factors with risk of acute myocardial infarction in 11119 cases and 13648 controls from 52 countries (the INTERHEART study): Case-control study. *Lancet (London, England), 364*(9438), 953–962. https://doi.org/10.1016/S0140-6736(04)17019-0
8. Sacco, R. L., Ellenberg, J. H., Mohr, J. P., Tatemichi, T. K., Hier, D. B., Price, T. R., & Wolf, P. A. (1989). Infarcts of undetermined cause: The NINCDS stroke data bank. *Annals of Neurology, 25*(4), 382–390. https://doi.org/10.1002/ana.410250410
9. Hassan, A., & Markus, H. S. (2000). Genetics and ischaemic stroke. *Brain: A Journal of Neurology, 123*(Pt 9), 1784–1812. https://doi.org/10.1093/brain/123.9.1784
10. Goldstein, L. B., Adams, R., Alberts, M. J., Appel, L. J., Brass, L. M., Bushnell, C. D., Culebras, A., Degraba, T. J., Gorelick, P. B., Guyton, J. R., Hart, R. G., Howard, G., Kelly-Hayes, M., Nixon, J. V., Sacco, R. L., American Heart Association/American Stroke Association Stroke Council, Atherosclerotic Peripheral Vascular Disease Interdisciplinary Working Group, Cardiovascular Nursing Council, Clinical Cardiology Council, Nutrition, Physical Activity, and Metabolism Council, et al. (2006). Primary prevention of ischemic stroke: A guideline from the American Heart Association/American Stroke Association Stroke Council: Cosponsored by the Atherosclerotic Peripheral Vascular Disease Interdisciplinary Working Group; Cardiovascular Nursing Council; Clinical Cardiology Council; Nutrition, Physical Activity, and Metabolism Council; and the Quality of Care and Outcomes Research Interdisciplinary Working Group: The American Academy of Neurology affirms the value of this guideline. *Stroke, 37*(6), 1583–1633. https://doi.org/10.1161/01.STR.0000223048.70103.F1
11. Corella, D., & Ordovas, J. M. (2005). Integration of environment and disease into 'omics' analysis. *Current Opinion in Molecular Therapeutics, 7*(6), 569–576.
12. de Luis, D. A., Izaola, O., Primo, D., & Pacheco, D. (2016). Effect of the rs10767664 variant of the brain-derived neurotrophic factor gene on weight change and cardiovascular risk factors in morbidly obese patients after biliopancreatic diversion surgery. *Journal of Nutrigenetics and Nutrigenomics, 9*(2–4), 116–122. https://doi.org/10.1159/000448102
13. Eaton, D. L., Bammler, T. K., & Kelly, E. J. (2001). Interindividual differences in response to chemoprotection against aflatoxin-induced hepatocarcinogenesis: Implications for human biotransformation enzyme polymorphisms. *Advances in Experimental Medicine and Biology, 500*, 559–576. https://doi.org/10.1007/978-1-4615-0667-6_85
14. Mutch, D. M., Wahli, W., & Williamson, G. (2005). Nutrigenomics and nutrigenetics: The emerging faces of nutrition. *FASEB Journal: Official Publication of the Federation of American Societies for Experimental Biology, 19*(12), 1602–1616. https://doi.org/10.1096/fj.05-3911rev
15. Raubenheimer, D., & Simpson, S. (1996). Meeting nutrient requirements: The roles of power and efficiency. In *Proceedings of the 9th International Symposium on Insect-Plant Relationships*.
16. Wickens, G. E. (2001). Human and animal nutrition. In *Economic botany: Principles and practices* (pp. 127–149). Springer.
17. Sénéchal, S., & Kussmann, M. (2011). Nutriproteomics: Technologies and applications for identification and quantification of biomarkers and ingredients. *Proceedings of the Nutrition Society, 70*(3), 351–364.
18. McKenna, G. (2021). *Nutrition and oral health*. Springer.
19. Alex, M. (2020). Role of nutrition in maintaining health. *International Journal of Physical Education, Sports and Health, 7*(4), 279–280.
20. Shen, J., Shan, J., Zhong, L., Liang, B., Zhang, D., Li, M., & Tang, H. (2022). Dietary phytochemicals that can extend longevity by regulation of metabolism. *Plant Foods for Human Nutrition, 77*(1), 12–19.

21. Schatz, B. R., Berlin, R. B., Jr., Schatz, B. R., & Berlin, R. B. (2011). Genomes for individual ability (Features). In *Healthcare infrastructure: Health systems for individuals and populations* (pp. 189–209).
22. Hengst, J., & Yun, J. (2007). Control and regulation of gene expression. In *xPharm: The comprehensive pharmacology reference* (pp. 1–5). Elsevier.
23. Simopoulos, A. P. (2002). Genetic variation and dietary response: Nutrigenetics/nutrigenomics. *Asia Pacific Journal of Clinical Nutrition, 11*, S117–S128.
24. Bessesen, D. H. (2006). Book review: Ultra-metabolism: The simple plan for automatic weight loss. *Obesity Management, 2*(6), 245–248.
25. Chadwick, R. (2004). Nutrigenomics, individualism and public health. *Proceedings of the Nutrition Society, 63*(1), 161–166.
26. Mathers, J. C. (2017). Nutrigenomics in the modern era. *Proceedings of the Nutrition Society, 76*(3), 265–275.
27. Paoloni-Giacobino, A., Grimble, R., & Pichard, C. (2003). Genetics and nutrition. *Clinical Nutrition, 22*(5), 429–435.
28. Trujillo, E., Davis, C., & Milner, J. (2006). Nutrigenomics, proteomics, metabolomics, and the practice of dietetics. *Journal of the American Dietetic Association, 106*(3), 403–413.
29. Busstra, M. C., Hartog, R., Kersten, S., & Muller, M. (2007). Design guidelines for the development of digital nutrigenomics learning material for heterogeneous target groups. *Advances in Physiology Education, 31*(1), 67–75.
30. Sharma, P., & Dwivedi, S. (2017). Nutrigenomics and nutrigenetics: New insight in disease prevention and cure. *Indian Journal of Clinical Biochemistry, 32*, 371–373.
31. Pavlović, M., Radlović, N., Berenji, K., Arsić, B., & Rokvić, Ž. (2020). Lactose intolerance in children and adults. *Medicinski časopis, 54*(3), 105–112.
32. Murgia, C., & Adamski, M. M. (2017). Translation of nutritional genomics into nutrition practice: The next step. *Nutrients, 9*(4), 366.
33. Fenech, M. (2013). A public health genomics approach to improving well-being based on DNA damage prevention using nutritional, life-style, environmental and psycho-social strategies. *Journal of the Australasian College of Nutritional and Environmental Medicine, 32*(1), 9–16.
34. Corella, D. (2018). *Nutrigenetics*. MDPI-Multidisciplinary Digital Publishing Institute.
35. American Heart Association. (2003). *Heart disease and stroke statistics—2004 update*. American Heart Association.
36. Samsa, G. P., Matchar, D. B., Goldstein, L., Bonito, A., Duncan, P. W., Lipscomb, J., Enarson, C., Witter, D., Venus, P., Paul, J. E., & Weinberger, M. (1998). Utilities for major stroke: Results from a survey of preferences among persons at increased risk for stroke. *American Heart Journal, 136*(4 Pt 1), 703–713. https://doi.org/10.1016/s0002-8703(98)70019-5
37. Spence, J. D. (2019). Nutrition and risk of stroke. *Nutrients, 11*(3), 647. https://doi.org/10.3390/nu11030647
38. Chiuve, S. E., Rexrode, K. M., Spiegelman, D., Logroscino, G., Manson, J. E., & Rimm, E. B. (2008). Primary prevention of stroke by healthy lifestyle. *Circulation, 118*(9), 947–954. https://doi.org/10.1161/CIRCULATIONAHA.108.781062
39. Bazzano, L. A., Serdula, M. K., & Liu, S. (2003). Dietary intake of fruits and vegetables and risk of cardiovascular disease. *Current Atherosclerosis Reports, 5*(6), 492–499. https://doi.org/10.1007/s11883-003-0040-z
40. Johnsen, S. P., Overvad, K., Stripp, C., Tjønneland, A., Husted, S. E., & Sørensen, H. T. (2003). Intake of fruit and vegetables and the risk of ischemic stroke in a cohort of Danish men and women. *The American Journal of Clinical Nutrition, 78*(1), 57–64. https://doi.org/10.1093/ajcn/78.1.57
41. Sauvaget, C., Nagano, J., Allen, N., & Kodama, K. (2003). Vegetable and fruit intake and stroke mortality in the Hiroshima/Nagasaki life span study. *Stroke, 34*(10), 2355–2360. https://doi.org/10.1161/01.STR.0000089293.29739.97
42. Steffen, L. M., Jacobs, D. R., Jr., Stevens, J., Shahar, E., Carithers, T., & Folsom, A. R. (2003). Associations of whole-grain, refined-grain, and fruit and vegetable consumption with

risks of all-cause mortality and incident coronary artery disease and ischemic stroke: The Atherosclerosis Risk in Communities (ARIC) Study. *The American Journal of Clinical Nutrition, 78*(3), 383–390. https://doi.org/10.1093/ajcn/78.3.383
43. Kubota, Y., Iso, H., Date, C., Kikuchi, S., Watanabe, Y., Wada, Y., Inaba, Y., Tamakoshi, A., & JACC Study Group. (2011). Dietary intakes of antioxidant vitamins and mortality from cardiovascular disease: The Japan Collaborative Cohort Study (JACC) study. *Stroke, 42*(6), 1665–1672. https://doi.org/10.1161/STROKEAHA.110.601526
44. Cook, N. R., Albert, C. M., Gaziano, J. M., Zaharris, E., MacFadyen, J., Danielson, E., Buring, J. E., & Manson, J. E. (2007). A randomized factorial trial of vitamins C and E and beta carotene in the secondary prevention of cardiovascular events in women: Results from the women's antioxidant cardiovascular study. *Archives of Internal Medicine, 167*(15), 1610–1618. https://doi.org/10.1001/archinte.167.15.1610
45. Verhaar, M. C., Stroes, E., & Rabelink, T. J. (2002). Folates and cardiovascular disease. *Arteriosclerosis, Thrombosis, and Vascular Biology, 22*(1), 6–13. https://doi.org/10.1161/hq0102.102190
46. Van der Schueren, B. J., Verstuyf, A., & Mathieu, C. (2012). Straight from D-Heart: Vitamin D status and cardiovascular disease. *Current Opinion in Lipidology, 23*(1), 17–23. https://doi.org/10.1097/MOL.0b013e32834d7357
47. Sun, Q., Pan, A., Hu, F. B., Manson, J. E., & Rexrode, K. M. (2012). 25-Hydroxyvitamin D levels and the risk of stroke: A prospective study and meta-analysis. *Stroke, 43*(6), 1470–1477. https://doi.org/10.1161/STROKEAHA.111.636910
48. Perry, I. J., & Beevers, D. G. (1992). Salt intake and stroke: A possible direct effect. *Journal of Human Hypertension, 6*(1), 23–25.
49. He, J., Ogden, L. G., Vupputuri, S., Bazzano, L. A., Loria, C., & Whelton, P. K. (1999). Dietary sodium intake and subsequent risk of cardiovascular disease in overweight adults. *JAMA, 282*(21), 2027–2034. https://doi.org/10.1001/jama.282.21.2027
50. Nagata, C., Takatsuka, N., Shimizu, N., & Shimizu, H. (2004). Sodium intake and risk of death from stroke in Japanese men and women. *Stroke, 35*(7), 1543–1547. https://doi.org/10.1161/01.STR.0000130425.50441.b0
51. Ascherio, A., Rimm, E. B., Hernán, M. A., Giovannucci, E. L., Kawachi, I., Stampfer, M. J., & Willett, W. C. (1998). Intake of potassium, magnesium, calcium, and fiber and risk of stroke among US men. *Circulation, 98*(12), 1198–1204. https://doi.org/10.1161/01.cir.98.12.1198
52. Khaw, K. T., & Barrett-Connor, E. (1987). Dietary potassium and stroke-associated mortality. A 12-year prospective population study. *The New England Journal of Medicine, 316*(5), 235–240. https://doi.org/10.1056/NEJM198701293160502
53. Tobian, L., Lange, J. M., Ulm, K. M., Wold, L. J., & Iwai, J. (1984). Potassium prevents death from strokes in hypertensive rats without lowering blood pressure. *Journal of Hypertension. Supplement: Official Journal of the International Society of Hypertension, 2*(3), S363–S366.
54. Kaluza, J., Wolk, A., & Larsson, S. C. (2012). Red meat consumption and risk of stroke: A meta-analysis of prospective studies. *Stroke, 43*(10), 2556–2560. https://doi.org/10.1161/STROKEAHA.112.663286
55. Larsson, S. C., Virtamo, J., & Wolk, A. (2012). Dietary fats and dietary cholesterol and risk of stroke in women. *Atherosclerosis, 221*(1), 282–286. https://doi.org/10.1016/j.atherosclerosis.2011.12.043
56. He, K., Merchant, A., Rimm, E. B., Rosner, B. A., Stampfer, M. J., Willett, W. C., & Ascherio, A. (2003). Dietary fat intake and risk of stroke in male US healthcare professionals: 14 year prospective cohort study. *BMJ (Clinical Research ed.), 327*(7418), 777–782. https://doi.org/10.1136/bmj.327.7418.777
57. Iso, H., Sato, S., Kitamura, A., Naito, Y., Shimamoto, T., & Komachi, Y. (2003). Fat and protein intakes and risk of intraparenchymal hemorrhage among middle-aged Japanese. *American Journal of Epidemiology, 157*(1), 32–39. https://doi.org/10.1093/aje/kwf166
58. Iso, H., Stampfer, M. J., Manson, J. E., Rexrode, K., Hu, F., Hennekens, C. H., Colditz, G. A., Speizer, F. E., & Willett, W. C. (2001). Prospective study of fat and protein intake and

risk of intraparenchymal hemorrhage in women. *Circulation, 103*(6), 856–863. https://doi.org/10.1161/01.cir.103.6.856
59. Yaemsiri, S., Sen, S., Tinker, L., Rosamond, W., Wassertheil-Smoller, S., & He, K. (2012). Trans fat, aspirin, and ischemic stroke in postmenopausal women. *Annals of Neurology, 72*(5), 704–715. https://doi.org/10.1002/ana.23555
60. Floßmann, E., Schulz, U. G., & Rothwell, P. M. (2004). Systematic review of methods and results of studies of the genetic epidemiology of ischemic stroke. *Stroke, 35*(1), 212–227.
61. Boehme, A. K., Esenwa, C., & Elkind, M. S. (2017). Stroke risk factors, genetics, and prevention. *Circulation Research, 120*(3), 472–495.
62. Meschia, J. F., Brown, R. D., Jr., Brott, T. G., Hardy, J., Atkinson, E. J., & O'Brien, P. C. (2001). Feasibility of an affected sibling pair study in ischemic stroke: Results of a 2-center family history registry. *Stroke, 32*(12), 2939–2941.
63. Polychronopoulos, P., Gioldasis, G., Ellul, J., Metallinos, I., Lekka, N., Paschalis, C., & Papapetropoulos, T. (2002). Family history of stroke in stroke types and subtypes. *Journal of the Neurological Sciences, 195*(2), 117–122.
64. Jerrard-Dunne, P., Cloud, G., Hassan, A., & Markus, H. S. (2003). Evaluating the genetic component of ischemic stroke subtypes: A family history study. *Stroke, 34*(6), 1364–1369.
65. Brass, L. M., Isaacsohn, J. L., Merikangas, K. R., & Robinette, C. D. (1992). A study of twins and stroke. *Stroke, 23*(2), 221–223.
66. Rubattu, S., Volpe, M., Kreutz, R., Ganten, U., Ganten, D., & Lindpaintner, K. (1996). Chromosomal mapping of quantitative trait loci contributing to stroke in a rat model of complex human disease. *Nature Genetics, 13*(4), 429–434.
67. Corella, D., & Ordovas, J. M. (2009). Nutrigenomics in cardiovascular medicine. *Circulation: Cardiovascular Genetics, 2*(6), 637–651.
68. Holliday, E. G., Traylor, M., Malik, R., Bevan, S., Falcone, G., Hopewell, J. C., Cheng, Y.-C., Cotlarciuc, I., Bis, J. C., & Boerwinkle, E. (2015). Genetic overlap between diagnostic subtypes of ischemic stroke. *Stroke, 46*(3), 615–619.
69. Ilinca, A., Samuelsson, S., Piccinelli, P., Soller, M., Kristoffersson, U., & Lindgren, A. G. (2019). A stroke gene panel for whole-exome sequencing. *European Journal of Human Genetics, 27*(2), 317–324.
70. Muqtadar, H., & Testai, F. D. (2012). Single gene disorders associated with stroke: A review and update on treatment options. *Current Treatment Options in Cardiovascular Medicine, 14*, 288–297.
71. Narayan, S., Gorman, G., Kalaria, R., Ford, G., & Chinnery, P. (2012). The minimum prevalence of CADASIL in northeast England. *Neurology, 78*(13), 1025–1027.
72. Razvi, S., Davidson, R., Bone, I., & Muir, K. (2005). The prevalence of cerebral autosomal dominant arteriopathy with subcortical infarcts and leucoencephalopathy (CADASIL) in the west of Scotland. *Journal of Neurology, Neurosurgery & Psychiatry, 76*(5), 739–741.
73. Rutten, J. W., Dauwerse, H. G., Gravesteijn, G., van Belzen, M. J., van der Grond, J., Polke, J. M., Bernal-Quiros, M., & Lesnik Oberstein, S. A. (2016). Archetypal NOTCH3 mutations frequent in public exome: Implications for CADASIL. *Annals of Clinical and Translational Neurology, 3*(11), 844–853.
74. Joutel, A., Vahedi, K., Corpechot, C., Troesch, A., Chabriat, H., Vayssière, C., Cruaud, C., Maciazek, J., Weissenbach, J., & Bousser, M.-G. (1997). Strong clustering and stereotyped nature of Notch3 mutations in CADASIL patients. *The Lancet, 350*(9090), 1511–1515.
75. Peters, N., Opherk, C., Bergmann, T., Castro, M., Herzog, J., & Dichgans, M. (2005). Spectrum of mutations in biopsy-proven CADASIL: Implications for diagnostic strategies. *Archives of Neurology, 62*(7), 1091–1094.
76. Chabriat, H., Joutel, A., Dichgans, M., Tournier-Lasserve, E., & Bousser, M.-G. (2009). Cadasil. *The Lancet Neurology, 8*(7), 643–653.
77. Guidetti, D., Casali, B., Mazzei, R., & Dotti, M. (2006). Cerebral autosomal dominant arteriopathy with subcortical infarcts and leukoencephalopathy. *Clinical and Experimental Hypertension, 28*(3–4), 271–277.

78. Chabriat, H., Vahedi, K., Bousser, M., Iba-Zizen, M., Joutel, A., Nibbio, A., Nagy, T., Lasserve, E. T., Krebs, M., & Julien, J. (1995). Clinical spectrum of CADASIL: A study of 7 families. *The Lancet, 346*(8980), 934–939.
79. Opherk, C., Peters, N., Herzog, J., Luedtke, R., & Dichgans, M. (2004). Long-term prognosis and causes of death in CADASIL: A retrospective study in 411 patients. *Brain, 127*(11), 2533–2539.
80. Razvi, S. S., & Bone, I. (2006). Single gene disorders causing ischaemic stroke. *Journal of Neurology, 253*, 685–700.
81. Bersano, A., Kraemer, M., Burlina, A., Mancuso, M., Finsterer, J., Sacco, S., Salvarani, C., Caputi, L., Chabriat, H., & Oberstein, S. L. (2021). Heritable and non-heritable uncommon causes of stroke. *Journal of Neurology, 268*, 2780–2807.
82. O'Sullivan, M., Jarosz, J., Martin, R., Deasy, N., Powell, J., & Markus, H. (2001). MRI hyperintensities of the temporal lobe and external capsule in patients with CADASIL. *Neurology, 56*(5), 628–634.
83. Joutel, A., Favrole, P., Labauge, P., Chabriat, H., Lescoat, C., Andreux, F., Domenga, V., Cécillon, M., Vahedi, K., & Ducros, A. (2001). Skin biopsy immunostaining with a Notch3 monoclonal antibody for CADASIL diagnosis. *The Lancet, 358*(9298), 2049–2051.
84. Fukutake, T. (2011). Cerebral autosomal recessive arteriopathy with subcortical infarcts and leukoencephalopathy (CARASIL): From discovery to gene identification. *Journal of Stroke and Cerebrovascular Diseases, 20*(2), 85–93.
85. Hara, K., Shiga, A., Fukutake, T., Nozaki, H., Miyashita, A., Yokoseki, A., Kawata, H., Koyama, A., Arima, K., & Takahashi, T. (2009). Association of HTRA1 mutations and familial ischemic cerebral small-vessel disease. *New England Journal of Medicine, 360*(17), 1729–1739.
86. Mendioroz, M., Fernández-Cadenas, I., del Río-Espinola, A., Rovira, A., Solé, E., Fernández-Figueras, M., García-Patos, V., Sastre-Garriga, J., Domingues-Montanari, S., & Álvarez-Sabín, J. (2010). A missense HTRA1 mutation expands CARASIL syndrome to the Caucasian population. *Neurology, 75*(22), 2033–2035.
87. Oide, T., Nakayama, H., Yanagawa, S., Ito, N., Ikeda, S. I., & Arima, K. (2008). Extensive loss of arterial medial smooth muscle cells and mural extracellular matrix in cerebral autosomal recessive arteriopathy with subcortical infarcts and leukoencephalopathy (CARASIL). *Neuropathology, 28*(2), 132–142.
88. Zheng, D., Xu, F., Gao, Y., Zhang, H., Han, S., & Bi, G. (2009). A Chinese pedigree of cerebral autosomal recessive arteriopathy with subcortical infarcts and leukoencephalopathy (CARASIL): Clinical and radiological features. *Journal of Clinical Neuroscience, 16*(6), 847–849.
89. Vinters, H. V. (1987). Cerebral amyloid angiopathy. A critical review. *Stroke, 18*(2), 311–324.
90. Viswanathan, A., & Greenberg, S. M. (2011). Cerebral amyloid angiopathy in the elderly. *Annals of Neurology, 70*(6), 871–880.
91. Greenberg, S. M. (1998). Cerebral amyloid angiopathy: Prospects for clinical diagnosis and treatment. *Neurology, 51*(3), 690–694.
92. Revesz, T., Holton, J. L., Lashley, T., Plant, G., Frangione, B., Rostagno, A., & Ghiso, J. (2009). Genetics and molecular pathogenesis of sporadic and hereditary cerebral amyloid angiopathies. *Acta Neuropathologica, 118*, 115–130.
93. Di Fede, G., Giaccone, G., & Tagliavini, F. (2013). Hereditary and sporadic beta-amyloidoses. *Frontiers in Bioscience-Landmark, 18*(4), 1202–1226.
94. Yamada, M. (2002). Risk factors for cerebral amyloid angiopathy in the elderly. *Annals of the New York Academy of Sciences, 977*(1), 37–44.
95. Knudsen, K. A., Rosand, J., Karluk, D., & Greenberg, S. M. (2001). Clinical diagnosis of cerebral amyloid angiopathy: Validation of the Boston criteria. *Neurology, 56*(4), 537–539.
96. Charidimou, A., Jaunmuktane, Z., Baron, J.-C., Burnell, M., Varlet, P., Peeters, A., Xuereb, J., Jäger, R., Brandner, S., & Werring, D. J. (2014). White matter perivascular spaces: An MRI marker in pathology-proven cerebral amyloid angiopathy? *Neurology, 82*(1), 57–62.

97. Coucke, P. J., Willaert, A., Wessels, M. W., Callewaert, B., Zoppi, N., De Backer, J., Fox, J. E., Mancini, G. M., Kambouris, M., & Gardella, R. (2006). Mutations in the facilitative glucose transporter GLUT10 alter angiogenesis and cause arterial tortuosity syndrome. *Nature Genetics, 38*(4), 452–457.
98. Richards, A., Van Den Maagdenberg, A. M., Jen, J. C., Kavanagh, D., Bertram, P., Spitzer, D., Liszewski, M. K., Barilla-LaBarca, M.-L., Terwindt, G. M., & Kasai, Y. (2007). C-terminal truncations in human 3′-5′ DNA exonuclease TREX1 cause autosomal dominant retinal vasculopathy with cerebral leukodystrophy. *Nature Genetics, 39*(9), 1068–1070.
99. Breedveld, G., De Coo, I., Lequin, M. H., Arts, W. F. M., Heutink, P., Gould, D. B., John, S. W., Oostra, B., & Mancini, G. M. (2006). Novel mutations in three families confirm a major role of COL4A1 in hereditary porencephaly. *Journal of Medical Genetics, 43*(6), 490–495.
100. Bernardes, T. P., Foresto, R. D., & Kirsztajn, G. M. (2020). Fabry disease: Genetics, pathology, and treatment. *Revista da Associação Médica Brasileira, 66*, s10–s16.
101. Garman, S. C., & Garboczi, D. N. (2004). The molecular defect leading to Fabry disease: Structure of human α-galactosidase. *Journal of Molecular Biology, 337*(2), 319–335.
102. Mehta, A., & Ginsberg, L. (2005). Natural history of the cerebrovascular complications of Fabry disease. *Acta Paediatrica, 94*, 24–27.
103. Cocozza, S., Russo, C., Pontillo, G., Pisani, A., & Brunetti, A. (2018). Neuroimaging in Fabry disease: Current knowledge and future directions. *Insights Into Imaging, 9*, 1077–1088.
104. Kato, G. J., Piel, F. B., Reid, C. D., Gaston, M. H., Ohene-Frempong, K., Krishnamurti, L., Smith, W. R., Panepinto, J. A., Weatherall, D. J., & Costa, F. F. (2018). Sickle cell disease. *Nature Reviews Disease Primers, 4*(1), 1–22.
105. Ohene-Frempong, K., Weiner, S. J., Sleeper, L. A., Miller, S. T., Embury, S., Moohr, J. W., Wethers, D. L., Pegelow, C. H., Gill, F. M., & the Cooperative Study of Sickle Cell Disease. (1998). Cerebrovascular accidents in sickle cell disease: Rates and risk factors. *Blood, The Journal of the American Society of Hematology, 91*(1), 288–294.
106. Martinez-Fernandez, E., Gil-Peralta, A., Garcia-Lozano, R., Chinchón, I., Aguilera, I., Fernández-López, O., Arenas, J., Campos, Y., & Bautista, J. (2001). Mitochondrial disease and stroke. *Stroke, 32*(11), 2507–2510.
107. El-Hattab, A. W., Adesina, A. M., Jones, J., & Scaglia, F. (2015). MELAS syndrome: Clinical manifestations, pathogenesis, and treatment options. *Molecular Genetics and Metabolism, 116*(1–2), 4–12.
108. Hirano, M., Ricci, E., Koenigsberger, M. R., Defendini, R., Pavlakis, S. G., DeVivo, D. C., DiMauro, S., & Rowland, L. P. (1992). MELAS: An original case and clinical criteria for diagnosis. *Neuromuscular Disorders, 2*(2), 125–135.
109. Yatsuga, S., Povalko, N., Nishioka, J., Katayama, K., Kakimoto, N., Matsuishi, T., Kakuma, T., Koga, Y., & Taro Matsuoka for MELAS Study Group in Japan. (2012). MELAS: A nationwide prospective cohort study of 96 patients in Japan. *Biochimica et Biophysica Acta (BBA)-General Subjects, 1820*(5), 619–624.
110. Koga, Y., Povalko, N., Nishioka, J., Katayama, K., Kakimoto, N., & Matsuishi, T. (2010). MELAS and L-arginine therapy: Pathophysiology of stroke-like episodes. *Annals of the New York Academy of Sciences, 1201*(1), 104–110.
111. Sproule, D. M., & Kaufmann, P. (2008). Mitochondrial encephalopathy, lactic acidosis, and strokelike episodes: Basic concepts, clinical phenotype, and therapeutic management of MELAS syndrome. *Annals of the New York Academy of Sciences, 1142*(1), 133–158.
112. Testai, F. D., & Gorelick, P. B. (2010). Inherited metabolic disorders and stroke part 1: Fabry disease and mitochondrial myopathy, encephalopathy, lactic acidosis, and strokelike episodes. *Archives of Neurology, 67*(1), 19–24.
113. Oderich, G. S., Panneton, J. M., Bower, T. C., Lindor, N. M., Cherry, K. J., Jr., Noel, A. A., Kalra, M., Sullivan, T., & Gloviczki, P. (2005). The spectrum, management and clinical outcome of Ehlers-Danlos syndrome type IV: A 30-year experience. *Journal of Vascular Surgery, 42*(1), 98–106.
114. Cañadas, V., Vilacosta, I., Bruna, I., & Fuster, V. (2010). Marfan syndrome. Part 1: Pathophysiology and diagnosis. *Nature Reviews Cardiology, 7*(5), 256–265.

115. Al-Mohaissen, M., Allanson, J. E., O'Connor, M. D., Veinot, J. P., Brandys, T. M., Maharajh, G., Dennie, C. J., & Beauchesne, L. M. (2012). Brachial artery occlusion in a young adult with an ACTA2 thoracic aortic aneurysm. *Vascular Medicine, 17*(5), 326–329.
116. Tangestani, H., Emamat, H., Ghalandari, H., & Shab-Bidar, S. (2020). Whole grains, dietary fibers and the human gut microbiota: A systematic review of existing literature. *Recent Patents on Food, Nutrition & Agriculture, 11*(3), 235–248.
117. Devkota, S., & Chang, E. B. (2015). Interactions between diet, bile acid metabolism, gut microbiota, and inflammatory bowel diseases. *Digestive Diseases, 33*(3), 351–356.
118. Wang, J., Zhang, H., He, J., & Xiong, X. (2022). The role of the gut microbiota in the development of ischemic stroke. *Frontiers in Immunology, 13*, 845243.
119. Yamashiro, K., Kurita, N., Urabe, T., & Hattori, N. (2021). Role of the gut microbiota in stroke pathogenesis and potential therapeutic implications. *Annals of Nutrition and Metabolism, 77*, 36–44.
120. Matsuura, J., Inoue, R., Takagi, T., Wada, S., Watanabe, A., Koizumi, T., Mukai, M., Mizuta, I., Naito, Y., & Mizuno, T. (2019). Analysis of gut microbiota in patients with cerebral autosomal dominant arteriopathy with subcortical infarcts and leukoencephalopathy (CADASIL). *Journal of Clinical Biochemistry and Nutrition, 65*(3), 240–244.
121. Ewida, A., Ahmed, R., Luo, A., Ghonim, H. T., & Anilkumar, A. C. (2021). Mitochondrial myopathy, encephalopathy, lactic acidosis and stroke-like episodes syndrome presenting with Anton-Babinski syndrome and concurrent occipital lobe seizures. *Cureus, 13*(1), e12908.
122. Arulappan, J., Karkada, S., Jayapal, S. K., & Seshan, V. (2019). Ketogenic diet-an evidence-based direction for seizure control. *International Journal of Nutrition, Pharmacology, Neurological Diseases, 9*(1), 37–40.
123. Masino, S. A., & Rho, J. M. (2019). Metabolism and epilepsy: Ketogenic diets as a homeostatic link. *Brain Research, 1703*, 26–30.
124. Cosma, J., Russo, A., Schino, S., Belli, M., Mango, R., Chiricolo, G., Martuscelli, E., & Mariano, E. G. (2022). Acute myocardial infarction in a patient with MELAS syndrome: A possible link? *Minerva Cardiology and Angiology, 71*, 374–380.
125. Paoli, A., Bianco, A., Damiani, E., & Bosco, G. (2014). Ketogenic diet in neuromuscular and neurodegenerative diseases. *BioMed Research International, 2014*, 474296.
126. Steriade, C., Andrade, D. M., Faghfoury, H., Tarnopolsky, M. A., & Tai, P. (2014). Mitochondrial encephalopathy with lactic acidosis and stroke-like episodes (MELAS) may respond to adjunctive ketogenic diet. *Pediatric Neurology, 50*(5), 498–502.

Pharmacognosy, Antioxidants, Natural Products and Stroke

13

Kamrul Hassan Suman, Saba Akram, Aroosa Faheem, and Ayesha Wadood

Abstract

Stroke is a leading cause of death and morbidity worldwide that occurs when brain blood flow is suddenly cut off, causing neural cell death. The field of pharmacognosy has a rich and extensive historical background in the realm of stroke prevention and therapy. Oxidative stress is a major contributor to the onset of chronic and acute brain diseases such as hemorrhagic and ischemic strokes. Antioxidants, both endogenous and exogenous, have anti-inflammatory, and neuroprotective properties that either prevent the generation of RNS (reactive nitrogen species) and ROS (reactive oxygen species) and or destroy them, hence safeguarding the brain against damage caused by free radicals. Natural substances rich in curcuminoids, carotenoids, polyphenols, and flavonoids found in vegetables, fruits, animals, and medicinal plants are gaining popularity for their potential to mitigate stroke damage. Despite the enormous potential of pharmacognosy, antioxidants, and natural compounds in treating stroke, it encounters a number of obstacles, such as a lack of extensive clinical studies and difficulties with standardisation, optimal dosage determination, and safety. Combining Ayurveda, TCM (Traditional Chinese Medicine) and Western pharmacology with cutting-edge tools like nanotechnology, AI, and multiomics to find new bioactive compounds could aid stroke patients in getting better care.

K. H. Suman (✉)
Molecular Biology and Interdisciplinary Life Sciences, New Mexico State University, New Mexico, USA
e-mail: khsuman@nmsu.edu

S. Akram · A. Wadood
Nuclear Institute for Agriculture and Biology College, Pakistan Institute of Engineering and Applied Sciences (NIAB-C, PIEAS), Faisalabad, Pakistan

A. Faheem
Department of Biochemistry, Quaid-i-Azam University, Islamabad, Pakistan

W. Mohamed, M. B. Mat-Nor (eds.), *Nutrition and Stroke*, Nutritional Neurosciences, https://doi.org/10.1007/978-981-97-6818-9_13

Keywords

Stroke · Oxidative stress · Free radical · Ayurveda · Traditional Chinese medicine · Antioxidants

13.1 Introduction

Stroke is a major neurological disorder marked by the occlusion or rupture of cerebral blood vessels, resulting in the impairment or cessation of blood supply to the brain [1]. It is the second leading cause of mortality worldwide and is typically classified as hemorrhagic or ischemic [2]. Ischemic strokes account for about 71% of all strokes, and their prevalence is reported to be as high as 85% in developed nations [3]. Ischemic stroke has a high fatality rate, hence exerting an adverse influence on the well-being of individuals and imposing substantial psychological and economic stress on their families [4]. The designated primary treatment for stroke is recombinant tissue plasminogen activator (rtPA), which should be injected intravenously within 4.5 h of an ischemic stroke recurrence. The limited time frame for therapeutic intervention leads to a small percentage (2–5%) of stroke patients obtaining this treatment, and within that group, only 50% achieve effective reperfusion of the brain [5]. Consequently, this disease presents significant clinical challenges that have yet to be adequately addressed. While the diagnosis and therapy of stroke pose significant difficulties, stroke-related biomarkers promise a vital role in the identification and prediction of stroke. The study of oxidative stress and inflammatory biomarkers in the saliva of stroke patients has attracted a great deal of interest among researchers [6]. Previous research has provided preliminary evidence supporting the potential use of oxidative stress as a viable therapeutic target and antioxidant intervention for the treatment of ischemic stroke [4]. Pathological conditions give rise to oxidative stress due to an inequilibrium between the generation and removal of free radicals. This radical imbalance is typically attributed to ROS and RNS in biological processes [7]. The consequences of an imbalance in ROS in stroke are significant and involve a range of pathophysiological processes, including inflammation, apoptosis, impairment of the blood-brain barrier (BBB), edoema development, and autophagy [8]. Therefore, oxidative stress is a key factor in the development of chronic and acute brain illnesses like traumatic brain injury, stroke, and neurological disorders [7]. The cell employs numerous antioxidant pathways either to avoid the formation of ROS or to eliminate them. Glutathione (GSH), catalase (CAT), superoxide dismutase (SOD), glutathione peroxidase (GSH-Px), heme oxygenase (HO), glutathione reductase (GR), and vitamins C and E are remarkable antioxidant molecules that can be synthesised endogenously or consumed from exogenous (nutritional) sources [9]. The consumption of antioxidants is associated with a decreased risk of stroke. Numerous epidemiological studies have provided evidence that dietary interventions aimed at modifying stroke risk factors, specifically hypertension, through increased amounts of fruit and vegetable consumption are associated with a decrease in the occurrence of strokes [10].

Traditional medicine has a significant role in stroke treatment as it promotes the production of antioxidant and nitration compounds while also exhibiting anti-inflammatory, anti-excitotoxic, and anti-apoptotic properties [1]. Due to the intricate nature of the various components, the task of devising viable medications using the "one drug, one target" approach proves exceedingly challenging, leading to detrimental consequences associated with the treatment and prevention of stroke [11]. TCM has been adopted to treat people with stroke-related symptoms and signs in China and East Asia for centuries. Recent progress in network pharmacology and multiomic has improved our understanding of the cellular targets and circuit controls in both conventional pharmaceuticals and TCM preparations. These advancements have led to the development of novel pharmacological substances and greater insights into TCM formulations for ischemic brain injury [12]. There is rising interest in the neuroprotective properties of natural compounds such as curcuminoids, lignans, stilbenoids, phenolic acids, and flavonoids derived from fruits, vegetables, animals, and medicinal plants. Their antioxidant properties have garnered significant attention, particularly for their capacity to lessen ischemic stroke damage [13].

This chapter provides an in-depth exploration of the intricate connection among pharmacognosy, antioxidants, natural compounds, and stroke. This study examines various aspects of stroke care, specifically highlighting the use of traditional medicinal practices like Ayurveda and TCM. We also summarise the beneficial impact of various natural and synthetic antioxidants that inhibition effects on ROS generation and oxidative stress during and after stroke incidence. Finally, we address existing challenges in the use of natural products, including standardisation, dosage, and safety in stroke, and propose future directions for the development of policies and medicinal products to improve the management of stroke.

13.2 Pharmacognosy in Stroke

Pharmacognosy, originating from the Greek words "pharmakon" (drug) and "gnosis" (knowledge), is the oldest scientific field and studies drugs derived from plants and animals. The procedure includes the verification of identity, quality assurance, and macroscopic and microscopic analysis of raw pharmaceutical components. The term "Pharmacognosy" was initially proposed by Seydler in 1815 in a publication titled "Analecta Pharmacognostica" [14]. According to the American Society of Pharmacognosy, pharmacognosy is the study of the physical, chemical, biochemical, and biological aspects of medications, potential drugs, and natural substances and the quest for new natural drugs. Modern pharmacognosy investigates plant, bacterial, fungal, and marine substances [14]. Western researchers classified hundreds of therapeutic plants during the golden age of Western herbology, 500 BC–200 AD. In mediaeval times, every home had an herb garden for cures. Herbology was losing popularity in the West by WWII with the rise of the pharmaceutical industry and its synthetic medications [15]. Even today, aspirin, atropine, belladonna, capsaicin, ephedrine, ergotamine, opium, and vinblastine are among the most common drugs with plant-based active ingredients. According to the WHO, the utilisation of herbal medicine for primary healthcare by approximately 80% of the global population highlights the considerable significance and benefits associated with these natural remedies [15]. Treatment for ischemic stroke involves the use of thrombolytics, anticoagulants, and antiplatelet medications. Aspirin and clopidogrel are utilized as first-line medications for the prevention of secondary strokes [16]. A variety of alternative medical practices, such as Ayurveda, TCM, and indigenous medicine, have been successfully treating stroke and related illnesses for centuries.

13.2.1 Ayurveda and Stroke Management

Ayurveda has had widespread popularity as a traditional medicine for centuries. The long-term viability of this system can be attributed to its profound comprehension of nature-based medicine, the interconnectedness between the human body and nature, and the influence of cosmic phenomena on living organisms [17]. Ayurvedic formulations may use metallic substances for their medicinal properties, referred to as rasausadhies. It is advisable to employ these metals subsequent to their purification, utilising minimal quantities, employing designated vehicles, and adhering to dietary, activity, and environmental recommendations [18]. Different researches suggested the use of ayurvedic treatment can cause a relief in stroke. Brahmi (*Bacopa monniera*) has been used for years as an anti-stroke agent. The plus maze test demonstrates that Brahmi reduces the size of infarcts in the ischemic brain and treats memory loss. In rodents injured by ischemia, Brahmi improved both muscular coordination and catalase activity. Nitrite, nitrate, and lipid peroxidation were also significantly decreased. Brahmi may protect the brain from ischemia, but further investigation is required [19]. Similarly, ashwagandha (*Withania somnifera*) and arjuna (*Terminalia arjuna*) have been used in alternative medicine for millennia for their comparable aphrodisiac, nerve-tonic, anti-inflammatory, and anticancer effects [20]. The concept of polyherbalism finds its roots in ancient Ayurvedic books such as the Caraka Samhita, Susrutha Samhitha, and Ashtanga Hrudaya [21]. Most ayurvedic formulations commonly incorporate many herbs due to the insufficiency of active phytochemical components within a single plant to achieve the intended therapeutic effects [22]. For the stroke treatment numerous ayurvedic therapies have been examined for their efficiency. Panchakarma is a widely used ayurvedic therapy for the treatment of stroke, encompassing a series of five processes aimed at purification and rejuvenation. Patients who have undergone Panchakarma therapy have shown significant improvements in motor function, cognitive function, and daily living activities after their stroke [23]. A study proposed the use of ayurvedic interventions as a potential therapeutic approach for managing Pakshaghata, a specific form of cerebral stroke. Abhyanga, Nadi Swedana, Nasya, Karpuradi oil, Maharasnadi Kwath, and Palsinuron pills were all part of the treatment plan that brought complete recovery to a 48-year-old patient who had suffered a stroke [24].

13.2.2 Traditional Chinese Medicine

Traditional Chinese Medicine (TCM) refers to a comprehensive system of medical knowledge that originated in China. The idea of "harmony between man and nature" is central to this discipline, and as such, it incorporates the processing of visual data, cognitive activity from the perspective of traits and interactions, the use of both internal and external natural resources, and the maintenance of homeostasis and homeostatic regulation [25]. TCM utilises several philosophical concepts, including "qi," "yin and yang," and the "five elements," in order to understand the internal organs [26]. TCM categorises anti-stroke medications into four groups, each based on their primary mechanism of action. Patients suffering from "channel collateral stroke" are typically prescribed the first class of drugs. Drugs closest to TCM philosophy are the most difficult to obtain for "modern" pharmaceutical study. The second group consists of anti-inflammatory drugs, such as aspirin, which has been shown to provide neuroprotective effects in mice against neurodegeneration produced by glutamate. TCM anti-rheumatic drugs might contain newly discovered chemicals that can block NF-B or other transcription factors linked to excitotoxicity, in a way that is similar to how aspirin works [27]. The third group is made up of blood-thinning drugs that have been shown to improve blood flow and control "Qi" plants, like *Rhizoma ligustici* and *Salviae miltiorrhizae radix*. These medications reduce arteriolar spasm, quicken blood flow, and augment blood volume; they also prevent further thrombosis from forming and dissolve existing thrombuses [27]. The fourth type, which is the least explored area of pharmacological TCM research, is neuroprotective drugs for stroke symptoms such as spasm, high fever, dizziness, and headaches. While only a small number of TCM compounds have been shown to have neuroprotective properties, some medications, such as ginsenosides, may provide promising leads for the isolation of neuroprotective molecules [28].

Rehabilitation and medical care play a key role in the TCM theory of disease prevention, diagnosis, and treatment. TCM relies heavily on biological and chemical products in addition to herbal, animal, mineral, and chemical therapies [29]. The therapy modalities of acupuncture, moxibustion, and tuina massage are encompassed within the framework of TCM [25]. TCM has been recognised as a prospective therapeutic approach for the management of stroke. Several therapeutic methods, including acupuncture and herbal medications, illustrate the comprehensive goals of rehabilitation based on Chinese medicine [30]. Chinese herbal medicine (CHM) is considered effective in treating stroke due to its diverse properties, such as antioxidant, anti-inflammatory, anti-apoptotic, neuroprotective, and vascular protective effects [31]. According to several studies, acupuncture may help stroke survivors with their dependence, general neurological deficits, and specific neurological impairments while they are recovering without having any negative effects [32]. The Xiaoxuming decoction (XXMD), a breakthrough in Chinese medicine, was used to treat schemic stroke. As the first-line wind-dispelling remedy recorded in Qianjin Yaofang Zhufeng, it was considered the most important recipe among effective therapies [33]. Stroke survivors who get acupuncture report a decrease in their neurological impairments. Using these methods for stroke management can help bridge the gap between traditional and Western medicine.

TCM provides promising outcomes for the treatment and prevention of stroke, especially when systematically integrated with modern digital screening techniques [34]. It provides new possibilities for exploring herbal medicine compounds with multiple mechanisms of action that can prevent and treat strokes. Data mining has emerged as a robust method for doing searches on Traditional Chinese Medicine (TCM) formulas. This technology utilises methodologies such as association rules, cluster analysis, sorting algorithms, and various other techniques [35]. Network pharmacology is an emerging discipline within bioinformatics that employs computational methods to search through digital chemical libraries in order to identify drugs with the ability to target multiple biological targets simultaneously. TCM has evolved this theory over millennia to treat complex illnesses. Virtual digital screening employs molecular docking, pharmacophores, and other screening models to rapidly investigate vast libraries of compounds in order to identify the key substance [36]. TCM drugs have been shown to be effective in the treatment of stroke with minimal side effects. International researchers working with Chinese scientists and physicians can benefit from combining Chinese TCM knowledge and experience with Western pharmacology in basic and clinical research towards better stroke therapy.

13.2.3 Antioxidants as Therapeutic Neuroprotectants for Stroke

The primary strategy for addressing brain damage is prevention, as the process of repairing brain damage is difficult. Stroke risk can be minimised by controlling blood pressure [37] or atherosclerosis through the consumption of antioxidants like flavonoids and vitamins [38]; however, inconsistent findings have been reported. Stroke causes both rapid and gradual neuronal death, and oxidative stress plays a role in both of these outcomes through necrosis and apoptosis. Thus, numerous antioxidant substances exhibiting diverse chemical compositions have been extensively investigated as possible mediators for safeguarding the brain against damage [10]. The effectiveness of both natural and synthetic antioxidants has been evaluated through preclinical and clinical research. Antioxidant strategies can be categorised into two approaches: augmenting endogenous antioxidants and administering exogenous antioxidants.

13.2.3.1 Endogenous Antioxidants

The nuclear factor erythroid 2-related factor 2 (Nrf2) serves as a crucial regulator of the body's internal antioxidant activity. The redox-sensitive transcription factor activates gene promoters comprising the antioxidant response element (ARE), which are responsible for detoxifying ROS [39]. The innate antioxidative defence mechanism consists of antioxidant enzymes, such as SOD, CAT, and GSH-Px, as well as nonenzymatic antioxidants such as thiol antioxidants (glutathione and α-lipoic acid), uric acid, and coenzyme Q10 (CoQ10). Together, they serve to safeguard against the deleterious consequences of oxidative stress [40].

CAT: The enzyme, predominantly found in peroxisomes, safeguards cells by reducing ROS through the breakdown of H_2O_2 into H_2O and O_2, facilitated by a cofactor consisting of Fe or Mn [41].

GSH-Px: The enzyme, mostly located in the mitochondrial matrix with a smaller presence in the cytoplasm, functions to decrease levels of ROS via the transformation of H_2O_2 into H_2O while simultaneously oxidising GSH to generate glutathione disulfide (GSSG) and H_2O [42].

CoQ10: Coenzyme Q10, or ubiquinol, is ubiquitously found in cellular membranes and exerts inhibitory effects on lipid peroxidation. It is also essential for reprocessing and regenerating antioxidants like vitamin E and C [43].

GSH: GSH, the predominant water-soluble endogenous antioxidant, possesses the ability to efficiently counteract ROS and holds a pivotal role in the metabolic process of xenobiotics [44].

Uric Acid: It is a strong antioxidant that is produced as a byproduct of the metabolism of purine nucleotides and neutralises a wide variety of free radicals such as superoxide ($O_2^{\bullet-}$), OH^-, $HClO^-$, NO_3^-, and lipid peroxides [40].

13.2.3.2 Exogenous Antioxidants

The human body is incapable of producing the complex antioxidants found in a variety of foods, such as vitamins C, E, and A, carotenoids, and polyphenols. By giving electrons to radicals, these antioxidants stop chain reactions from happening, leaving behind safe byproducts [45]. Carotenes, including α-carotene, lycopene, and β-carotene, as well as xanthophylls such as fucoxanthin, lutein, astaxanthin, zeaxanthin, capsanthin, and canthaxanthin, are classified as carotenoids. Carrots, sweet potatoes, papayas, mangos, tomatoes, and oranges are only few of the many popular natural crops that contain these ingredients [46]. Carotenes are powerful antioxidants and neuroprotectants due to their capacity to neutralise free radicals and absorb singlet oxygen. Consequently, they play a pivotal role in impeding lipid oxidation and suppressing apoptosis by mitigating oxidative stress [47]. Ascorbic acid transforms free radicals into stable molecules within the cytoplasm, and the lipid-soluble vitamins E in the form of tocopherols and tocotrienols safeguard membrane lipids by stopping oxidative and inflammatory lipid degradation [40]. By scavenging or binding free radicals and boosting the activity of the body's own endogenous antioxidants, polyphenols neutralise a wide range of free radicals [48]. In the aged and ischemic brain, anthocyanins have considerable anti-oxidative stress-protective capabilities [10], including the highest level of permeation through the BBB [49]. The blueberries' anthocyanin concentration, which is 25 times higher than that of strawberries and raspberries, demonstrates their superior impact on brain oxidative stress [49]. Resveratrol, a polyphenol compound, has demonstrated efficacy in mitigating the danger of cardiovascular disease and averting the development of atherosclerosis. It has strong neuroprotective benefits against ischemia/reperfusion by reducing thrombosis [50].

13.2.3.3 Synthetic Antioxidants

Researchers have successfully developed artificial antioxidants, altered natural antioxidants, and combined them with other potent chemicals to make them more effective than natural antioxidants in terms of fighting free radicals, body absorption, and resistance to metabolism breakdown [51]. Synthetic antioxidant compounds have shown promise in recent investigations for their potential to combat oxidative stress and various pathways in neurodegenerative disorders. A decrease in infarct volume in both an in vitro hypoxia and an in vivo stroke model was observed when the drug 5-amino-1-phenyl-1H-pyrazole-4-carbonitrile (APPC) was combined with lipoic acid, referred to as UPEI-800 [52]. In a case study of stroke, the combination of ALA and 3-n-butylphthalide (NBP) called Dlx-23 was found to protect neurons from dying by stopping mitochondrial degeneration and restoring redox balance [53]. Synthetic edaravone demonstrates neuroprotective and antioxidant properties, effectively delaying the progression of diseases such as ischemic strokes and amyotrophic lateral sclerosis (ALS) by mitigating lipid peroxidation and minimising cell membrane damage caused by oxidative stress [54]. Table 13.1 provides a comprehensive overview of the most commonly used antioxidants in the treatment of stroke.

13.2.4 Antioxidant Mechanisms for Stroke Oxidative Stress Regulation

In physiological homeostasis, the brain's different physiological processes consistently produce ROS, which are neutralised by antioxidant systems within the brain. However, ROS generation after a stroke exceeds the capabilities of the body's endogenous antioxidant defences [7]. The following are the basic mechanisms by which antioxidants function:

13.2.4.1 Prevention of Free Radical Generation

This strategy involves the targeted inhibition of enzymes responsible for producing ROS, which are implicated in the promotion of illness. NADPH oxidases (NOXs) play a significant role in the creation of ROS during cerebral ischaemia or reperfusion injury. Studies using rat models of experimental stroke have shown that the administration of apocynin can effectively reduce cerebral ischaemia [65], indicating the vital role of NOX in neurological damage. Since apocynin functions as a Rho kinase inhibitor, it's likely to serve as a nonspecific antioxidant by blocking membrane migration of the p47phox component and NOX group assembly [66]. With pharmacological suppression of NOX using VAS2870, reduction of infarct size, oxidative damage, neuronal apoptosis, and BBB permeability is achieved [67]. Proteolytic conversion of xanthine dehydrogenase to xanthine oxidase (XO) occurs during ischemia. Ischemia also causes ATP to be lost inside the cell, which leads to more hypoxanthine and xanthine, which are both used by the enzyme XO to make ROS [68]. It has been reported that both allopurinol and its metabolite, oxypurinol, can scavenge hydroxyl radicals [69]. The enzyme myeloperoxidase (MPO) plays a crucial role in the manifestation of neurological inflammation and oxidative stress

Table 13.1 Summary of the most prevalent antioxidant-containing substances used to treat stroke

Antioxidant	Compound type	Stroke type	Outcome	Reference
Sulforaphane	Natural	Ischemic, Hemorrhagic	NOX inhibition, HO-1 ↑	[55]
Quercetin	Natural	Ischemic	Infarct size reduction, MPO ↓, SOD ↑, CAT ↑	[56]
Resveratrol	Natural	Hemorrhagic, Ischemic	Decrease infract volume and neuronal damage, MMPs ↓	[57]
Mitoquinone	Synthetic	Ischemic	Restoration of complex I function and oxygen utilisation	[7]
ML090, ML171	Synthetic	Ischemic	Suppression of NOX1, NOX4, and NOX5	[58]
Epigallocatechin gallate (EGCG)	Natural	Ischemic	MMP-9 ↓, MMP-2 ↓	[59]
Dauricine	Natural	Ischemic	MPO ↓	[7]
Allopurinol	Synthetic	Ischemic	Decreased stroke volume and brain edoema	[7]
Melatonin	Natural	Hemorrhagic, Ischemic	Improvement of brain edema, apoptosis, neurological impairments, and BBB dysfunction	[60]
Edaravone	Synthetic	Ischemic	Boost in functional results and suppression of lipid peroxidation	[61]
Neuroglobin	Natural	Ischemic	Better neurological results, less oxidative stress due to hypoxia	[7]
Pyrroloquinoline quinone (PQQ)	Natural	Hemorrhagic	Retardation of progression of cerebral edoema, ROS ↓	[62]
Trigonelline	Natural	Ischemic	Lower risk of infarction, suppression of MPO, and GSH	[63]
Vitamin E (α-tocopherol)	Natural	Hemorrhagic, Ischemic	Suppress ROS chain reaction	[64]

HO-1–heme oxygenase 1, ↑–increase, ↓–decrease, MMPs–matrix metalloproteinases

in many diseases, including stroke. Reduced brain lust and enhanced neurological results can be achieved by inhibiting MPO using substances such as curcumin, quercetin, resveratrol, eriodictyol, trigonelline, and dauricine [56].

13.2.4.2 Free Radical Scavengers

Compounds possessing the capacity to eliminate free radicals have been synthesised for alleviating cerebral ischemic stroke. Nevertheless, getting these molecules from the laboratory into human therapeutic trials has gone mainly unsatisfactorily. U-74006F (tirilazad mesylate) is a lipid peroxidation inhibitor that is now in clinical trials. Substantial pre-clinical research began on this chemical in the mid-1990s, and the results showed that it decreased the size of infarcts in rats that had suffered from transitory focal ischaemia but had not undergone permanent occlusion [70]. Another medicine that showed promise in the lab but failed to deliver in the clinic is NXY-059, a spin trap compound. Numerous pre-clinical investigations have substantiated the neuroprotective properties of NXY-059, in diminishing the size of infarcts and facilitating neurological recuperation in various stroke models, encompassing both rodents and primates other than humans [71]. Since 2001, Edaravone, a compound with antioxidant properties, has received approval for use in Japan. In a study of both acute and chronic cerebral ischaemia, edaravone decreased the size of infarcts in patients with small-vessel occlusion strokes within a year. However, the resulting neurological improvements faded rapidly after a year [72].

13.2.4.3 Free Radical Degradation

ROS play a critical role in the development of lesions, as shown in animal experimental models of stroke by therapies targeted at reducing oxidative stress by augmenting SOD, an antioxidant enzyme. SOD reduces the reactivity of $O_2^{\bullet -}$ by splitting it into H_2O_2 and O_2 via catalysis, whereas CAT and GPx aid in the removal of H_2O_2 byproducts, thereby enhancing the general performance of SOD [73]. Multiple investigations in rat models of localised ischaemia have found that normobaric oxygen (NBO) decreases the volume of infarction and neurological impairments and increases the window of effective reperfusion [74]. In addition, previous studies have demonstrated that the administration of both NBO and ethanol in combination has neuroprotective abilities following ischemia/reperfusion (I/R) injury in rats [75].

13.3 Natural Products in Stroke Cure

The potential of using natural products for stroke treatment and management has been extensively investigated. These products come from a variety of sources, including plants, animals, fruits, and herbs.

13.3.1 Ginkgo

Stroke prevention is aided by the anti-inflammatory, neuroprotective, and anticancer characteristics discovered in the plant's leaf extract of the ginkgo tree (*Ginkgo biloba*). Flavonoids and terpene trilactones, the primary active components in the extracts, regulate nuclear factor-B signalling to influence immunological activity. The terpene trilactones section contains the potent platelet activating factor inhibitors bilobalide and seven ginkgolides (A, B, C, J, K, L, and M). Flavonoids and terpene trilactones regulate nuclear factor-β signalling to influence immunological activity. One of the antioxidant functions of leaf extract is to stop lipid peroxidation by getting rid of ROS [76]. Ginkgo extract contains flavonoids that guard against oxidative stress, which is associated with damage to the heart and brain from ischemia/reperfusion (I/R). Lipid peroxidation can be slowed down after I/R damage by turning on Nrf2 and making more heme oxygenase-1 (HO-1). New research shows that ischemia preconditioning (IPC) can help protect against tissue infarction and keep the heart and brain working properly by turning on Nrf2 [77].

13.3.2 Ginseng

Active chemical components of the medicinal herb ginseng (*Panax ginseng),* known as panaxosides (also called ginsenosides), have been employed for centuries in East Asian traditional medicine [78]. The positive effects of ginseng extract on atherosclerosis patients suggest that it may be useful in delaying the onset of an ischemic stroke in patients with severe atherosclerosis [79]. According to the literature, major and minor ginsenosides like Rg3, Rg5, and Rh2 may have a neuroprotective effect, primarily by lowering excitotoxicity, Ca^{2+} excess, mitochondrial dysfunction, and inflammation. By increasing BBB permeability, it can also promote angiogenesis and neurogenesis and have anti-apoptotic, anti-oxidative, anti-pyroptotic, and anti-autophagic effects [80].

13.3.3 Brahmi

Brahmi (*Bacopa monnieri*) is associated with numerous health benefits, particularly in relation to mental and cognitive well-being, which can be attributed to its primary active ingredients, including bacosides (A and B), bacopasides I and II, betulinic acid, bacogenin, quercetin, ebelin lactone, stigmasterol, asiatic acid, -sitosterol, and loliolide. They influence neurotransmission, neurogenesis, the neuroendocrine system, and inflammatory processes [81]. Ebelin lactone had the strongest binding energy, the ability to cross the blood-brain barrier, and the strongest binding affinity for M1 and 5-HT2A, which are receptors that play a role in cognition. Bacopasides A and X increased the binding of D1 receptors to M1 and 5-hydroxytryptamine receptors, which are involved in memory and thinking [82]. In a different study, brahmi extract lessens the cognitive decline caused by trimethyltin (TMT) in mice. It does this by protecting hippocampal neurons from TMT damage and promoting neurodegeneration in the dentate gyrus. It protects the brain from aluminium, glutamate, and nitric oxide [83], suggesting that brahmi could be useful for the prevention and treatment of stroke.

13.3.4 Blueberries

According to [84], berries are considered a superfood due to their high content of active phytochemicals, particularly polyphenols. Blueberries have been recognised as potential functional foods and nutraceutical ingredients due to their ability to inhibit the activity of NADPH oxidase through their circulating polyphenolic metabolites. This inhibition leads to an improvement in nitric oxide bioavailability and endothelial-dependent vasodilation, which has positive effects on cardiovascular wellness [85].

13.3.5 Citrus Fruits

Stroke risk reduction has been linked to eating citrus fruits, including oranges, lemons, and grapefruit, which are rich in vitamin C and flavonoids. Eriodyctiol (ERD), a naturally occurring flavonoid, has a diverse range of pharmacological properties and is abundantly present in fruits, vegetables, and medicinal herbs. Reports suggest that ERD's vascular and neuroprotective activities contribute to maintaining homeostasis in living things [86].

13.3.6 Pomegranate

Pomegranate, when incorporated into a healthy diet, may help prevent cardiovascular problems like heart attacks and strokes. This is due to the fact that pomegranates have been shown to improve heart function by thinning blood, increasing cardiac blood flow, reducing blood pressure, clearing artery plaque, and lowering bad cholesterol while elevating good cholesterol [87]. A study conducted by Li et al. [77] found that pomegranate supplements improved neuroplasticity and repair after an ischemic stroke. These improvements were especially obvious in memory, focus, and neurological function. Fisetin, a pomegranate flavonoid, activates the Nrf2-ARE pathway, a vital cellular defence against oxidative damage. After fisetin treatment, ROS levels dropped and antioxidant enzyme levels rose. Fisetin's ability to improve brain cell antioxidant capacity via the Nrf2-ARE pathway may contribute to its neuroprotective qualities.

13.3.7 Krill Oil

As the primary food source for the vast majority of marine organisms, Antarctic Ocean krill (*Euphausia superba*) plays an important role in marine ecosystems [88]. It contains easily digestible ω-3 polyunsaturated fatty acids, including eicosapentaenoic acid (EPA) and docosahexaenoic acid (DHA). Clinical investigations have demonstrated significant metabolic, vascular, anti-inflammatory, neuroprotective, and chemopreventive advantages associated with the consumption of krill oil [89].

13.3.7.1 Bee Propolis

Propolis has garnered attention as a promising natural product due to its several advantageous characteristics, rendering it a viable option for consideration [90]. According to research, propolis's antioxidant, anti-inflammatory, immunomodulatory, antihypertensive, antiangiogenic, and antiatherosclerotic qualities are usually linked to its active ingredients especially its phenolic components, including quercetin, chrysin, luteolin, and pinocembrin [91]. Propolis' phenolic chemicals are linked to its antioxidant properties, which protect multiple tissues from IR injury (including the liver, heart, colon, and brain) by reducing cyclooxygenase, ROS, and nitric oxide (NO) productions [92].

13.4 Natural Products in Stroke Prevention

Natural products contain several active chemicals that exhibit a wide range of biological actions, including reducing the risk of strokes. Natural products, such as food and dietary supplements, may serve a crucial role in the prevention of strokes. The following studies demonstrate the function of natural products in the prevention of stroke.

13.4.1 Dietary Fiber

Data taken from various cohort studies suggests reduced risk of stroke with increased dietary fiber intake. These studies suggest that there is a significant decrease in stroke risk when dietary fibre intake exceeds 25 g per day. A daily increase of 4 g of soluble fibre reduces the risk by up to 6% [93]. Larsson et al. [94] also reported an important correlation between the consumption of insoluble fibre and a reduced risk of total stroke. Cereals, fruits and vegetables are a good source of dietary fibers. The cohort studies suggested that cereal fiber intake has a relative risk (RR) of 0.66 for total stroke, RR 0.51 for hemorrhagic stroke and RR 0.80 for ischemic stroke [95]. Another study found a link between stroke and fruit and vegetable fibre with RRs of 0.87 and 0.92, respectively. To reduce the incidence of hemorrhagic and ischemic stroke, it is recommended that women and men consume 21–25 g and 30–38 g of fibre daily [96].

13.4.2 Turmeric

Metabolic syndrome, which affects 20–25% of adult population worldwide, significantly increases stroke risk. The curcumin in turmeric has been demonstrated to effectively regulate metabolic syndrome [97]. According to Ray et al. [98], regular turmeric has about 2–4% curcumin, and lakadong turmeric possesses about 7–10% curcumin. High levels of cholesterol are a major cause of stroke, curcuminoids decreases the bad cholesterol (LDL) and increases vitamin E levels which is an antioxidant and neuroprotective [99].

13.4.3 ω-3 Fatty Acids

EPA and DHA are two prominent instances of ω-3 polyunsaturated fatty acids, which exhibit potent anti-inflammatory characteristics, impede platelet aggregation, and reduce the size of atherosclerotic plaques [100]. The consumption of fish and ω-3 fatty acids has been associated with a potential decrease in the occurrence of strokes and mortality rates. According to the meta-analysis conducted by Larsson et al. [94], women and adults with body mass indexes (BMIs) below 24 kg/m^2 had a reduced risk of stroke when incorporating fish into their diet. Lean fish provide fewer omega-3 fatty acids than fatty fish, yet a recent meta-analysis indicated that both types of fish reduce stroke risk [101].

13.4.4 Dark Chocolate (Cocoa)

The flavonoids found in cocoa are beneficial in lowering the incidence of heart disease and stroke [102]. Depending on the dose, cocoa decreased flow-mediated dilation, pulse wave velocity, endothelin-1, and blood pressure tracking. The results show that cocoa may be a good food to include in our diet to lower the risk of heart disease and stroke because it has few calories [102]. Regular consumption of cocoa flavanol reduces uric acid and trimethylamine N-oxide (TMAO) levels and increases flow-mediated dilation, thereby enhancing cardiovascular health. Similarly, these effects are associated with blood polyphenol levels. TMAO concentrations were observed to be negatively correlated with polyphenol levels following the intervention [103]. The suggested daily intake of chocolate differs based on the type, dietary plans, and other individual targets. Dark chocolate with a cocoa content of 70% or higher is widely regarded as being more healthful compared to milk chocolate. The potential health advantages of this substance may be attributed to its antioxidant characteristics. It is advisable to incorporate a small portion of dark chocolate, weighing approximately 28 g, into one's daily meals [104].

13.4.5 Green Tea

Green tea (*Camellia sinensis*) contains epigallocatechin-3-gallate (EGCG), which stimulates the formation of neural stem cells and progenitor cells as well as brain growth and repair. It prevents pyroptosis by polarising microglia to M2 and inhibiting NOD-like Receptor Protein-3 (NLRP3), thereby inhibiting the pyroptotic process. Thus, EGCG may have neuroprotective properties. These two methods provide assistance to individuals who have experienced ischemic stroke [105]. Green tea's high levels of catechin derivatives, such as epigallocatechin, gallocatechin, epicatechin gallate, epicatechin, and epigallocatechin-3-gallate, make it an energising beverage. Green tea may be beneficial for treating stroke due to its antioxidants, anti-inflammatory properties, anti-apoptotic properties, and neuroprotective properties [106].

13.5 Issues and Limitations

Herbal medications and dietary supplements have garnered interest as a potential adjunct treatment for stroke. The utilisation of natural products for stroke treatment poses several challenges, such as the need for regulation, appropriate dosage determination, and ensuring safety. Thus, ensuring the quality, efficacy, and safety of herbal treatments is crucial in both developed and developing countries [107]. Herbal medicines must be standardised to ensure quality due to natural variability. The necessity for standardisation in this field is further highlighted by the fact that the biochemical diversity of herbal products is in part attributable to the large variation in secondary metabolites among plant components such as stems, roots, barks, and leaves [108]. In addition, conducting feasibility studies presents numerous challenges. Placebo-controlled studies on herbal medicine pose ethical challenges due to the complexities involved in replicating sensory attributes, potential conflicts with holistic therapeutic ideologies, and challenges in determining optimal dosages based on a wide range of literature sources. Thus, therapeutic outcomes may vary when formulations fail quality control [109]. The phytochemical makeup of herbal products is intricate, posing challenges for quality control in order to guarantee safety, effectiveness, and consistency across different batches [110]. Furthermore, a number of issues need to be resolved before natural therapies for the treatment of ischemic stroke may move from the realm of animal research into human trials. To further boost the likelihood of clinical translation, research should be undertaken on older animals rather than young adult animals. Research into the precise mechanisms by which herbal remedies mediate their neuroprotective advantages against stroke should be conducted [111]. There is no universal consensus on how to regulate dietary supplements and herbal medicines [111], but regulatory authorities can play an important role in addressing the issue of uniformity. Compliance with good manufacturing practices and good agricultural and extraction practices can be facilitated by international standards, DNA barcoding, and omics-based methods [112, 113], despite the difficulties that these approaches present.

13.6 Conclusion with Future Perspectives

Stroke remains a significant global health problem that affects countless lives and burdens our healthcare systems. Hence, a comprehensive understanding of the fundamental roles played by pharmacognosy, antioxidants, and natural compounds is of utmost importance in the domain of stroke care. The standard therapy method uses rtPA, which can only be given during a certain time window, and only a small portion of stroke patients can receive these advantages. Natural products have bioactive compounds that may help lower oxidative stress and inflammation during and after a stroke, which would improve the healing process for patients. Nonetheless, there are additional possibilities for additional investigation in this sector. The integration of traditional knowledge with modern scientific research presents a viable pathway for the better prevention and treatment of strokes. To make stroke management more effective by addressing the problems that come up when antioxidants and natural products are used to treat stroke, the following guidelines can be devised:

- Comprehensive clinical investigations are necessary to evaluate the efficacy and durability of antioxidants and natural products in the management of stroke. It is imperative that this research encompass a diverse range of patient demographics and subtypes of stroke.
- Technological advancement and our growing understanding of the molecular mechanisms underlying stroke encourage the development of neuroprotective drugs. The discovery of bioactive substances found in nature with therapeutic promise in the treatment of stroke may be aided by cutting-edge approaches like nanotechnology, artificial intelligence (AI), and multiomics (i.e., metabolomics and proteomics).
- Conducting in-depth studies on relevant biomarkers linked to stroke is of utmost importance. The discovery of dependable biomarkers has great potential for enhancing personalised treatment approaches in the fields of stroke prevention and control, specifically in relation to natural antioxidants.
- The examination of the collective effects of multiple natural substances and antioxidants exhibits potential for augmenting neuroprotective results, hence broadening the spectrum of therapy alternatives for stroke.
- There is also a common phenomenon in which people are unconcerned or uninterested in the possibility of using natural ingredients as medical remedies for stroke. Future efforts must also include encouraging the public about the potential benefits of antioxidants and organic products as safeguards against stroke.

By acknowledging and promoting diverse perspectives and interdisciplinary collaboration, it is feasible to improve the effectiveness of stroke management interventions, thus improving the overall welfare of individuals who are susceptible to or have experienced a stroke.Author's ContributionConceptualization (KHS); Writing and Editing (KHS, SA, AF, and AW); Review (KHS).

Funding No funding was available for this research.

Declaration of Competing Interest All authors approved the final manuscript and declare that there are no known competing interests.

References

1. Xu, M., Wu, R. X., Li, X. L., Zeng, Y. S., Liang, J. Y., Fu, K., Liang, Y., & Wang, Z. (2022). Traditional medicine in China for ischemic stroke: Bioactive components, pharmacology, and mechanisms. *Journal of Integrative Neuroscience, 21*(1), 026. https://doi.org/10.31083/j.jin2101026

2. Zhou, M., Wang, H., Zeng, X., Yin, P., Zhu, J., Chen, W., Li, X., Wang, L., Wang, L., Liu, Y., Liu, J., Zhang, M., Qi, J., Yu, S., Afshin, A., Gakidou, E., Glenn, S., Krish, V. S., Miller-Petrie, M. K., et al. (2019). Mortality, morbidity, and risk factors in China and its provinces, 1990–2017: A systematic analysis for the Global Burden of Disease Study 2017. *The Lancet, 394*(10204), 1145–1158. https://doi.org/10.1016/s0140-6736(19)30427-1
3. GBD 2016 Lifetime Risk of Stroke Collaborators. (2018). Global, regional, and country-specific lifetime risks of stroke, 1990 and 2016. *New England Journal of Medicine, 379*(25), 2429–2437. https://doi.org/10.1056/nejmoa1804492
4. Zhou, Y., Zhang, S., & Fan, X. (2021). Role of polyphenols as antioxidant supplementation in ischemic stroke. *Oxidative Medicine and Cellular Longevity, 2021*, 1–19. https://doi.org/10.1155/2021/5471347
5. Adeoye, O., Hornung, R., Khatri, P., & Kleindorfer, D. (2011). Recombinant tissue-type plasminogen activator use for ischemic stroke in the United States: A doubling of treatment rates over the course of 5 years. *Stroke, 42*(7), 1952–1955.
6. Maciejczyk, M., Bielas, M., Zalewska, A., & Gerreth, K. (2021). Salivary biomarkers of oxidative stress and inflammation in stroke patients: From basic research to clinical practice. *Oxidative Medicine and Cellular Longevity, 2021*, 1–22. https://doi.org/10.1155/2021/5545330
7. Jelinek, M., Jurajda, M., & Duris, K. (2021). Oxidative stress in the brain: Basic concepts and treatment strategies in stroke. *Antioxidants, 10*(12), 1886. https://doi.org/10.3390/antiox10121886
8. Orellana-Urzúa, S., Claps, G., & Rodrigo, R. (2021). Improvement of a novel proposal for antioxidant treatment against brain damage occurring in ischemic stroke patients. *CNS & Neurological Disorders - Drug Targets, 20*(1), 3–21. https://doi.org/10.2174/1871527319666200910153431
9. Hishiki, T., Yamamoto, T., Morikawa, T., Kubo, A., Kajimura, M., & Suematsu, M. (2012). Carbon monoxide: Impact on remethylation/transsulfuration metabolism and its pathophysiologic implications. *Journal of Molecular Medicine, 90*(3), 245–254. https://doi.org/10.1007/s00109-012-0875-2
10. Weber, J., Slemmer, J., Shacka, J., & Sweeney, M. (2008). Antioxidants and free radical scavengers for the treatment of stroke, traumatic brain injury and aging. *Current Medicinal Chemistry, 15*(4), 404–414. https://doi.org/10.2174/092986708783497337
11. Chen, H. S., Qi, S. H., & Shen, J. G. (2017). One-compound-multi-target: Combination prospect of natural compounds with thrombolytic therapy in acute ischemic stroke. *Current Neuropharmacology, 15*(1), 134–156. https://doi.org/10.2174/1570159x14666160620102055
12. Chen, H., He, Y., Chen, S., Qi, S., & Shen, J. (2020). Therapeutic targets of oxidative/nitrosative stress and neuroinflammation in ischemic stroke: Applications for natural product efficacy with omics and systemic biology. *Pharmacological Research, 158*, 104877. https://doi.org/10.1016/j.phrs.2020.104877
13. Xu, H., Wang, E., Chen, F., Xiao, J., & Wang, M. (2021). Neuroprotective phytochemicals in experimental ischemic stroke: Mechanisms and potential clinical applications. *Oxidative Medicine and Cellular Longevity, 2021*, 1–45. https://doi.org/10.1155/2021/6687386
14. Sarker, S. D. (2012). Pharmacognosy in modern pharmacy curricula. *Pharmacognosy Magazine, 8*(30), 91.
15. Shashikiran, N. (2016). Pharmacognosy. *Journal of Indian Society of Pedodontics and Preventive Dentistry, 34*(2), 103–104.
16. Wassef, A., & Butcher, K. (2016). Novel oral anticoagulant management issues for the stroke clinician. *International Journal of Stroke, 11*(7), 759–767.
17. Jaiswal, Y. S., & Williams, L. L. (2017). A glimpse of Ayurveda–The forgotten history and principles of Indian traditional medicine. *Journal of Traditional and Complementary Medicine, 7*(1), 50–53.
18. Parasuraman, S., Thing, G. S., & Dhanaraj, S. A. (2014). Polyherbal formulation: Concept of ayurveda. *Pharmacognosy Reviews, 8*(16), 73.

19. Mathur, D., Goyal, K., Koul, V., & Anand, A. (2016). The molecular links of re-emerging therapy: A review of evidence of Brahmi (Bacopa monniera). *Frontiers in Pharmacology, 7*, 44.
20. Kulkarni, S., & Dhir, A. (2008). Withania somnifera: An Indian ginseng. *Progress in Neuro-Psychopharmacology and Biological Psychiatry, 32*(5), 1093–1105.
21. Rajini, P., & Muralidhara, M. (2023). Therapeutic efficacy of ayurvedic polyherbal formulations (PHF): Interactive mechanisms and broad-spectrum activities against neurological disorders. In *Ayurvedic herbal preparations in neurological disorders* (pp. 89–111). Elsevier.
22. Parasuraman, S., Kumar, E., Kumar, A., & Emerson, S. (2010). Free radical scavenging property and diuretic effect of triglize, a polyherbal formulation in experimental models. *Journal of Pharmacology & Pharmacotherapeutics, 1*(1), 38.
23. Poluboiartsev, I., & George, A. (2023). Ayurvedic medicine in treating stroke complications: A literature review. *Journal of Kazakhstan-Russian Medical University, 39*(1), 99–102.
24. Mirza, S., Sawarkar, G., & Sawarkar, P. (2021). Management of cerebro-vascular stroke (Pakshaghata) with the successful intervention of Ayurveda: A case report. *Drugs and Cell Therapies in Hematology, 10*(1), 3433–3439.
25. Ma, D., Wang, S., Shi, Y., Ni, S., Tang, M., & Xu, A. (2021). The development of traditional Chinese medicine. *Journal of Traditional Chinese Medical Sciences, 8*, S1–S9.
26. Raphals, L. (2015). Chinese philosophy and Chinese medicine. In E. N. Zalta (Ed.), *The Stanford encyclopedia of philosophy*. Retrieved from http://plato.stanford.edu/archives/sum2015/entries/chinesephil- medicine/
27. Gong, X., & Sucher, N. J. (1999). Stroke therapy in traditional Chinese medicine (TCM): Prospects for drug discovery and development. *Trends in Pharmacological Sciences, 20*(5), 191–196.
28. Garcia, J. H., Lassen, N. A., Weiller, C., Sperling, B., & Nakagawara, J. (1996). Ischemic stroke and incomplete infarction. *Stroke, 27*(4), 761–765.
29. Bensky, D., Clavey, S., & Stõger, E. (2004). Materia medica. *Chinese Herbal Medicine*, 3–6.
30. Junhua, Z., Menniti-Ippolito, F., Xiumei, G., Firenzuoli, F., Boli, Z., Massari, M., et al. (2009). Complex traditional Chinese medicine for poststroke motor dysfunction: A systematic review. *Stroke, 40*(8), 2797–2804.
31. Gaire, B. P. (2018). Herbal medicine in ischemic stroke: Challenges and prospective. *Chinese Journal of Integrative Medicine, 24*, 243–246.
32. Hung, C. Y.-f., Wu, X.-y., Chung, V. C.-h., Tang, E. C.-h., Wu, J. C.-y., & Lau, A. Y.-l. (2019). Overview of systematic reviews with meta-analyses on acupuncture in post-stroke cognitive impairment and depression management. *Integrative Medicine Research, 8*(3), 145–159.
33. Zhang, Q., Wang, Y., Chen, A., Huang, X., Dong, Q., Li, Z., et al. (2021). Xiaoxuming decoction: A traditional herbal recipe for stroke with emerging therapeutic mechanisms. *Frontiers in Pharmacology, 12*, 802381.
34. Liang, H., Ruan, H., Ouyang, Q., & Lai, L. (2016). Herb-target interaction network analysis helps to disclose molecular mechanism of traditional Chinese medicine. *Scientific Reports, 6*(1), 36767.
35. Guo, J., Shang, E., Zhao, J., Fan, X., Duan, J., Qian, D., et al. (2014). Data mining and frequency analysis for licorice as a "Two-Face" herb in Chinese Formulae based on Chinese Formulae Database. *Phytomedicine, 21*(11), 1281–1286.
36. Lionta, E., Spyrou, G., Vassilatis, K., & D., & Cournia, Z. (2014). Structure-based virtual screening for drug discovery: Principles, applications and recent advances. *Current Topics in Medicinal Chemistry, 14*(16), 1923–1938.
37. Houston, M. C. (2007). Treatment of hypertension with nutraceuticals, vitamins, antioxidants and minerals. *Expert Review of Cardiovascular Therapy, 5*(4), 681–691. https://doi.org/10.1586/14779072.5.4.681
38. Riccioni, G., Bucciarelli, T., Mancini, B., Di Ilio, C., Capra, V., & D'Orazio, N. (2007). The role of the antioxidant vitamin supplementation in the prevention of cardiovascular diseases. *Expert Opinion on Investigational Drugs, 16*(1), 25–32. https://doi.org/10.1517/13543784.16.1.25

39. Vasconcelos, A. R., dos Santos, N. B., Scavone, C., & Munhoz, C. D. (2019). Nrf2/ARE pathway modulation by dietary energy regulation in neurological disorders. *Frontiers in Pharmacology, 10*. https://doi.org/10.3389/fphar.2019.00033
40. Ashok, A., Andrabi, S. S., Mansoor, S., Kuang, Y., Kwon, B. K., & Labhasetwar, V. (2022). Antioxidant therapy in oxidative stress-induced neurodegenerative diseases: Role of nanoparticle-based drug delivery systems in clinical translation. *Antioxidants, 11*(2), 408. https://doi.org/10.3390/antiox11020408
41. Nandi, A., Yan, L. J., Jana, C. K., & Das, N. (2019). Role of catalase in oxidative stress- and age-associated degenerative diseases. *Oxidative Medicine and Cellular Longevity, 2019*, 1–19. https://doi.org/10.1155/2019/9613090
42. Lubos, E., Loscalzo, J., & Handy, D. E. (2011). Glutathione peroxidase-1 in health and disease: From molecular mechanisms to therapeutic opportunities. *Antioxidants & Redox Signaling, 15*(7), 1957–1997. https://doi.org/10.1089/ars.2010.3586
43. Yang, X., Zhang, Y., Xu, H., Luo, X., Yu, J., Liu, J., & Chuen-Chung, R. (2015). Neuroprotection of coenzyme Q10 in neurodegenerative diseases. *Current Topics in Medicinal Chemistry, 16*(8), 858–866. https://doi.org/10.2174/1568026615666150827095252
44. Lushchak, V. I. (2012). Glutathione homeostasis and functions: Potential targets for medical interventions. *Journal of Amino Acids, 2012*, 1–26. https://doi.org/10.1155/2012/736837
45. Sharifi-Rad, M., Anil Kumar, N. V., Zucca, P., Varoni, E. M., Dini, L., Panzarini, E., Rajkovic, J., Tsouh Fokou, P. V., Azzini, E., Peluso, I., Prakash Mishra, A., Nigam, M., El Rayess, Y., Beyrouthy, M. E., Polito, L., Iriti, M., Martins, N., Martorell, M., Docea, A. O., et al. (2020). Lifestyle, oxidative stress, and antioxidants: Back and forth in the pathophysiology of chronic diseases. *Frontiers in Physiology, 11*. https://doi.org/10.3389/fphys.2020.00694
46. Chen, P., Li, L., Gao, Y., Xie, Z., Zhang, Y., Pan, Z., et al. (2019). β-Carotene provides neuroprotection after experimental traumatic brain injury via the Nrf2-ARE pathway. *Journal of Integrative Neuroscience, 18*(2), 153–161.
47. Manochkumar, J., Doss, C. G. P., El-Seedi, H. R., Efferth, T., & Ramamoorthy, S. (2021). The neuroprotective potential of carotenoids in vitro and in vivo. *Phytomedicine, 91*, 153676. https://doi.org/10.1016/j.phymed.2021.153676
48. Ma, H., Johnson, S., Liu, W., DaSilva, N., Meschwitz, S., Dain, J., & Seeram, N. (2018). Evaluation of polyphenol anthocyanin-enriched extracts of blackberry, black raspberry, blueberry, cranberry, red raspberry, and strawberry for free radical scavenging, reactive carbonyl species trapping, anti-glycation, anti-β-amyloid aggregation, and microglial neuroprotective effects. *International Journal of Molecular Sciences, 19*(2), 461. https://doi.org/10.3390/ijms19020461
49. Galli, R. L., Shukitt-Hale, B. A. R. B. A. R. A., Youdim, K. A., & Joseph, J. A. (2002). Fruit polyphenolics and brain aging: Nutritional interventions targeting age-related neuronal and behavioral deficits. *Annals of the New York Academy of Sciences, 959*(1), 128–132.
50. Inoue, H., Jiang, X. F., Katayama, T., Osada, S., Umesono, K., & Namura, S. (2003, December). Brain protection by resveratrol and fenofibrate against stroke requires peroxisome proliferator-activated receptor α in mice. *Neuroscience Letters, 352*(3), 203–206. https://doi.org/10.1016/j.neulet.2003.09.001
51. Lourenço, S. C., Moldão-Martins, M., & Alves, V. D. (2019). Antioxidants of natural plant origins: From sources to food industry applications. *Molecules, 24*(22), 4132. https://doi.org/10.3390/molecules24224132
52. Saleh, T. M., Saleh, M. C., Connell, B. J., Kucukkaya, I., & Abd-El-Aziz, A. S. (2017). A novel synthetic chemical entity (UPEI-800) is neuroprotective in vitro and in an in vivo rat model of oxidative stress. *Clinical and Experimental Pharmacology and Physiology, 44*(10), 993–1000. https://doi.org/10.1111/1440-1681.12785
53. Uppakara, K., Jamornwan, S., Duan, L. X., Yue, K. R., Sunrat, C., Dent, E. W., Wan, S. B., & Saengsawang, W. (2020). Novel α-lipoic acid/3-n-butylphthalide conjugate enhances protective effects against oxidative stress and 6-OHDA induced neuronal damage. *ACS Chemical Neuroscience, 11*(11), 1634–1642. https://doi.org/10.1021/acschemneuro.0c00105

54. Watanabe, T., Tahara, M., & Todo, S. (2008). The novel antioxidant edaravone: From bench to bedside. *Cardiovascular Therapeutics, 26*(2), 101–114. https://doi.org/10.1111/j.1527-3466.2008.00041.x
55. Guerrero-Beltrán, C. E., Calderón-Oliver, M., Pedraza-Chaverri, J., & Chirino, Y. I. (2012). Protective effect of sulforaphane against oxidative stress: Recent advances. *Experimental and Toxicologic Pathology, 64*(5), 503–508. https://doi.org/10.1016/j.etp.2010.11.005
56. Chen, S., Chen, H., Du, Q., & Shen, J. (2020). Targeting Myeloperoxidase (MPO) mediated oxidative stress and inflammation for reducing brain ischemia injury: Potential application of natural compounds. *Frontiers in Physiology, 11*. https://doi.org/10.3389/fphys.2020.00433
57. Lei, J., Tu, X., Wang, Y., Tu, D., & Shi, S. (2019). Resveratrol downregulates the TLR4 signaling pathway to reduce brain damage in a rat model of focal cerebral ischemia. *Experimental and Therapeutic Medicine*. https://doi.org/10.3892/etm.2019.7324
58. Casas, A. I., Kleikers, P. W., Geuss, E., Langhauser, F., Adler, T., Busch, D. H., Gailus-Durner, V., de Angelis, M. H., Egea, J., Lopez, M. G., Kleinschnitz, C., & Schmidt, H. H. (2019). Calcium-dependent blood-brain barrier breakdown by NOX5 limits postreperfusion benefit in stroke. *Journal of Clinical Investigation, 129*(4), 1772–1778. https://doi.org/10.1172/jci124283
59. Wang, X. H., & You, Y. P. (2017). Epigallocatechin gallate extends therapeutic window of recombinant tissue plasminogen activator treatment for brain ischemic stroke: A randomized double-blind and placebo-controlled trial. *Clinical Neuropharmacology, 40*(1), 24–28. https://doi.org/10.1097/wnf.0000000000000197
60. Wang, Z., Ma, C., Meng, C. J., Zhu, G. Q., Sun, X. B., Huo, L., Zhang, J., Liu, H. X., He, W. C., Shen, X. M., Shu, Z., & Chen, G. (2012). Melatonin activates the Nrf2-ARE pathway when it protects against early brain injury in a subarachnoid hemorrhage model. *Journal of Pineal Research, 53*(2), 129–137. https://doi.org/10.1111/j.1600-079x.2012.00978.x
61. Otomo, E. (2003). Effect of a novel free radical scavenger, edaravone (MCI-186), on acute brain infarction-randomized, placebo-controlled, double-blind study at multicenters. *Cerebrovascular Diseases, 15*, 222–229.
62. Lu, H., Shen, J., Song, X., Ge, J., Cai, R., Dai, A., & Jiang, Z. (2015). Protective effect of Pyrroloquinoline Quinone (PQQ) in rat model of intracerebral hemorrhage. *Cellular and Molecular Neurobiology, 35*(7), 921–930. https://doi.org/10.1007/s10571-015-0187-5
63. Pravalika, K., Sarmah, D., Kaur, H., Vats, K., Saraf, J., Wanve, M., Kalia, K., Borah, A., Yavagal, D. R., Dave, K. R., & Bhattacharya, P. (2019). Trigonelline therapy confers neuroprotection by reduced glutathione mediated myeloperoxidase expression in animal model of ischemic stroke. *Life Sciences, 216*, 49–58. https://doi.org/10.1016/j.lfs.2018.11.014
64. Loh, H. C., Lim, R., Lee, K. W., Ooi, C. Y., Chuan, D. R., Looi, I., Kah Hay, Y., & Abdul Karim Khan, N. (2021). Effects of vitamin E on stroke: A systematic review with meta-analysis and trial sequential analysis. *Stroke and Vascular Neurology, 6*(1), 109–120. https://doi.org/10.1136/svn-2020-000519
65. Genovese, T., Mazzon, E., Paterniti, I., Esposito, E., Bramanti, P., & Cuzzocrea, S. (2011). Modulation of NADPH oxidase activation in cerebral ischemia/reperfusion injury in rats. *Brain Research, 1372*, 92–102. https://doi.org/10.1016/j.brainres.2010.11.088
66. Schluter, T., Steinbach, A. C., Steffen, A., Rettig, R., & Grisk, O. (2008). Apocynin-induced vasodilation involves Rho kinase inhibition but not NADPH oxidase inhibition. *Cardiovascular Research, 80*(2), 271–279. https://doi.org/10.1093/cvr/cvn185
67. Kleinschnitz, C., Grund, H., Wingler, K., Armitage, M. E., Jones, E., Mittal, M., Barit, D., Schwarz, T., Geis, C., Kraft, P., Barthel, K., Schuhmann, M. K., Herrmann, A. M., Meuth, S. G., Stoll, G., Meurer, S., Schrewe, A., Becker, L., Gailus-Durner, V., et al. (2010). Post-stroke inhibition of induced NADPH oxidase type 4 prevents oxidative stress and neurodegeneration. *PLoS Biology, 8*(9), e1000479. https://doi.org/10.1371/journal.pbio.1000479
68. Reynolds, A., Laurie, C., Mosley, R. L., & Gendelman, H. E. (2007). Oxidative stress and the pathogenesis of neurodegenerative disorders. *International Review of Neurobiology, 82*, 297–325.

69. Tsutsumi, Z., Moriwaki, Y., Takahashi, S., Ka, T., & Yamamoto, T. (2004). Oxidized low-density lipoprotein autoantibodies in patients with primary gout: Effect of urate-lowering therapy. *Clinica Chimica Acta, 339*(1–2), 117–122. https://doi.org/10.1016/j.cccn.2003.09.019
70. Park, C. K., & Hall, E. D. (1994). Dose-response analysis of the effect of 21-aminosteroid tirilazad mesylate (U-74006F) upon neurological outcome and ischemic brain damage in permanent focal cerebral ischemia. *Brain Research, 645*(1–2), 157–163. https://doi.org/10.1016/0006-8993(94)91649-7
71. Marshall, J. W., Cummings, R. M., Bowes, L. J., Ridley, R. M., & Green, A. R. (2003). Functional and histological evidence for the protective effect of NXY-059 in a primate model of stroke when given 4 hours after occlusion. *Stroke, 34*(9), 2228–2233. https://doi.org/10.1161/01.str.0000087790.79851.a8
72. Nakase, T., Yoshioka, S., & Suzuki, A. (2011). Free radical scavenger, edaravone, reduces the lesion size of lacunar infarction in human brain ischemic stroke. *BMC Neurology, 11*(1). https://doi.org/10.1186/1471-2377-11-39
73. Warner, D. S., Sheng, H., & Batinić-Haberle, I. (2004). Oxidants, antioxidants and the ischemic brain. *Journal of Experimental Biology, 207*(18), 3221–3231. https://doi.org/10.1242/jeb.01022
74. Esposito, E., Mandeville, E. T., Hayakawa, K., Singhal, A. B., & Lo, E. H. (2013). Effects of normobaric oxygen on the progression of focal cerebral ischemia in rats. *Experimental Neurology, 249*, 33–38. https://doi.org/10.1016/j.expneurol.2013.08.005
75. Geng, X., Parmar, S., Li, X., Peng, C., Ji, X., Chakraborty, T., Li, W. A., Du, H., Tan, X., & ling, F., Guthikonda, M., Rafols, J. A., & Ding, Y. (2013). Reduced apoptosis by combining normobaric oxygenation with ethanol in transient ischemic stroke. *Brain Research, 1531*, 17–24. https://doi.org/10.1016/j.brainres.2013.07.051
76. Lan, W., Wu, X., Wu, Y., & Zhang, H. (2023). Evaluating the role of Ginkgo biloba extract in the secondary prevention of acute ischemic stroke with cerebral microbleeds by Quantitative Susceptibility Mapping (QSM). *International Journal of Neuroscience*, 1–10. Advance online publication. https://doi.org/10.1080/00207454.2023.2268264
77. Li, Y., Zhu, X., Wang, K., Zhu, L., Murray, M., & Zhou, F. (2022). The potential of Ginkgo biloba in the treatment of human diseases and the relationship to Nrf2–mediated antioxidant protection. *Journal of Pharmacy and Pharmacology, 74*(12), 1689–1699.
78. Jin, M., Kim, K.-M., Lim, C., Cho, S., & Kim, Y. K. (2022). Neuroprotective effects of Korean White ginseng and Red ginseng in an ischemic stroke mouse model. *Journal of Ginseng Research, 46*(2), 275–282.
79. Kwon, B., Song, Y., Kim, J.-G., Lee, D., Lee, S.-H., Cho, Y.-K., et al. (2022). Preventive effects of ginseng against atherosclerosis and subsequent ischemic stroke: A randomized controlled trial (PEGASUS trial). *Journal of Ginseng Research, 46*(4), 585–591.
80. Zhao, A., Liu, N., Yao, M., Zhang, Y., Yao, Z., Feng, Y., et al. (2022). A review of neuroprotective effects and mechanisms of ginsenosides from Panax ginseng in treating ischemic stroke. *Frontiers in Pharmacology, 13*, 946752.
81. Vinod, A., Sathianarayanan, S., Babu, A. E., Sadanandan, P., Venu, A. K., & Venkidasamy, B. (2022). Bacopa monnieri for disorders affecting brain: Current perspectives. *Current Topics in Medicinal Chemistry, 22*(23), 1909–1929.
82. Ramasamy, S., Chin, S. P., Sukumaran, S. D., Buckle, M. J. C., Kiew, L. V., & Chung, L. Y. (2015). In silico and in vitro analysis of bacoside A aglycones and its derivatives as the constituents responsible for the cognitive effects of Bacopa monnieri. *PLoS One, 10*(5), e0126565.
83. Pham, H. T. N., Phan, S. V., Tran, H. N., Phi, X. T., Le, X. T., Nguyen, K. M., et al. (2019). Bacopa monnieri (L.) ameliorates cognitive deficits caused in a trimethyltin-induced neurotoxicity model mice. *Biological and Pharmaceutical Bulletin, 42*(8), 1384–1393.
84. Manful, C. F., Hameed, A., & Thomas, R. H. (2023). Berries. In *Functional foods and their implications for health promotion* (pp. 161–217). Elsevier.

85. Onuh, J. O., Dawkins, N. L., & Aluko, R. E. (2023). Cardiovascular disease protective properties of blueberry polyphenols (Vaccinium corymbosum): A concise review. *Food Production, Processing and Nutrition, 5*(1), 27.
86. Guo, S., Xing, N., Xiang, G., Zhang, Y., & Wang, S. (2023). Eriodictyol: A review of its pharmacological activities and molecular mechanisms related to ischemic stroke. *Food & Function, 14*(10), 4949.
87. Hasan, S., Khatri, N., Rahman, Z. N., Menezes, A. A., Martini, J., Shehjar, F., et al. (2023). Neuroprotective potential of flavonoids in brain disorders. *Brain Sciences, 13*(9), 1258.
88. Colletti, A., Cravotto, G., Citi, V., Martelli, A., Testai, L., & Cicero, A. F. (2021). Advances in technologies for highly active omega-3 fatty acids from krill oil: Clinical applications. *Marine Drugs, 19*(6), 306.
89. Yamagata, K. (2023). Dietary lipids: The effect of docosahexaenoic acid on stroke-related neuronal damage. In *Diet and nutrition in neurological disorders* (pp. 937–953). Elsevier.
90. Kim, J. H., Kismali, G., & Gupta, S. C. (2018). *Natural products for the prevention and treatment of chronic inflammatory diseases: Integrating traditional medicine into modern chronic diseases care*. Hindawi.
91. Vatansever, H. S., Sorkun, K., Gurhan, S. I. D., Ozdal-Kurt, F., Turkoz, E., Gencay, O., & Salih, B. (2010). Propolis from Turkey induces apoptosis through activating caspases in human breast carcinoma cell lines. *Acta Histochemica, 112*(6), 546–556.
92. Zullkiflee, N., Taha, H., & Usman, A. (2022). Propolis: Its role and efficacy in human health and diseases. *Molecules, 27*(18), 6120.
93. Threapleton, D. E., Greenwood, D. C., Evans, C. E., Cleghorn, C. L., Nykjaer, C., Woodhead, C., et al. (2013). Dietary fiber intake and risk of first stroke: A systematic review and meta-analysis. *Stroke, 44*(5), 1360–1368.
94. Larsson, S. C., Männistö, S., Virtanen, M. J., Kontto, J., Albanes, D., & Virtamo, J. (2009). Dietary fiber and fiberrich food intake in relation to risk of stroke in male smokers. *European Journal of Clinical Nutrition, 63*(8), 1016–1024.
95. Oh, K., Hu, F. B., Cho, E., Rexrode, K. M., Stampfer, M. J., Manson, J. E., et al. (2005). Carbohydrate intake, glycemic index, glycemic load, and dietary fiber in relation to risk of stroke in women. *American Journal of Epidemiology, 161*(2), 161–169.
96. Zello, G. A. (2006). Dietary Reference Intakes for the macronutrients and energy: Considerations for physical activity. *Applied Physiology, Nutrition, and Metabolism, 31*(1), 74–79.
97. Prasad, S., & Aggarwal, B. B. (2011). Turmeric, the golden spice. In *Herbal medicine: Biomolecular and clinical aspects* (2nd ed.). CRC Press.
98. Ray, M. K., Santra, D. K., Mishra, P. K., & Das, S. (2023). Indigenous Lakadong turmeric of Meghalaya and its future prospects. *Journal of Applied Biology and Biotechnology, 11*(5), 133–142.
99. Mughal, M. H. (2019). Turmeric polyphenols: A comprehensive review. *Integrative Food, Nutrition and Metabolism, 6*, 1–6.
100. Ueno, Y., Miyamoto, N., Yamashiro, K., Tanaka, R., & Hattori, N. (2019). Omega-3 polyunsaturated fatty acids and stroke burden. *International Journal of Molecular Sciences, 20*(22), 5549.
101. Zhang, Y., Tuomilehto, J., Jousilahti, P., Wang, Y., Antikainen, R., & Hu, G. (2011). Lifestyle factors on the risks of ischemic and hemorrhagic stroke. *Archives of Internal Medicine, 171*(20), 1811–1818.
102. Grassi, D., Desideri, G., Necozione, S., Ruggieri, F., Blumberg, J. B., Stornello, M., & Ferri, C. (2012). Protective effects of flavanol-rich dark chocolate on endothelial function and wave reflection during acute hyperglycemia. *Hypertension, 60*(3), 827–832.
103. García-Cordero, J., Martinez, A., Blanco-Valverde, C., Pino, A., Puertas-Martín, V., San Román, R., & de Pascual-Teresa, S. (2023). Regular consumption of cocoa and red berries as a strategy to improve cardiovascular biomarkers via modulation of microbiota metabolism in healthy aging adults. *Nutrients, 15*(10), 2299.

104. Shin, J.-H., Kim, C.-S., Cha, L., Kim, S., Lee, S., Chae, S., et al. (2022). Consumption of 85% cocoa dark chocolate improves mood in association with gut microbial changes in healthy adults: A randomized controlled trial. *The Journal of Nutritional Biochemistry, 99*, 108854.
105. Putri, I. S., Robbani, T. N., Divamillenia, D., Pratiwi, O. G., & I'tishom, R. (2022). Potential neurogenesis and neuroprotective effects of epigallocatechin-3-gallate (EGCG) in green tea (Camellia sinensis) through microglia M2 induction process and NLRP3 inhibition as an innovation for ischemic stroke adjuvant therapy: A review. *International Journal of Research Publications, 92*(1), 8–8.
106. Azami, S., & Forouzanfar, F. (2023). Therapeutic potentialities of green tea (Camellia sinensis) in ischemic stroke: Biochemical and molecular evidence. *Metabolic Brain Disease, 39*(2), 347–357.
107. Amer, H. M., & Mohammad, A. A. (2022). Medicinal plants and their validation challenges in traditional Egyptian medicine. *Journal of Applied Pharmaceutical Science, 12*(3), 023–033.
108. Kushwaha, A., Pathak, D., Saha, S., Kumar, S., Singh, K., & Saha, S. (2022). Standardization of some ayurvedic formulation used in Covid-19. *YMER Digital, 21*(12), 2756–2766.
109. Husain, G. M., Khan, M. A., Urooj, M., & Kazmi, M. H. (2020). Pharmacodynamic evaluation: Herbal medicine. In *Drug discovery and evaluation: Methods in clinical pharmacology* (pp. 483–497). Springer.
110. Mukherjee, P. K., Banerjee, S., Gupta, B. D., & Kar, A. (2022). Evidence-based validation of herbal medicine: Translational approach. In *Evidence-based validation of herbal medicine* (pp. 1–41). Elsevier.
111. Tao, T., Liu, M., Chen, M., Luo, Y., Wang, C., Xu, T., et al. (2020). Natural medicine in neuroprotection for ischemic stroke: Challenges and prospective. *Pharmacology & Therapeutics, 216*, 107695.
112. Clark, A. M., Zorn, K. M., Lingerfelt, M. A., & Ekins, S. (2018). Developing next generation tools for computational toxicology. In *Computational toxicology: Risk assessment for chemicals* (pp. 363–387). Wiley.
113. Thakkar, S., Anklam, E., Xu, A., Ulberth, F., Li, J., Li, B., et al. (2020). Regulatory landscape of dietary supplements and herbal medicines from a global perspective. *Regulatory Toxicology and Pharmacology, 114*, 104647.

14 Enteral Access and Dysphagia in Stroke

Farah Nadia Mohd Hanafiah

Abstract

There is a fourfold increase in mortality when a patient is diagnosed of having post-stroke dysphagia. Pneumonia, malnutrition and dehydration contributes towards the high mortality associated with post-stroke dysphagia. Screening and assessment for dysphagia must be done using tools such as the water-swallow test, videofluoroscopic swallowing study and fibreoptic endoscopic evaluation of swallowing. The focus in managing post-stroke dysphagia is to prevent aspiration and to improve the swallowing mechanism. Several methods have been developed such as dietary modification along with the use of enteral access, maneuvers that facilitate swallowing, neurostimulation as well as rehabilitation techniques aiming to improve swallowing. Currently, the ongoing treatment of post-stroke dysphagia demonstrate potential benefit however further study and recommendations are required to improve the quality of life of patient with post-stroke dysphagia.

Keywords

Post-stroke dysphagia · Enteral access · Dysphagia screening assessment

14.1 Introduction

Dysphagia is quite a common presentation of stroke and the incidence range from 29% to 81% depending on location and lesion size [1, 2]. Dysphagia can be defined as difficulty in swallowing and eating. Other symptoms of dysphagia include

F. N. M. Hanafiah (✉)
Department of Anaesthesiology and Intensive Care, International Islamic University Malaysia, Kuala Lumpur, Malaysia
e-mail: drfarahnadia@iium.edu.my

W. Mohamed, M. B. Mat-Nor (eds.), *Nutrition and Stroke*, Nutritional Neurosciences, https://doi.org/10.1007/978-981-97-6818-9_14

repetitive swallowing, slurred speech, hoarseness of voice, coughing, choking, weight loss, food sticking, and recurrent pneumonia. Post-stroke dysphagia is highly associated with respiratory complications. Likewise, there is also an increased risk of malnutrition and dehydration in these patients [3]. Furthermore, the mortality of patients with post-stroke dysphagia rises 4 times fold as compared to patients of stroke without dysphagia [2]. Although in the majority of patients, there will be a rapid recovery within 7 days, persistent dysphagia up to 6 months post-stroke may happen [3]. These patients with persistent dysphagia would require alternative means of enteral feeding.

14.2 Pathophysiology of Dysphagia in Stroke

Multiple muscle groups are involved in the swallowing process from the propulsion of food or liquid bolus that start from the mouth into the stomach while simultaneously protecting the airway. Swallowing can be characterized into subconscious swallowing which will happen once every minutes, reflexive swallowing which is part of the airway protective mechanism stimulated by a drop of food into the pharynx and lastly, volitional swallowing [4]. The volitional swallowing involves a complex swallowing network in the cerebral cortex which includes the cingulate gyrus, prefrontal gyrus, insula, somatosensory cortex, and precuneus regions. Any lesions related to stroke in any of these areas can disrupt the swallowing network leading to dysphagia. Swallowing abnormalities can develop centrally due to any damage in brain stem and cortex and it can also happen peripherally when it involves damage to the nerve or muscles which are involved with swallowing [4].

14.3 Risk Factor of Post-Stroke Dysphagia

Identifying dysphagia is important to anticipate and prevent the complications related to dysphagia and for treatment. The risk factors associated with dysphagia include age, comorbidities, previous history of cerebrovascular disease, and the location of current infarction [5]. The body function progressively weaken with increasing age, and this contributes to impaired oral and maxillomandibular function leading to dysphagia. Furthermore, the elderly are associated with higher risk of developing swallowing dysfunction as a result of degeneration of advanced cranial nerve function and abnormal swallowing reflex [5]. There is also higher incidence of post-stroke dysphagia with comorbidities of hypertension and diabetes mellitus. The presence of brain stem lesion is also an independent risk factor for dysphagia as it is involved with the swallowing process. A more severe stroke as assessed by NIHSS is another independent risk factor for developing dysphagia. An NIHSS ≥ 12 predicts that there will be persistent dysphagia more than 14 days [5]. The risk factors of developing post-stroke dysphagia are listed in Table 14.1.

Table 14.1 Risk factors of developing post-stroke dysphagia

Older age
Greater NIHSS
Hypertension
Diabetes Mellitus
Greater lesion volume
Subcortical vs cortical involvement
Brainstem involvement
Corticobulbar tract involvement
Presence of dysarthria
Presence of dysphonia
Cognitive impairment or dementia

NIHSS National Institute of Health Stroke Scale

14.4 Complications of Post-Stroke Dysphagia

Post-stroke dysphagia can lead to increase mortality, unfavorable outcomes, longer hospitalization, and increased dependency. In addition, there is a heightened risk of developing malnutrition, dehydration, reduced quality of life, as well as aspiration pneumonia [3, 4]. In general, all-cause of dysphagia can lead to lead to increase hospital stays by 2.8 days, increased cost by 34% as well as doubling of the in-hospital mortality [6].

The patients with post-stroke dysphagia tend to have a longer oral passage time and there is a delay with pharyngeal swallowing which is predisposed to aspiration. Furthermore, as patients tend to be kept nil-by-mouth, they will continue to aspirate saliva and they will have persistent reflux. Fluid and diet modification are parts of the management of post-stroke dysphagia and both are associated with dehydration. Dehydration itself contributes to poorer outcomes as it is associated with increased mortality, and higher healthcare costs as there is an increased risk of developing urinary tract infections, constipation, and falls. In an observational study done, the risk of developing aspiration pneumonia, dehydration, UTI, and constipation is higher when patients have dysphagia regardless of hemiparesis status as compared to patients without dysphagia.

Pneumonia occurred in around 10% of hospitalized stroke patients [3]. The incidence of pneumonia can increase up to 40% in elderly patients with severe stroke who develops post-stroke dysphagia [3]. Pneumonia most often occurs within a week of presentation of stroke as firstly, the immune system is suppressed due to the acute event and secondly, due to the presence of dysphagia. Patients developing pneumonia has a longer length of hospitalization with an increased risk of mortality.

14.5 Screening and Assessment of Dysphagia

Dysphagia screening must be done with patients with stroke. One of the widely used screening tools is the water-swallow test. In this test, the patients need to drink a 50–90 ml of water and if there is the presence of clinical signs of aspiration such as cough, voice and stridor, the test is considered positive [1]. Subsequently, the patient will be kept nil-by-mouth and proceed to another diagnostic test. The gagging swallowing screen (GUSS) evaluates the ability of the patient to swallow liquid, semi-solid solid, and solid boluses with increasing volume. If there are any signs of clinical aspiration, the test is terminated, and the patient is then graded from no dysphagia to severe dysphagia. Specific diet recommendations can then be suggested for patients based on the severity of dysphagia.

Clinical swallow examination (CSE) is a method of assessing dysphagia by using videofluoroscopic swallowing study (VFSS) [7]. Different texture of food and different maneuvers are introduced to improve efficacy and safety of swallowing. Specific parameters such as oral transit time, pharyngeal transit time, or laryngeal vestibule closure time can then be determined. VFSS gives a rating of swallowing function by aggregating several single items to a sum of scores. In a retrospective observational study of 499 patients, VFSS was done and removal of nasogastric tube and start of oral diet was suggested in 199 patients. Only 5 patients developed pneumonia during follow up which demonstrated that safety had been adequately assessed by VFSS. However, it is expensive to perform as it requires a radiology suite and several different staff. In addition, it involves radiation exposure, and it is not readily repeatable [7]. Additional procedures can also be performed such as cough reflex test, swallow-provocation test, or peak flow measurement which enable testing for aspiration.

Fibreoptic endoscopic evaluation of swallowing (FEES) is another method of assessing swallowing where a flexible nasolaryngoscope is initially introduced through the nose into the pharynx and subsequently dyed food and drinks are introduced. The pharyngeal swallow can then be assessed and any signs of impairment of the oral and oesophageal stages of deglutition can then be detected. FEES is advantageous as it can be done bedside which facilitates severely impaired bedridden patients. In addition, the efficacy of clearing mechanism in the form of coughing and throat clearing can also be tested which will inform the physician regarding pooling and residue in the hypopharynx allowing detection of aspiration [7]. The advantages of FEES are it is repeatable and safe, but the downside is it is not available in all hospitals.

14.6 Management of Post-stroke Dysphagia

The primary aim of dysphagia management is to reduce aspiration as well as managing the swallowing difficulties. The therapeutic intervention for post-stroke dysphagia includes dietary and nutritional interventions, behavioral interventions with altering posture, and maneuvers to facilitate swallowing and rehabilitation techniques.

14.7 Dietary Interventions

The cornerstone of dietary management of post-stroke dysphagia is through using texture-modified foods and thickened liquid. The principle of this approach is that by modifying food and liquids, it is easier and safer for stroke patients to swallow. However, the evidence of modifying food and liquids is still inconclusive. A meta-analysis included 6 RCTs and 3 non-randomized controlled trials (NRCTs), and dietary modification which included combination of texture-modified food and liquid thickening demonstrated lower risk of developing pneumonia. However, in the largest RCT done for dysphagia of more than 500 patients, there is no increased in the incidence of pneumonia between using thickened liquid or normal liquid with chin-down postures [1].

14.8 Behavioral Interventions

The most widespread treatment of dysphagia comprises exercise and maneuvers ranging from combined or isolated, direct or indirect, and swallowing or non-s swallowing tasks. The European Stroke Organization and European Society for Swallowing Disorder guidelines recommend the use of behavioral swallowing exercises for swallowing rehabilitation. Rehabilitation exercises that can be performed include shaker head lift in patients with upper oesophageal sphincter dysfunction, Masako maneuver that can strengthen the base of the tongue and pharyngeal wall movement and expiratory and submental muscle strength training. Other maneuvers that can be used is chin-down posture that is used in patients with premature spillage and Mendelsohn maneuver in patients with laryngeal excursion [1].

14.9 Rehabilitation Techniques

Rehabilitation techniques focused on improving both the strength and endurance of the patients. The aim is to increase isometric pressure in specific parts of swallowing and can be done via oral and lingual exercise. Neurostimulation therapy such as pharyngeal electrical stimulation, neuromuscular electrical stimulation, and transcranial direct current stimulation has been employed as part of the rehabilitation technique and has been shown to lower aspiration risk, pharyngeal residue, hospital length of stay and improved swallowing performance [1].

14.10 Neurostimulation

Increased cortical activities along with neural repair mechanism have a part in improving swallowing process in post-stroke dysphagia. Neurostimulation can promote reorganization of the cortex which can accelerate stroke recovery thus benefitting the patients with post-stroke dysphagia [6]. A systematic review by Cheng et al. demonstrated that neurostimulation whether peripheral or central stimulation are superior to conventional treatment [8].

14.11 Peripheral Stimulation

14.11.1 Pharyngeal Electrical Stimulation

A transnasal catheter with electrodes transmits low-amplitude electrical impulses into the pharynx. Pharyngeal electrical stimulation is postulated to increase activity over the brain center for swallowing thus promoting cortical reorganization of the swallowing motor complex. Previous studies have shown mixed results with the exception of tracheostomized patients with stroke where there is an evidence of significant improvement in dysphagia using PES. This might be due to the fact that the patient with a more severe stroke showed better response to PES as compared to the less severe patients [6].

14.11.2 Neuromuscular Electrical Stimulation

Transcutaneous surface electrodes are applied to transmit electrical impulses which can generate muscle contractions in suprahyoid and infrahyoid muscles. Stimulation of the suprahyoid muscle will strengthen the muscle allowing elevation of the hololaryngeal complex for airway protection during swallowing. The infrahyoid stimulation depresses the hololaryngeal complex allowing swallowing to happen against resistance. NMES has demonstrated positive results in treating post-stroke dysphagia, however the optimal electrode placement, frequency, and intensity of stimulation are still unclear [6, 7].

14.12 Central Stimulation

14.12.1 Repetitive Transcranial Magnetic Stimulation

Placing a copper coil on the scalp allows delivery of magnetic stimulation which results in changes in electrical activity on the neocortex in repetitive transcranial magnetic stimulation (rTMS) [4]. Two meta-analyses by Pisegna et al. and Liao et al. evaluated 11 studies which concluded that there is a benefit in terms of improving swallowing function in patient with post-stroke dysphagia by using rTMS along with traditional dysphagia therapy. However, there is no consensus length of treatment, specific stimulation parameters to be used, and number of pulses per each treatment [6].

14.12.2 Transcranial Direct Current Stimulation

Transcranial direct current stimulation (tDCS) applies low-intensity electrical current (1–2 mA) through electrodes placed on the scalp. There is depolarization of the resting membrane potential by anodal tCDS leading to increased neuronal excitability. A cathodal tDCS works oppositely to hyperpolarize the resting membrane potential causing reduced neuronal excitability. Previous studies have been inconclusive however recent studies such as those by Ahn et al. and Li et al. demonstrated that tDCS improves swallowing functions in chronic stroke patients. Further research is required as information regarding the placement of electrodes, intensity, and duration of parameters of stimulation, timing post-stroke as well and type of stroke patients are still lacking [6].

14.12.3 Paired Associative Stimulation

Both peripheral and central are stimulated promoting excitation of the pharyngeal motor cortex in paired associative stimulation (PAS). A study by Hamdy et al. demonstrated that there is a short term benefit in swallowing function by applying PAS to post-stroke dysphagia patients leads as well as reduced aspirations in chronic stroke patients. Furthermore, repetitive stimulation and administration over a short period of approximately 10 min can induce excitability in previously unresponsive patients. However, there is no recent research to advocate the effectiveness of PAS as a modality for post-stroke dysphagia [6, 7].

14.13 Enteral Access for Post-stroke Dysphagia

All the stroke patients are at risk of developing malnutrition. The possible causes range from reduced consciousness, reduced mobility, depression, poor oral hygiene, and post-stroke dysphagia. Malnutrition have a severe impact on the stroke patient's life as it worsens the functional status and reduces the quality of life.

The most natural way in which nutrients are being absorbed is through enteral nutrition. Enteral nutrition is the only way to allow the administration of nutrition, and hydration as well as for the administration of medications. Enteral nutrition can be administered with nasogastric (NG), nasoduodenal, or nasojejunal (NJ) tube feeding, percutaneous endoscopic gastrostomy (PEG), or percutaneous endoscopic jejunostomy (PEJ) A NG tube should be done as the first choice of the patients who receive enteral nutrition within 2 weeks. The recommendation for placement of NG tube ranges from within 24 h to 3 days and up to 7 days. However, NG tube enteral feeding is only preferred for short-term use. If enteral nutrition is expected to last for more than 14 days, a nasal bridle tube can be applied in cases where NG tube is intolerable or requires frequent replacement. In patients that are unable to tolerate NG tube, gastrostomy feeding (PEG or PEJ) can be considered more so if the patients are unable to tolerate feeding over 2–3 weeks after the onset of stroke. The delay in placing PEG or PEJ is because half of the patients with dysphagia improved after 2 weeks [9]. A gastrostomy tube is a better choice than NG tube for patients with post-stroke dysphagia that requires long-term enteral nutrition.

NJ tube may be the best choice for patients with gastroesophageal reflux and aspiration [10]. In stroke patients with poor prognosis, the presence of advance directive as well as ethical issues must be taken into consideration, and when in doubt, feeding via NG tube would be the better choice as it is reversible. A comparison study done between NG tube and PEG demonstrated that PEG is associated with lower treatment failures, higher delivery of feed and reduced incidence of gastrointestinal bleed. However, the study showed no difference in terms of death and dependency in the study [11]. In addition, there is a lower risk of tube dislodgement and improved quality of life with a PEG tube as compared to NG tube [12]. The European Society of Gastrointestinal Endoscopy (ESGE) guidelines recommended the use of gastric route for primary options in patients requiring enteral nutrition as it is more physiological and well tolerated. Furthermore, it requires less expertise on top of that it can be bedside thus reducing the delay in starting enteral nutrition. However, if the patients are unable to tolerate gastric feeding despite adding prokinetic agents, jejunal feeding would be the next option. There are no significant differences in terms of mortality, infection, gastrointestinal complications as well as hospital length stay between a jejunal feed or gastric feed [12].

14.13.1 Contraindications of Nasogastric or Nasojejunal Tube and Percutaneous Enteral Tube Insertion

The absolute contraindication for NG and NJ tube insertion includes the presence of mechanical obstruction in the gastrointestinal tract, active peritonitis, severe coagulopathy, and ongoing bowel ischemia [12]. The contraindications for a "blind" NG and NJ tube insertions are anatomical deformities such as facial fracture, skull fracture with CSF leakage, high cervical fracture, upper digestive obstruction, and recent oronasal surgery [12].

14.13.1.1 Contraindications for Percutaneous Enteral Tube Insertion

All the absolute contraindications for NG and NJ tube apply to percutaneous tube insertion. In addition, the relative contraindications also include recent gastrointestinal bleeding with a high risk of rebleeding due to peptic ulcer disease as well as hemodynamic and respiratory instability. The insertion of a percutaneous enteral tube can be challenging in a patient with ascitic fluid. In addition, the healing process of the stomach tract might be impaired as well as there is a higher risk of developing bacterial peritonitis. In patients who require ventriculoperitoneal shunt (VPS) insertion, there is an increased risk of VPS infection if PEG is inserted first. Thus, VPS must be inserted first before percutaneous tube insertion. In patients with concomitant abdominal wall defects such as the presence of stomas, and surgical scars as well as in the presence of adhesions, there is an increased risk of intrabdominal infections. The placement of these tubes must be 2 cm distance from any previous surgery site to avoid potential scar tissues and adhesions intrabdominally [12].

14.13.2 Complications of Enteral Access

The complications of enteral access include nasal trauma, displacement, or poor positioning which can lead to increased risk of aspiration. In a study that determined the incidence and consequences of pharyngeal misplacement using fibreoptics endoscopic, 5% of the total patients had pharyngeal misplacement of NG tube which was associated with worsening dysphagia [9]. NG tube insertion is also associated with traumatic lesions such as nasal trauma, nosebleed, and pressure areas. In a small prospective study evaluating the effect of NG tube insertion, difficult NG tube insertion is associated with hypoxemia [6]. Furthermore, the enteral access can be blocked and require change. Patients might also have gastrointestinal complications and metabolic disorders which ultimately result in poor quality of life.

14.14 Quality of Life with Post-Stroke Dysphagia

In general, the impact of post-stroke dysphagia is very profound and detrimental. There is an inverse proportion between severity of post-stroke dysphagia and the quality of life of the patients. This is mainly due to difficulty in meeting a fundamental physiological need which is eating leading to a high risk of malnutrition [13]. In a recent retrospective study comparing the quality of life of stroke patients between the tube-feeding group and oral-feeding group, the tube-feeding group scored lower than the oral-feeding group in terms of mental health, fear, and desire to eat. The quality of sleep is also affected, and this is possibly due to tube maintenance during sleep [14]. The ongoing recent treatment of post-stroke dysphagia demonstrated potential benefit however further study and recommendations are required to improve the quality of life of patient with post-stroke dysphagia.

References

1. Dziewas, R., Michou, E., Trapl-Grundschober, M., Lal, A., Arsava, E. M., Bath, P. M., et al. (2021). European Stroke Organisation and European Society for Swallowing Disorders guideline for the diagnosis and treatment of post-stroke dysphagia. *European Stroke Journal, 6*(3), LXXXIX–CXV.
2. Banda, K. J., Chu, H., Kang, X. L., Liu, D., Pien, L. C., Jen, H. J., et al. (2022). Prevalence of dysphagia and risk of pneumonia and mortality in acute stroke patients: A meta-analysis. *BMC Geriatrics, 22*(1), 420.
3. Cohen, D. L., Roffe, C., Beavan, J., Blackett, B., Fairfield, C. A., Hamdy, S., et al. (2016). Post-stroke dysphagia: A review and design considerations for future trials. *International Journal of Stroke, 11*(4), 399–411.
4. Shaker, R., & Geenen, J. E. (2011). Management of dysphagia in stroke patients. *Gastroenterology & Hepatology, 7*(5), 308.

5. Yang, C., & Pan, Y. (2022). Risk factors of dysphagia in patients with ischemic stroke: A meta-analysis and systematic review. *PLoS One, 17*(6), e0270096.2.
6. Jones, C. A., Colletti, C. M., & Ding, M. C. (2020). Post-stroke dysphagia: Recent insights and unanswered questions. *Current Neurology and Neuroscience Reports, 20*, 1–12.
7. González-Fernández, M., Ottenstein, L., Atanelov, L., & Christian, A. B. (2013). Dysphagia after stroke: An overview. *Current Physical Medicine and Rehabilitation Reports, 1*, 187–196.
8. Cheng, I., Sasegbon, A., & Hamdy, S. (2021). Effects of neurostimulation on poststroke dysphagia: A synthesis of current evidence from randomized controlled trials. *Neuromodulation: Technology at the Neural Interface, 24*(8), 1388–1401.
9. Reddy, K. M., Lee, P., Gor, P. J., Cheesman, A., Al-Hammadi, N., Westrich, D. J., & Taylor, J. (2022). Timing of percutaneous endoscopic gastrostomy tube placement in post-stroke patients does not impact mortality, complications, or outcomes. *World Journal of Gastrointestinal Pharmacology and Therapeutics, 13*(5), 77.
10. Gong, L., Wang, Y., & Shi, J. (2021). Enteral nutrition management in stroke patients: A narrative review. *Annals of Palliative Medicine, 10*(10), 111911202–111911202.
11. Geeganage, C., Beavan, J., Ellender, S., & Bath, P. M. (2012). Interventions for dysphagia and nutritional support in acute and subacute stroke. *Cochrane Database of Systematic Reviews, 10*, CD000323.
12. Itkin, M., DeLegge, M. H., Fang, J. C., McClave, S. A., Kundu, S., d'Othee, B. J., et al. (2011). Multidisciplinary practical guidelines for gastrointestinal access for enteral nutrition and decompression from the Society of Interventional Radiology and American Gastroenterological Association (AGA) Institute, with endorsement by Canadian Interventional Radiological Association (CIRA) and Cardiovascular and Interventional Radiological Society of Europe (CIRSE). *Journal of Vascular and Interventional Radiology, 22*(8), 1089–1106.
13. Al Rjoob, M., Hassan, N. F. H. N., Aziz, M. A. A., Zakaria, M. N., & Mustafar, M. F. B. M. (2022). Quality of life in stroke patients with dysphagia: Asystematic review. *La Tunisie Medicale, 100*(10), 664.
14. Kim, D. Y., Park, H. S., Park, S. W., & Kim, J. H. (2020). The impact of dysphagia on quality of life in stroke patients. *Medicine, 99*(34), e21795.

Ethical Guidelines for Nutritional Management of Stroke

15

Azarinah Izaham

Abstract

The adherence to key ethical principles in providing nutritional management in stroke patients is essential to ensure high-quality patient-centred care. Respect for autonomy is an important aspect especially in necessitating informed consent, addressing concerns, and involving patients in shared decision-making regarding nutritional interventions in consideration of patients' values and preferences.

Beneficence principal require healthcare providers to promote well-being through comprehensive nutritional assessments, evidence-based individualised nutritional management, judicious selection feeding routes, continuous monitoring avoiding harms, and management of any occurrence of complications.

The importance of non-maleficence avoiding harm through prevention of under or overfeeding, early identification and management of complications and diligent management of feeding tubes to avoid complications.

Justice is important in advocating for equal access and opportunities to nutritional screening, support resources, feeding assistance and supplementation, irrespective of socioeconomic status.

Advocating these ethical doctrine of respect for autonomy, beneficence, non-maleficence and justice, healthcare provider and team can deliver ethical, high-quality nutritional management that optimizes outcomes while preserving patient dignity, well-being, and equitable access to care for all stroke patients.

A. Izaham (✉)
Department of Anaesthesiology & Intensive Care, Faculty of Medicine, Universiti Kebangsaan Malaysia, Kuala Lumpur, Malaysia
e-mail: azaizaham@ppukm.ukm.edu.my; azaizaham@ukm.edu.my

W. Mohamed, M. B. Mat-Nor (eds.), *Nutrition and Stroke*, Nutritional Neurosciences, https://doi.org/10.1007/978-981-97-6818-9_15

Keywords

Ethical aspects · Stroke · Autonomy · Beneficence · Non-maleficence · Justice

15.1 Introduction

Post-stroke malnutrition is associated with poor functional recovery and outcome. It also have a higher rate of mortality, longer hospital stays, and morbidity. Many patients face difficulties meeting their nutritional needs due to effects of the stroke itself or other factors [1]. Therefore, a proper nutritional plan is crucial in recovery and preventing complications after a stroke. This highlights the need for evidence-based nutritional management as a key component of stroke care.

The development of clear ethical guidelines for healthcare professionals involved in providing nutritional support to stroke patients is vital to deliver high-quality, patient-centred care. A multidisciplinary nutritional support team including doctors, nurses, dietitians, speech-language pathologists and other providers is typically involved in assessing patients' needs, recommending interventions, and monitoring progress. Adhering to ethical principles helps coordinate care across specialties and ensures patients' values and preferences are respected.

Research indicates post-stroke malnutrition is common, affecting up to 60% of patients. Contributing factors include dysphagia which can impair ability to swallow safely, diminished appetite, metabolic changes increasing nutritional needs, and limited mobility affecting food access [2]. Timely nutritional management can help counter these issues and impacts patient outcomes. Adequate protein and caloric intake aids wound healing, immunity, and rehabilitation. Dysphagia management reduces aspiration pneumonia risk. Micronutrient deficiencies are also addressed.

The indication, therapeutic goals, and ethical framework that should guide clinicians in providing optimal nutritional support for stroke patients. A focus on sound ethical reasoning and decision-making processes can promote achievement of positive patient-centred outcomes. Practical considerations for nutritional assessment, individualised diet planning and patient/caregiver education will also be discussed.

15.2 Indication and Definition

Nutritional management is indicated for all stroke patients due to the high risk of malnutrition following a stroke. Stroke can directly impact the ability to consume, digest and absorb nutrients through effects such as dysphagia, gastrointestinal changes and altered metabolic states [1]. Indirect factors like reduced mobility, depression and functional disability can also contribute to poor nutrition. Timely assessment and management can help counter these effects.

The European Society for Clinical Nutrition and Metabolism (ESPEN) guidelines define nutritional management in stroke care as "the provision of an adequate amount and composition of nutrients to stroke patients in order to prevent and/or treat malnutrition and its consequences." [3] The key goal is meeting increased nutritional requirements associated with the stroke recovery process and preventing complications of malnutrition.

Proper nutritional support is crucial for stroke patients dealing with various complications Table 15.1. For those experiencing dysphagia or difficulty swallowing safely, often arising from neurological damage, texture-modified diets or enteral feeding may be necessary to ensure adequate nutrient intake. Patients with reduced consciousness or cognitive impairments may require assistance as their ability to communicate needs and self-feed could be compromised. Additionally, stroke can increase resting energy expenditure and protein needs due to hypermetabolism, necessitating adjustments to nutrient targets. Unintentional weight loss, a frequent occurrence after a stroke, signifies inadequate intake and may warrant oral supplementation or enteral/parenteral feeding if the weight loss persists. Evaluating a patient's nutritional status by assessing dietary intake, laboratory results, and body composition is vital, as nutrition support can help replenish any deficiencies identified. Addressing these specific indications through appropriate nutritional interventions is crucial for stroke recovery and rehabilitation.

Appropriate nutritional management starts with screening for risk factors using validated tools, followed by a comprehensive nutritional assessment if risk is identified. This helps determine the optimal nutrition interventions for each patient. The assessment of dysphagia after stroke is crucial prior to the decision of the route of the nutrition [2, 3].

The screening for dysphagia should be conducted early, within 24–48 h of stroke onset, before starting any oral food or fluids. This helps identify patients at risk who require further swallow evaluation. Usually a formal swallowing assessment is typically done by a speech-language pathologist using clinical and instrumental techniques.

The aim of the non-invasive bedside assessment clinical swallowing test are to evaluate the functions of the oral, pharyngeal, and laryngeal in relation to swallowing and also to test the cranial nerve involved by overserving the anatomy, sign of weakness or any asymmetrical movement during swallowing (Table 15.2). Different

Table 15.1 Specific indications for nutritional support

Indication	Effect
Dysphagia	Impaired ability to swallow safely often occurs due to neurological damage. Texture modified diets or enteral feeding may be required
Reduced consciousness/ cognitive impairments	Can affect ability to communicate needs and self-feed. Assistance may be required
Hypermetabolism	Stroke increases resting energy expenditure and protein needs. Nutrient targets may need adjustment
Unintentional weight loss	Frequent after a stroke and indicates inadequate intake. Oral supplementation or enteral/parenteral feeding may be considered if weight loss continues
Poor nutritional status	Assessed by evaluating intake, labs, intake, body composition. Nutrition support helps replenish deficiencies

texture and consistency of foods and fluid used to examine the ability to swallow safely and the cough reflex was also been assess at the same setting which act as protective mechanism against aspiration and at the end of the test the quality of the voice is assess. A systematically evaluation of these components, will guide the healthcare professionals can obtain valuable insights into the patient's swallowing function and develop appropriate intervention strategies.

The instrumental assessments provide a more comprehensive visual analysis of swallowing physiology. The fibreoptic endoscopic evaluation of swallowing (FEES) identification of any potential deficits or abnormalities by inserting of a flexible endoscope through the nasal cavity and direct visualization of the pharyngeal and laryngeal structures can be observe during the swallowing process. Video fluoroscopic swallow study (VFSS) is another technique with the same aim also known as a modified barium swallow. The process involve utilizing fluoroscopy to capture real-time X-ray images of the oral, pharyngeal, and oesophageal phases of swallowing. These offers a detailed assessment of the swallowing mechanism and can detect issues such as aspiration or residue.

The results of the assessment will guide the appropriate diet texture, liquid thickness and route to ensure adequate calories intake and safe method. Diet may be gradually been step up as swallow function improves.

15.3 Therapeutic Goals

The therapeutic goal of nutritional management in stroke is to improve patients' overall health and well-being which include providing adequate nourishment to prevent malnutrition, and addressing any specific dietary requirements or restrictions. Therefore, in order to achieve the goals a personalized nutritional plan tailored to individual needs is crucial for supporting recovery and improving overall prognosis

Table 15.2 Clinical swallow evaluation

1. Observation of anatomy for asymmetry and weakness
2. Testing of cranial nerves involved in swallowing
3. Trials of food and liquid consistency to check ability to safely swallow different textures
4. Cough reflex testing
5. Voice quality assessment after swallows

[3–7]. The nutritional management in stroke care are twofold by providing adequate nutrients, and preventing complications of malnutrition.

The more specific therapeutic goals are maintaining adequate caloric and protein intake.

Preventing aspiration and its complications, good hydration, correcting micronutrient deficiencies, supporting neurological recovery, improving gastrointestinal tolerance and promoting overall functional status (Table 15.3).

Maintaining appropriate caloric intake is essential, as stroke often increases resting energy expenditure with typically targeted at 25–30 kcal/kg body weight, are needed to prevent catabolism and undesirable weight loss. Protein intake, estimated at 1.2–1.5 g/kg body weight daily, is another vital objective as protein plays a crucial role in for stroke recovery such as in immunity and muscle preservation [4].

Aspiration and its complications, such as aspiration pneumonia, is a known complication in stroke patients with dysphagia. The diet texture and liquid modification can help reduce this risk. Adequate hydration is also a priority, supports wound healing and metabolism.

Micronutrient deficiencies, is common in stroke patient and correcting it is one the specific goal. The deficiencies in vitamins and minerals warrant supplementation to replenish levels. Supporting neurological recovery, by the certain nutrients like DHA omega-3 s and antioxidants, which support brain health and neurological function, is an important for positive outcome [4].

Gastrointestinal intolerance such as constipation and diarrhoea are frequent occurrences post-stroke [4, 7, 8]. A proper nutritional plan can help improve gastrointestinal function and promoting overall functional status needed for rehabilitation. Finally, the improvement in quality of life is a crucial goal, with a specific attention to food enjoyment, preferences and patient satisfaction. In order achieve it, these specific therapeutic goals through comprehensive nutritional management plan, healthcare providers can ensure stroke patients' recovery and improve their overall well-being.

Table 15.3 Goals of nutritional management in stroke care

1. Maintaining appropriate caloric intake
2. Ensuring adequate protein intake
3. Preventing aspiration and its complications
4. Maintaining hydration
5. Correcting micronutrient deficiencies
6. Supporting neurological recovery
7. Improving gastrointestinal tolerance
8. Promoting overall functional status
9. Preventing adverse events
10. Optimising quality of life

Ongoing monitoring of nutritional status, weight trends, laboratory markers and achievement of therapeutic goals allows for adjustments to the nutrition care plan to optimise outcomes.

15.4 Ethical Framework

An ethical framework provides guidance to healthcare professionals involved in stroke nutrition management, helping ensure decision-making aligns with principles of sound medical ethics. Key components includes; respect for autonomy, beneficence, non-maleficence and justice [9].

Respect for patient autonomy is a fundamental ethical principle in nutritional management. Healthcare providers must obtain informed consent before initiating any nutrition support interventions, thoroughly explaining the available options, potential risks, and anticipated benefits. Patients have the right to refuse recommended nutritional support, and it is crucial to address their concerns and provide comprehensive education to facilitate informed decision-making. Religious, cultural, and personal preferences and values surrounding food should be carefully considered and integrated into the care plan. Promoting shared decision-making is essential, actively involving patients and their families in discussions about goals of care and proposed nutrition interventions. By upholding autonomy, healthcare teams can ensure that nutritional management aligns with patients' values, beliefs, and wishes, fostering a collaborative and patient-centred approach to care. Respectful dialogue and a commitment to preserving patient autonomy are vital for building trust and delivering ethical, high-quality nutritional care [9–11].

Beneficence is a core principle in medical ethics that guides practitioners to act as they believe is in the best interest of the patient [9]. Therefore in order to apply beneficence in the ethics of nutrition in stroke patient, the health care provider must consider a few aspects. Firstly, assessment of the patient and plan for individualized nutrition prescription plan based on evidence-based guidelines is essential to meet each patient's specific needs. Follow by establishing the least invasive route of feeding should be used during the initiation of feeding establishment. Continuous monitoring is essential to avoid under feeding or overfeeding. If nutritional plan causes any complication, such as vomiting or aspiration, it has to handle promptly. A vigilant management of feeding tube handling must be practice to avoid dislodgment, discomfort, bleeding or any other type of its complications. According to the principal of beneficence, a maximum benefit of nutrition management plan with minimal risk must be applied to improved the overall quality of care [9–12]. Avoiding harm is the main principal of non-maleficence. Therefore, in the nutritional plan management of stroke patient the crucial step is monitoring. Similar to the principal of management discuss in beneficence have to be followed.

The principle of justice, is ensures fair and equitable access to nutritional management plan for all stroke patients [9, 12]. Therefore, advocating for sufficient resources to enable comprehensive malnutrition screening for every stroke patient, regardless of their circumstances. An important principal that the access to appropriate nutritional support should not be influenced by socioeconomic status or other factors that could lead to disparities in care. By promoting justice in nutritional management, providing equal opportunities for patients to receive improved food textures, oral nutritional supplements, or feeding assistance when needed, without any discrimination. Effort towards eliminating unjust barriers and ensuring that all stroke patients have the chance to receive the nutritional interventions is important for their recovery and overall well-being. Addressing inequities and promoting fair allocation of resources is crucial to delivering ethical, high-quality nutritional care that upholds the principle of justice for all individuals.

Practical application of these principles along with good communication, shared decision-making, and patient advocacy helps safeguard ethical delivery of nutrition care. Ongoing analysis of complex cases from an ethical perspective promotes highly principled practice (Table 15.4).

15.5 Respect for Autonomy

Respecting patient autonomy is crucial in stroke nutrition management. It involves involving patients in decision-making regarding their nutritional care. Healthcare professionals should provide information about available options, risks, benefits,

Table 15.4 Key component of ethical framework in stroke care

Respect for Autonomy
1. Obtain informed consent before initiating nutrition support interventions and explain the options, risks and benefits
2. Respect right to refuse recommended nutritional support by addressing the concerns and provide education
3. Consider religious, cultural and personal preferences and values around food when planning care
4. Promote shared decision-making around goals of care and nutrition interventions
Beneficence
1. Complete thorough nutritional assessment and regularly monitor status
2. Design an individualized nutrition prescription utilizing evidence-based guidelines
3. Select least invasive feeding route able to meet nutritional goals
4. Monitor benefits and any potential harms of nutrition interventions
5. Avoid under or overfeeding which can worsen outcomes
6. Monitor for and manage any complications of nutrition support like aspiration, diarrhoea
7. Taking care when handling feeding tubes to avoid discomfort, dislodgement
Non-Maleficence
1. Avoid under or overfeeding which can worsen outcomes
2. Monitor for and manage any complications of nutrition support like aspiration, diarrhea
3. Take care when handling feeding tubes to avoid discomfort, dislodgement
Justice
1. Advocate for resources to provide malnutrition screening for all stroke patients
2. Strive for equitable access to nutritional support, regardless of socioeconomic status
3. Allow equal access to improved food textures, supplements, feeding assistance

and involve patients in creating a personalized nutrition plan that aligns with their preferences. This empowers patients to actively participate in their own care and make informed choices [9, 13].

In stroke nutrition management, respecting autonomy means that recognizing and upholding the patient's right to accept or refuse nutrition support interventions like tube feeding after being informed of risks/benefits. This respects their capacity for self-determination. Involving the patient in shared decision-making around nutritional goals and therapies, rather than making paternalistic decisions on their behalf. This values their perspectives.

Obtaining informed consent before initiating nutrition support therapies. This ensures the patient understands and authorizes the intervention. Accommodating patient preferences regarding food textures, supplements etc. as much as possible within safety limits. This demonstrates respect for their values.

Providing adequate information and education to patients/families to empower their participation in nutrition decisions. This promotes autonomous decision-making. Being sensitive to cultural, religious or personal food choices and accommodating them if feasible. This shows respect for diversity [12, 14].

Overall, respect for autonomy requires care providers to recognize stroke patients as equal partners whose choices and preferences matter in determining their nutritional care. Honouring their right to decide demonstrates this core ethical principle.

Examples to help explain respect for autonomy in stroke nutrition management, a patient with dysphagia may prefer to eat orally rather than receive enteral feeding despite risks. After explaining the risks of aspiration, the speech pathologist should respect the patient's choice if they have capacity to make informed decisions.

In a patient wants to continue eating only soft, pureed foods despite improving swallow function. Their preferences should be accommodated. If a patient expresses feeling overwhelmed by nutrition decisions after stroke. The dietitian should take time to simplify information and actively involve the patient in setting goals and priorities. The health care provider should not make assumptions about goals of care. Preferences regarding artificial nutrition at end-of-life should be directly discussed with the patient/family.

Staff should take care to offer food/fluids in a respectful manner, honouring the dignity of stroke patients who need assistance eating. The patient food preferences related to culture, taste or intolerances should be incorporated into nutrition plans when possible.

The key is recognizing nutrition decisions post-stroke can be complex and emotional. Respecting patient perspectives, providing explanatory information, and actively involving patients in shared decision-making helps uphold autonomy. While safety is also important, patient wishes generally take priority when competent.

15.6 Beneficence

The principle of beneficence emphasizes acting in the best interest of the patient's health and well-being. In stroke nutrition management, the dietary modifications, use of nutritional supplements, or other interventions aimed at optimizing nutritional status and promoting recovery must based on evidence-based recommendations according to individual needs [9–14].

The principle of beneficence can be applied by conducting comprehensive nutritional assessments at regular intervals using tools by evaluating few parameters such as weight, calory and fluid intake and specific laboratory investigation for example the level of trace elements. These assessment allows identification of risks and deficiencies requiring further monitoring and intervention. Specific plan nutritional management plan according to the patients' needs and goal with adequate calories, protein and micronutrients will provide optimal nutritional status and balances risks and benefits.

Monitoring intake and nutritional markers to ensure adequacy and make timely adjustments to the care plan as needed. This maximizes benefits. Cooperating across disciplines like nurses, dietitians, and speech pathology to coordinate nutritional care. This promotes a unified approach. Advocating for resources to manage malnutrition including supplements, feeding assistants, and community referrals. This expands access.

Educating patients and caregivers on the importance of proper nutrition and how to follow dietary advice. This empowers involvement in care. Considering quality of life factors like food enjoyment and socialization opportunities around meals. This contributes to overall wellbeing. Reporting substandard nutritional practices that compromise patient health and safety. This enhances quality of care.

The key is focusing both short and long-term nutrition interventions toward restoring and maintaining the patient's health in alignment with clinical standards and their goals. Ongoing re-evaluation and improvement of care is needed to fully enact the principle of beneficence.

15.6.1 Short-Term Nutritional Interventions

The short-term nutritional interventions that align with the principle of beneficence in acute stroke care, includes screening for dysphagia within 24 h of admission to identify aspiration risk and need for modified diet textures and this will prevents pneumonia [4].

Initiating enteral feeding within 24–48 h if oral intake is unsafe or inadequate. This prevents catabolism from developing. Assessing hydration status and providing IV fluids if dehydration is present. This stabilizes fluid balance. Ordering lab work to check electrolytes, glucose, vitamin levels and supplementing deficiencies. This will corrects acute imbalances.

Consulting speech pathology for swallow evaluations and diet texture recommendations. This will facilitates safe oral feeding. Involving a hospital dietitian to determine calorie and protein needs for wound healing and supporting the clinical recovery. Discussing goals of care and use of feeding tubes with patient and family. This is an informs decision-making. Educating patient/family on modified diet, proper feeding techniques to prevents complications. Managing gastrointestinal issues like constipation proactively to reduces discomfort. Facilitating early mobility and feeding assistance, so that it will maintains independence.

The focus is taking quick action to maintain nutrition, hydrate, correct deficits, implement feeding adaptations, and prevent adverse events in the acute phase. This provides stabilization until longer term nutrition plans can be made. Ongoing monitoring helps ensure interventions produce intended benefits.

15.6.2 Longer-Term Nutritional Interventions

As longer-term nutritional interventions that uphold the principle of beneficence during stroke recovery, continuing modified diets if dysphagia persists, with ongoing swallowing assessments to upgrade texture as appropriate to facilitates safe oral feeding [10, 11].

Establishing tube feeding regimens to meet full calorie, protein, fluid needs for patients unable to obtain adequate oral intake to prevent malnutrition. Consulting dietitians to adjust nutrition prescriptions as needs change during recovery and rehabilitation for optimizes nutrition.

Coordinating community referrals like meal delivery or outpatient dietitian follow-up after hospital discharge to provides ongoing support. Recommending adaptive equipment like built-up utensils, cups with handles, and plate guards to aid independent eating to promotes functionality. Screening for secondary nutritional issues like poor dentition, gastrointestinal disorders, that emerge post-discharge to allows timely intervention.

Providing nutrition education for lifestyle factors like heart healthy diets, hydration, avoiding alcohol/drug interactions to supports long-term health. Discussing goals like dietary management of risk factors to prevent recurrent stroke to motivates behavioural change. Monitoring weight, laboratory, intake at follow-up visits and adjusting interventions as needed. This ensures continued adequacy.

The focus is maintaining nutritional support, adapting to evolving needs, coordinating transitions of care, and promoting patient/family education and engagement in the long-term. This achieves ongoing nutritional wellbeing.

15.7 Non-Maleficence

Non-maleficence highlights the importance of avoiding harm to patients during nutritional interventions for stroke management. Healthcare professionals should carefully consider potential risks or adverse effects when implementing specific dietary plans or interventions [9, 14]. This includes monitoring for any side effects or complications that may arise from dietary changes or the use of nutritional supplements.

Non-maleficence refers to the ethical obligation of healthcare providers to avoid causing harm or injury to patients during the course of treatment and care. In the context of stroke nutrition, it means that refraining from providing inadequate or excessive nutrition that can physically harm the patient. Under or overfeeding contradicts clinical standards of care. Avoiding actions that could aspiration like improper diet textures or poor feeding techniques. Aspiration pneumonia would be a preventable harm. Not using inappropriate medications or therapies that interact poorly with nutrition recommendations. Malnutrition or side effects may ensue. Abstaining from restraining patients unnecessarily during meals or forcing food intake beyond tolerance. This can causes distress.

By not allowing commercial interests to influence nutrition prescriptions in ways that undermine health. The patient's wellbeing takes priority. Identifying risks associated with nutrition support like tube complications and taking steps to avoid them and prevention of harm is key. Upholding safety and sanitation protocols to prevent infectious or mechanical complications of feeding tubes or dietary intake. Committing to regular monitoring and evaluation of the patient's response to nutritional interventions. This will allows early intervention for adverse events. Reporting substandard nutritional care practices that place patients at risk for harm and this is accountability to promote safety [4].

In essence, non-maleficence means carefully considering how nutritional care may produce avoidable harm and taking reasonable steps to prevent it. The emphasis is on safe, evidence-based nutrition support aligned with patient goals.

The principle of non-maleficence can be applied to nutritional management of stroke patients by carefully calculating calorie and nutrient needs to prevent complications of under or overfeeding. Underfeeding can worsen wounds and immunity. Overfeeding can cause metabolic issues. Monitoring for signs of aspiration during oral intake such as coughing or change in voice. Stopping the meal and reassessing helps prevent pneumonia [4, 10, 11].

Using proper sanitation protocols when handling feeding tubes to avoid infection. Securing tubes properly prevents dislodgement. Watching for diarrhoea, cramping or bloating and adjusting rate/formula if feeding intolerance occurs. This prevents worsening gastrointestinal issues.

Being aware of potential food/drug interactions and adjusting medications or diet to avoid reduced drug efficacy or increased side effects. Taking care not to force-feed stroke patients with swallowing difficulties. Proceeding slowly and allowing breaks prevents distress. Avoiding restraining patients' arms during meals unless essential for safety to allows a more functional eating experience [12].

By reviewing bloodwork regularly to detect refeeding syndrome in malnourished patients and adjusting intake to avoid it. Selecting enteral feeding tubes appropriate to patient size and anatomy to minimize discomfort and to replace with another plan when needed to maintain integrity. Reporting and addressing lapses in oral care, feeding assistance or safety checks that could jeopardize patients.

The focus is vigilance in monitoring for and promptly managing any potential harms of nutritional support through individualized plans, safe protocols and quality oversight. This upholds the principle of non-maleficence.

15.8 Justice

Justice ensures fairness and equity in resource allocation when it comes to stroke nutrition management. Healthcare professionals should strive for equal access to appropriate nutritional support based on individual needs, without any form of discrimination [10, 14]. Besides socioeconomical status, other important aspects in nutritional management plan to ensure justice are cultural preferences and religious beliefs and restriction of the patients.

Advocating for vulnerable patient populations to receive appropriate meal assistance, feeding aids, and dietary counselling. Striving to overcome social and economic barriers patients may face in procuring nutrition supplements, transportation to clinics, or obtaining prescribed modified food textures.

Treating patients equitably by basing nutrition recommendations on clinical characteristics and avoiding biases. Distributing healthcare resources and staff time needed for nutritional management according to actual patient needs and priorities. Ensuring health care provider have adequate training in evidence-based stroke nutrition protocols to standardize and optimize care [12].

Referring malnourished stroke patients to community nutrition programs regardless of ability to pay if needed. Reporting nutritional policies or practices that discriminate against subgroups of patients to promote meaningful change. Including family and community perspectives when devising programs and policies around stroke nutrition. Advocating for research on stroke nutrition that addresses disparities and aims to improve outcomes for diverse populations [12].

The key focus is identifying and removing potential barriers to optimal nutritional care across patient demographics. This upholds fairness and embodies the ethical principle of justice.

The principle of justice as it applies to nutritional management of stroke patients, the health care providers should ensure all patients receive basic screening for nutritional risk, regardless of socioeconomic status. Missing malnutrition in disadvantaged groups due to lack of resources is unjust.

The health care provider should advocate for equitable insurance coverage and reimbursement for nutritional therapies like supplements and feeding tubes. Financial barriers undermine care. The hospitals should invest in adequate professional translation services so nutrition education and counselling can be delivered in the patient's language. Language barriers should not impede care. When budget limitations arise, cuts should not disproportionately target resources needed for nutritional care like feeding assistants or multidisciplinary nutrition teams. This would worsen disparities.

The health care provider should take time to understand patients' cultural food traditions, practices or religious fasting and make accommodations. Patients should not have to choose between beliefs and health. The health care provider should be proactively assess for food insecurity and collaborate with social work to address economic limitations, not just medical nutrition needs in isolation. Access to food is a justice issue.

The policy makers need to engage community representatives on stroke nutrition guidelines to incorporate diverse perspectives and needs. Top-down approaches alone are insufficient. Research agendas should prioritize studying nutritional disparities in underserved stroke populations. Justice requires evidence on eliminating inequities.

The hospitals should invest in patient navigation services and community health workers to connect vulnerable patients to nutrition resources upon discharge. This promotes equity. The focus should move beyond treating all patients equally to treating them equitably according to need. Eliminating disparities requires analyzing root causes and tackling injustice.

15.9 Practical Considerations

In addition to the ethical framework, healthcare professionals should also consider practical aspects of stroke nutrition management. This includes assessing the patient's nutritional status, addressing any specific dietary requirements or restrictions, and providing education and support to both the patient and their caregivers [10–13].

15.9.1 Nutritional Assessment

In order to have a comprehensive nutritional management plan, a comprehensive nutritional assessment is essential, which includes evaluating the patient's dietary intake, weight, height, biochemical markers, laboratory tests (albumin, prealbumin, electrolytes) that will indicate any specific nutritional deficiencies or requirements, functional ability such as chewing and swallowing, grip strength and mobility. A validated nutritional screening tools have to use to identify signs of malnutrition needing intervention.

15.9.2 Individualised Nutrition Management Plan

The individulised nutritional management plan is based on the nutritional assessment, the specific nutritional needs, dietary preferences, and any swallowing difficulties or other challenges they may face. The calorie and protein targets should be calculated based on weight, body mass index and clinical factors like wounds, tube feeding needs. Micronutrient levels are evaluated and deficiencies corrected through supplements if needed. The modifications is included the texture of food, use of thickened liquids, or the use of enteral or parenteral nutrition if necessary [12].

15.9.3 Education and Support

Education and support are crucial components of stroke nutrition management. Healthcare professionals should provide patients and their caregivers with information about the importance of proper nutrition, dietary modifications, and strategies to address swallowing difficulties. This empowers them to make informed choices and ensures compliance with the nutrition plan [4, 12].

Teach meals preparation techniques like added seasonings, minced meats, gravy to improve taste of texture modified foods. Providing hands-on feeding assistance training for caregivers supporting stroke patients. The explain signs of feeding intolerance like vomiting, diarrhoea requiring provider notification must also been thought. Reviewing strategies to track accurate oral intake and weight. Refer families to community nutrition resources post-discharge and assist with food stamps applications.

The focus is individualizing care through comprehensive assessment, setting nutrition targets, and equipping patients and caregivers with knowledge and support. This facilitates optimal recovery.

15.10 Conclusion

Proper nutrition is a vital yet complex component of recovery and wellbeing for stroke patients. Malnutrition can arise due to effects of the stroke itself as well as from secondary impairments in functional ability. Regular screening and assessment to identify risks, defiance, specific and tailored to the patients' nutritional support reduce malnutrition and its negative outcome.

An ethical framework guided nutritional decision-making in stroke care that upholds principles of autonomy, beneficence, non-maleficence and justice. Patients have a right to participate in determining their nutritional plan in alignment with personal values and preferences. The health care provider are ethically obligated to plan for evidence-based management plan and interventions aiming on optimizing nutritional status and avoiding complication. Specific consideration in addressing any disparities in access, resources or quality of nutritional care across patient populations is essential.

Integrating a sound ethical principals to patient nutritional management plan in the clinical decision, the stroke care teams can collaborate with patients and their families to provide a holistic nutrition management that is patient-centred, equitable, adapted to evolving needs, and focused on the highest standards of care. However, to strengthen nutritional care through expanded research, improved quality metrics, better clinician training, and heightened public awareness are needed. Continuous progress in strong ethics, the nutritional status and overall recovery of stroke survivors can be significantly enhanced.

References

1. Gomez-Roig, M. D., Miret-Cuesta, M., & Tur, J. A. (2020). Management of post-stroke dysphagia, aspiration, and malnutrition. *Nutrients, 12*(7), 2155.
2. Holas, M. A., DePippo, K., & Reding, M. J. (1994). Aspiration and relative risk of medical complications following stroke. *Archives of Neurology, 51*, 1051–1053. https://doi.org/10.1001/archneur.1994.00540220099020
3. Burgos, R., Bretón, I., Cereda, E., et al. (2018). ESPEN guideline clinical nutrition in neurology. *Clinical Nutrition, 37*(1), 354–396.
4. Lakhan, S. E., Makay, A. B., Maski, L., et al. (2020). Nutritional therapies for acute stroke: A systematic review. *Nutrients, 12*(7), 2096.
5. Bally, M. R., Yildirim, P. Z. B., Bounoure, L., et al. (2016). Nutritional support and outcomes in malnourished medical inpatients: A systematic review and meta-analysis. *JAMA Internal Medicine, 176*(1), 43–53.
6. Foley, N., Finestone, H., & Woodbury, M. G. (2012). Energy and protein intakes of acute stroke patients housed on a dedicated stroke unit: How close are we to meeting evidence-based targets? *Journal of the Academy of Nutrition and Dietetics, 112*(4), 550–555.
7. Wirth, R., Smoliner, C., Jäger, M., Warnecke, T., Leischker, A. H., & Dziewas, R. (2013). Guideline clinical nutrition in patients with stroke. *Experimental & Translational Stroke Medicine, 5*(1), 14.
8. Finestone, R., & Hillel, M. (2008). Malnutrition and dehydration in the elderly stroke patient. *The Open Nursing Journal, 2*, 58–61.
9. Geppert, C. M. A., et al. (2020). Ethical issues in the nutritional support of patients with dementia near the end of life. *American Journal of Hospice and Palliative Medicine®, 37*(8), 614–620. https://doi.org/10.1177/1049909120910703
10. Mitchell, S. L. (2015). Financial incentives to place feeding tubes in nursing home residents with advanced dementia. *Journal of the American Geriatrics Society, 63*(1), 129–131.
11. Smith, L. E., et al. (2017). Decision making about enteral tube feeding for older adults with dementia. *American Journal of Alzheimer's Disease & Other Dementias, 32*(6), 305–310.
12. Orrevall, Y. (2015). Nutritional support at the end of life. *Nutrients, 7*(3), 1672–1680. https://doi.org/10.3390/nu7031672
13. Plonk, W. M., & Arnold, C. S. (2005). Terminal care: The last weeks of life. *Journal of Palliative Medicine, 8*(5), 1042–1054. https://doi.org/10.1089/jpm.2005.8.1042
14. Crary, M. A., et al. (2013). Ethical challenges in dysphagia management: Making good choices. *Seminars in Speech and Language, 34*(4), 214–224. https://doi.org/10.1055/s-0033-1353447

16 Nutritional Interventions in Clinical Practice Guidelines for Stroke

Iqbalmunauwir Ab-Rashid

Abstract

This chapter explores on the nutritional interventions in the Clinical Practice Guidelines in stroke patients, with focus on the post stroke period. It examines the available interventions that has been well practiced worldwide and described further the components and importance of the interventions for recovery in stroke patients. The importance of the interventions themselves cannot be obscured by the importance of education and counselling on the interventions. Following through the chapter, readers will be exposed on the possible barriers and facilitators for the implementations of the nutritional interventions for the betterment of the post stroke patients care. At the end of the chapters, writer open up several areas of opportunities for further research aiming to look into better interventions or even to overcome all the barriers for implementation of such interventions.

Keywords

Mediterranean · Hypertension · DASH · Nutrition · Stroke

16.1 Introduction

Malnutrition among post stroke patients is associated with more complications and poor functional status, and eventually increased the mortality rate [1]. Stroke causes reduced conscious state and possible dysphagia which are the most common problems causes reduced nutritional intake [1, 2]. Other factors associated with

I. Ab-Rashid (✉)
Department of Anaesthesiology and Intensive Care, Kulliyyah of Medicine, International Islamic University Malaysia, Kuala Lumpur, Malaysia
e-mail: iqbal@iium.edu.my

W. Mohamed, M. B. Mat-Nor (eds.), *Nutrition and Stroke*, Nutritional Neurosciences, https://doi.org/10.1007/978-981-97-6818-9_16

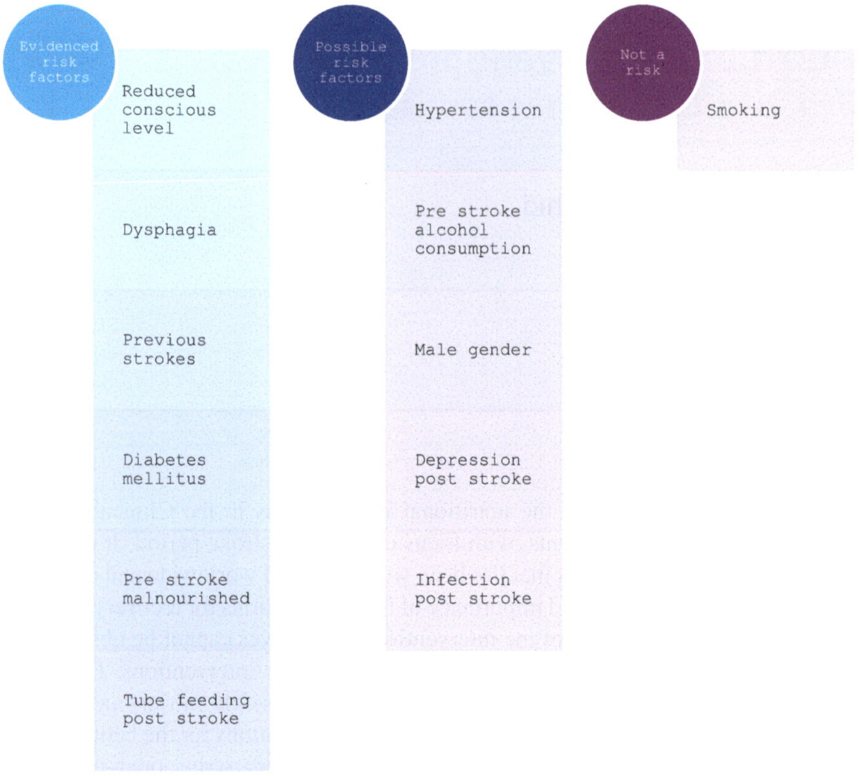

Fig. 16.1 Factors at risk for malnutrition in post stroke patients [2]

malnutrition in peri stroke period include previous stroke state, premorbid diabetes mellitus, malnutrition on stroke presentation and use of tube feeding in post stroke period [2]. Smoking is not considered as a risk factors [2]. Hypertension, pre stroke alcohol consumption, male gender and depression in post stroke period have vague impact on the malnutrition in post stoke patients [2]. Infections, including pneumonia require further analysis in determining association of risk for malnutrition [1] (Fig. 16.1).

16.2 Nutrition to Reduce Risk of Stroke

Evidence on dietary interventions to reduce risk of stroke is limited and the recommendations are mainly taken from increased risk of cardiovascular diseases (CVD) and coronary heart diseases (CHD) patients [3]. Mediterranean diets with added olive oil or nuts as supplements are generally reduces risk of CVD and CHD [4].

16.3 Nutritional Intervention in Post Stroke Patients

Previous studies on nutrition observed reduced risk of stroke from increased consumptions of fish [5], fruits and vegetables [3], fibers [6], adherence towards Mediterranean diet [7], and Dietary Approaches to Stop Hypertension (DASH) diet [8]. The optimum diet for post stroke patients should include plant-based diet including vegetables and fruits, counting for more than four servings per day, that come from various sources, and fish oil of two portions per week (including salmon, trout, herring, pilchards, sardine or fresh tuna) [9].

Consumption of high potassium and low salt are related to reduced incidence of stroke [3]. Reduction of salt intake is important in post stroke patient to reduce the CVD risk [20]. This includes avoiding adding salt on table, using no or reduce salt in cooking and restrain from high salt containing foods, such as processed meat, stock cubes, cheese, snack or pre-prepared soup [9].

Collective recommendations were similarly found in all American, UK and Ireland, and Malaysian Clinical Practice Guidelines in the Management of Stroke, produced by the American Heart Association (AHA) and American Stroke Association (ASA), Royal College of Physician of UK (RCPUK) and Ireland (RCPI), and Malaysian Society of Neuroscience (MSN), respectively. The recommendation includes implementation of Mediterranean type diet in post stroke care [3, 9, 10]. All guidelines recommend consumption of low salt diet, with focus on following DASH diet [3, 9, 10]. Table 16.1 summarize the recommendations from the guidelines.

16.3.1 Mediterranean Diet

Mediterranean diet reflects to the eatery pattern of communities in the Mediterranean region, and this refers back to early 1960s-time frame [7]. It is not merely the description of specific component on micro- and macro-nutrients, but more of description of historical and cultural lifestyle and eating habits [7]. Multiple studies had described the elements in the Mediterranean diet, which includes the following (Fig. 16.2):

- Large quantities of plant food intake, comprising of good variance of vegetables and fruits, breads of whole grain ingredients and cereals, seeds, nuts, and beans.
- Use of fresh own grown plants.
- Use of unprocessed foods.
- Use of honey or concentrated sugars taken at reduced quantities and frequency per week.
- Use of plant-based oil such as olive oil in cooking and act as fat source.
- Consumption of more fishes, and low red meat.
- Reduce intake of red wine.
- Low to moderate use of cheese or yogurt.

Table 16.1 Summary of recommendations in nutritional interventions from Malaysian, American, and UK and Ireland CPGs

No	Recommendations	Malaysian CPG [10] (by MSN)		American CPG [3] (by AHA and ASA)		UK and Ireland CPG [9] (by RCPUK and RCPI)
		COR	LOE	COR	LOE	2023 vs 2016 guidelines
1	To implement Mediterranean diet with focus on monounsaturated fats, plant-based diet and increase consumption of fish	2	B	2a	B-R	These are recommended in latest 2023 guideline similar to 2016 guidelines.
2	In patient with co-morbid of hypertension, sodium restriction is recommended	–	–	2a	B-R	
3	Adherence to DASH diet with low saturated fats and high fruits and vegetables, to reduce BP	1	A	–	–	
4	It is not recommended to supplement diet with vitamins A, B, C, D or E, or Calcium	–	–	–	–	
5	Reduction of alcohol consumption reduce risk of stroke	2	B	1	B-NR	
6	In diabetic patients, target HbA1C should be strictly less than 6.5–7.0%	1	A	1	A	
7	Strict glucose level control by multidisciplinary approach	1	A	1	C-EO	

Mediterranean diet in general reduce stroke incidence in high risk CVD patients with Hazard ratio (HR) of 0.60 (95% CI, 0.45–0.80) [6]. While collective primary end point of Myocardial Infarction (MI), stroke and cardiovascular death, showed adjusted HRs of 0.72 (95% CI, 0.54–0.95) and 0.69 (95% CI, 0.53–0.91) for Mediterranean diet with supplement of olive oil and nuts, respectively, comparatively to the control diet [6].

Reduction and replacement of saturated fats with mono- and poly-unsaturated fats are encouraged in post stroke patients [9]. This include consumption of low-fat dairy products, use if vegetable oils in replacement of butter and ghee, and limiting red meat and processed meat [9].

16.3.2 Dietary Approach to Stop Hypertension (DASH)

Back to early 2000, a thorough research under sponsor of US National Health Institute (NIH), was successfully proof that certain modification in diet is an alternative to classic pharmacological method to treat hypertension [8]. Apart from reduction of blood pressure, the proposed modification of diet also reduces the lipoprotein-cholesterol of low-density family (LDL-C). The described diet also

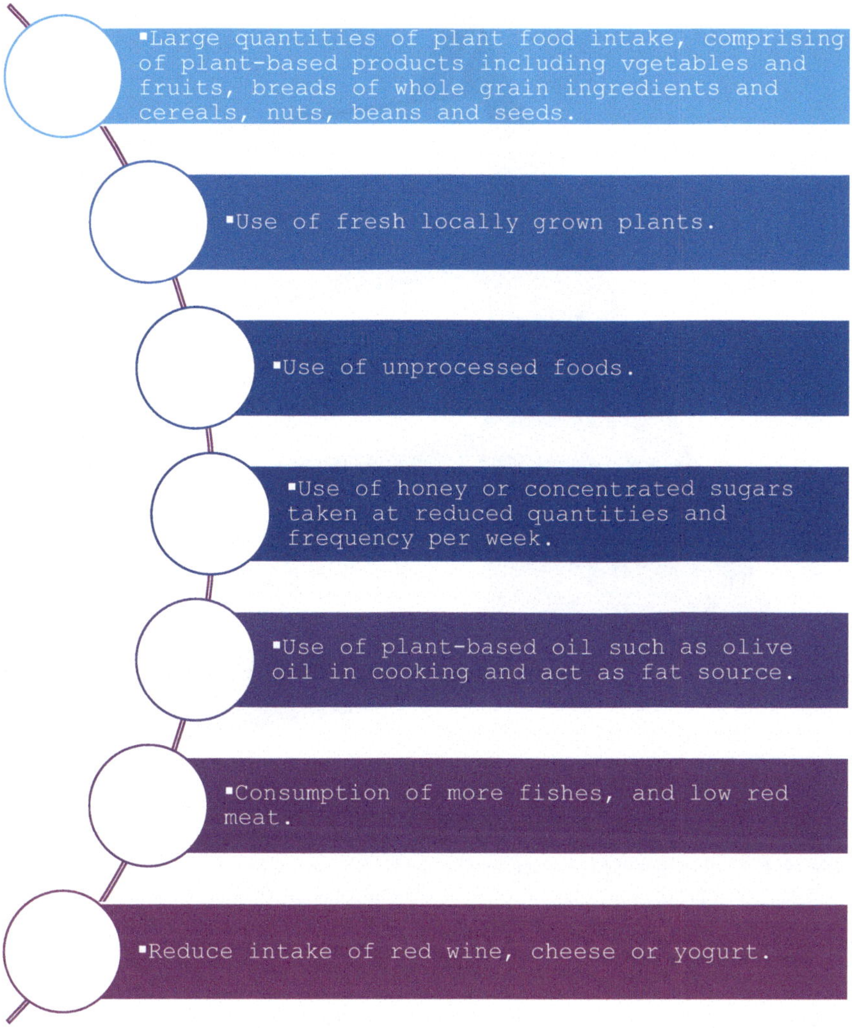

Fig. 16.2 Summary of Mediterranean diet recommendations

corresponded to reduce of diabetes mellitus risk and complications and cardiovascular related death [8, 11].

This diet modification was termed as Dietary Approach to Stop Hypertension or DASH in short. DASH diet is similar to Mediterranean diet (Table 16.2) with focus on increasing intake of vegetables, fruits, low-fat diary, nuts, whole grains type, and beans; reducing consumption of sweets based products, beverages with simple sugars, added sugars, and processed meats; and minimize intake of total or saturated fat and cholesterol [8] (Fig. 16.3).

Consumption on DASH diet does demonstrate reduce risk of CVDs, CHD and stroke [12]. The reduction of Relative Risk (RR) are calculated at 0.80 (95% CI, 0.76–0.85), 0.79 (95% CI, 0.71–0.88) and 0.81 (95% CI, 0.72–0.92), respectively

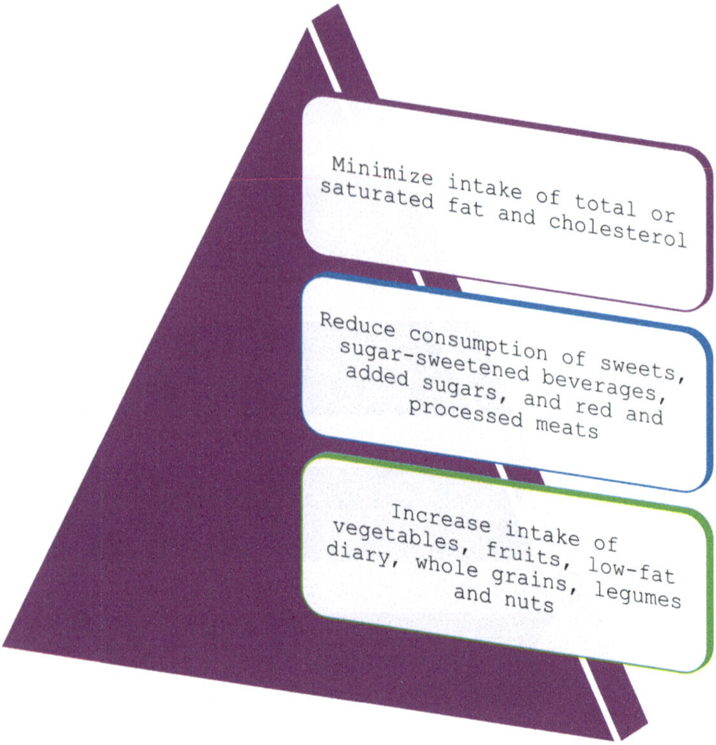

Fig. 16.3 Summary of DASH diet components

Table 16.2 Comparison of Mediterranean diet and DASH diet [3, 8]

Components	Mediterranean diet	DASH diet
Fat	High monounsaturated/saturated fat ratio	Reduce saturated fat and cholesterol
	Use of olive oil as main source	
Sugar, sweets	Discourage soda, sweets, commercial bakery	Reduce sweets, sugar-sweetened beverages, added sugars
Plant-based food	Increase fruits, vegetables and legumes	Emphasize on intake of vegetables, fruits and legumes
Grains and cereal	Large consumptions	Focus on whole grains
Fish	Increase intake	–
Red meat	Reduce consumption	Limit intake
Dairy products	Moderate intake	Emphasize use of low-fat products

[8, 12]. Thus the findings showed the significant of implementation of DASH diet in post stroke patients, which reduce the complications and risk of recurrent stroke.

Apart from that, DASH diet consumptions affect level of blood pressure, cholesterol, glucose control, and body weight. SBP reduced by Mean Difference (MD) of −5.20 mmHg (95% CI, −7.00 to −3.40 mmHg), while DBP reduced by MD of −2.60 mmHg (95% CI, −3.50 to −1.70 mmHg) [12]. Following DASH diet, cholesterol level showed reduction by MD = −0.20 mmol/L (95% CI, −0.31 to −0.10 mmol/L) and MD = −0.10 mmol/L (95% CI, −0.20 to −0.01 mmol/L) for Total-C and LDL-C, respectively. HDL-C showed no difference [12]. As for glucose control, there was a reduction of HbA1C by MD = −0.53% (95% CI, −0.62 to −0.43%) [12]. Body weight has been proved to reduce by MD = − 1.42 kg (95% CI, −2.03 to −0.82 kg).

16.3.3 Other Interventions of Interest

Alteration in alcohol consumption habits help in reduction of stroke recurrence, as stroke risk has been postulated to high alcohol consumption [21]. This is described as more than 4 shots in a day or 14 shots in a week, and more than 3 shots in a day or 7 shots in a week, for men and women, respectively [3]. Level of consumption of alcohol for recurrent stroke was not well defined. However, consumption of more than 60 g/day or 4 drinks per day has been associated with recurrence of stroke in 90 days [3, 13]. This similar to RR described as 1.69 (95% CI, 1.34–2.15) in a meta-analysis with similar level of alcohol consumption.

Glucose control should be considered as part of the intervention in stroke patient. Aim of HbA1C level of less than 6.5–7.0% should be the target for patient post stroke with diabetes mellitus [3]. Separately, the strict daily glucose control should be implemented to reduce risk of recurrent stroke. Diabetes is associated with increased risk of first ischaemic stroke (RR 1.5–3.7%) [3]. Diabetes mellitus has independently described to increase risk of stroke up to 1.8- to nearly sixfolds [10]. The HR is increased in patients with diabetic and high HbA1C, up to 2.15 (95% CI, 1.76–2.13) [14].

16.3.4 Nutritional Counseling and Education

The nutrition interventions themselves is of important in improving the outcome of post stroke patients. But what is more important is the education and counselling for patient and care-givers for the discussed interventions [15]. Studies have proven that counselling by the professionals such as dietician, or nutritionist has the bigger impact on the patients care [15]. The earlier counselling or education is the best time period for opportunity to gather knowledge on the patients care [16]. But careful assessment of readiness of patients to accept the knowledge need to be looked into carefully to avoid overwhelming of the situation and eventually might cause disappointment and depression [16].

16.4 Challenges, Barriers and Future Directions

16.4.1 Barriers and Facilitators in Implementing Nutritional Interventions

By understanding the challenges and barriers for implementation of nutritional interventions, we would be able to plan to encounter those problems. Studies were done to look into this challenges and barriers. Three main themes had been identified as the barriers, including changes in the pre and post stroke situation, impact of environmental support, and inspiration and motivation to changes [17] (Fig. 16.4).

For each themes, certain sub-themes had been recognized, for easier management of each problems and improvement in the clinical practice of stroke patients. For changes in post stroke situation, patients had such limitation in transportation

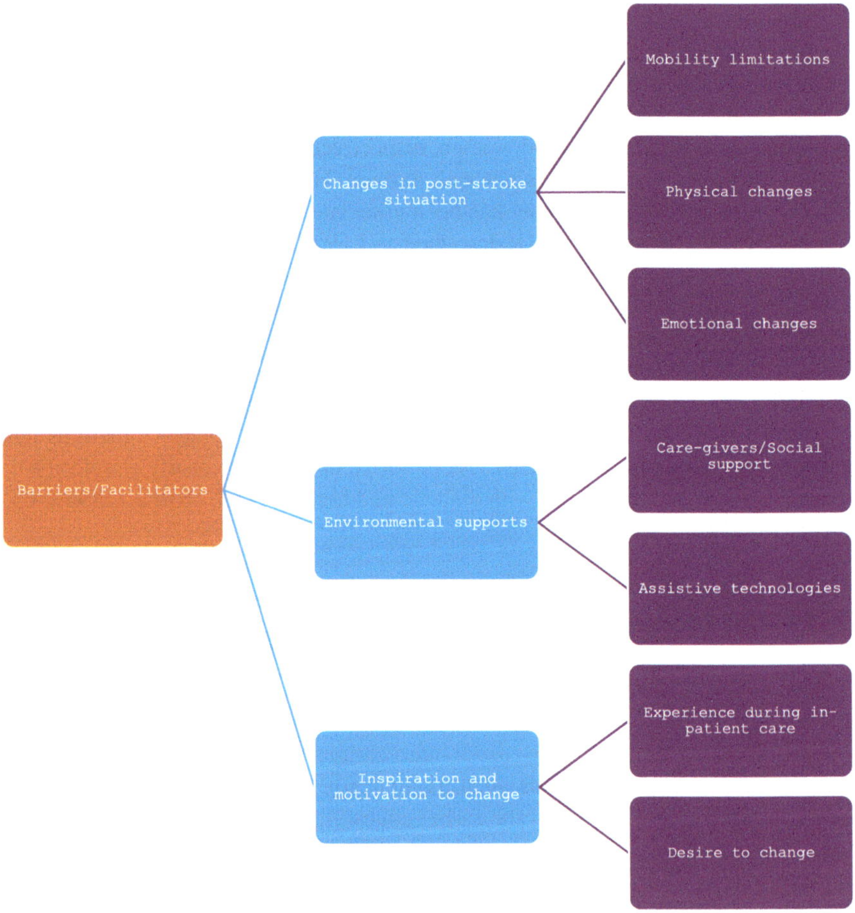

Fig. 16.4 Barriers or facilitators in implementing nutritional interventions in post-stroke patients

and mobility [17]. This include difficulties in food preparation from the very beginning. Patients report difficulties in driving for groceries, buying the products, preparation in kitchen, cooking and eventually eating [18]. These may be attributed by the body structure changes, physiological disturbances [18], and changes in vision, flavour, and taste buds [17]. Negative feelings do affect the changes in the post stroke situation, including the fear feeling on driving and cooking, dissatisfaction, and disappointment [18].

The environmental support could be both either barriers or facilitators, and can be described in wide discussion in term of stroke patients' care. The care-givers support are mostly considered as facilitators towards better nutritional interventions in post stroke patients [17, 18]. This include facilitation in transportation, preparation of food, or even during the serving and eating [18]. Unfortunately, the full assistance might not be considered at all time as facilitator, as patients do feel dependency situation, which further lead to frustration, discouragement and depression [18]. Some patients depends on the assistive technologies to overcome the limitations, including walker and motor-powered cart [18].

As of inspiration and motivation to change, study observed the desire to change will definitely exhibit dietary changes in post stroke situation [17]. This might be closely related with the experience during the in-patients care in health care facilities with involvement of all the related health care professionals, including, but not limited to, doctors, nurses, dieticians, and rehabilitation personnel.

16.4.2 Areas for Future Research and Development in Stroke Nutrition

Post stroke patients are at high risk of recurrence of stroke and malnutrition from poor dietary intake. There are area for exploration for better outcome in stroke patients. One of the area is support group involving the patients and care-givers for better understanding and cross-check mechanism for better dietary intake [19]. One of the initiative was Healthy Eating and Lifestyle after Stroke (HEaLS) initiative which is a multi-disciplinary approach involving all parties in care of stroke patients. The study showed improvement in stroke patients care after the initiatives, and proposed better outcome of stroke patients [19].

16.5 Chapter Summary

Nutritional interventions are important in both primary and secondary prevention of stroke. Some risk factors with malnutrition have been identified in post stroke patients, which include, but not limited to, reduce conscious level and dysphagia. The two most well described nutritional interventions include Mediterranean diet and DASH diet. Both has similar recommendations with focus on increasing plant-based consumption, reduction of processed red meat, and limit of sweet based products and salt intake. Strict reduction on alcohol and management of good glycemic control are among other interventions that need to be managed together. All these interventions are meant to improve outcome in post stroke patient and eventually reduce the risk of recurrence stroke. Some barriers and facilitators have been identified for better care of stroke patients. Area of exploration are widely open to facilitate improvement in post stroke patients' care.

References

1. Sato, Y., Yoshimura, Y., & Abe, T. (2021). Nutrition in the first week after stroke is associated with discharge to home. *Nutrition, 13*, 943. https://doi.org/10.3390/nu13030943
2. Chen, N., Li, Y., Fang, J., Lu, Q., & He, L. (2019). Risk factors for malnutrition in stroke patients: A meta-analysis. *Clinical Nutrition, 38*(1), 127–135. https://doi.org/10.1016/j.clnu.2017.12.014
3. Kleindorfer, D. O., Towfighi, A., Chaturvedi, S., Cockroft, K. M., Gutierrez, J., et al. (2021). 2021 guideline for the prevention of stroke in patients with stroke and transient ischemic attack: A guideline from the American Heart Association/American Stroke Association. *Stroke, 52*, e364–e467. https://doi.org/10.1161/STR.0000000000000375
4. Estruch, R., Ros, E., Salas-Salvadó, J., Covas, M. I., Corella, D., Arós, F., Gómez-Gracia, E., Ruiz-Gutiérrez, V., Fiol, M., Lapetra, J., et al. (2018). PREDIMED study investigators. Primary prevention of cardiovascular disease with a Mediterranean diet supplemented with extra-virgin olive oil or nuts. *New England Journal of Medicine, 378*, e34. https://doi.org/10.1056/NEJMoa1800389
5. Zhao, W., Tang, H., Yang, X., Luo, X., Wang, X., Shao, C., & He, J. (2018). Fish consumption and stroke risk: A meta-analysis of prospective cohort studies. *Journal of Stroke and Cerebrovascular Diseases, 28*, 604–611. https://doi.org/10.1016/j.jstrokecerebrovasdis.2018.10.036
6. Tong, T. Y. N., Appleby, P. N., Key, T. J., Dahm, C. C., Overvad, K., Olsen, A., Tjønneland, A., Katzke, V., Kühn, T., Boeing, H., et al. (2020). The associations of major foods and fibre with risks of ischaemic and haemorrhagic stroke: A prospective study of 418 329 participants in the EPIC cohort across nine European countries. *European Heart Journal, 41*, 2632–2640. https://doi.org/10.1093/eurheartj/ehaa007
7. Rees, K., Takeda, A., Martin, N., Ellis, L., Wijesekara, D., Vepa, A., Das, A., Hartley, L., & Stranges, S. (2019). Mediterranean-style diet for the primary and secondary prevention of cardiovascular disease. *Cochrane Database of Systematic Reviews, 3*, CD009825. https://doi.org/10.1002/14651858.CD009825.pub3
8. Chiavaroli, L., Viguiliouk, E., Nishi, S. K., Mejia, B. S., Rahelić, D., Kahleová, H., Salas-Salvadó, J., Kendall, C. W., & Sievenpiper, J. L. (2019). DASH dietary pattern and cardiometabolic outcomes: An umbrella review of systematic reviews and meta-analyses. *Nutrients, 11*, 338. https://doi.org/10.3390/nu11020338
9. Royal College of Physicians (UK and Ireland). (2023). National Clinical Guidelines for Stroke for the United Kingdom and Ireland (2023 ed.). Retrieved on 1 May 2024 from http://www.strokeguideline.org
10. Malaysian Society of Neuroscience. (2020). Clinical Practice Guidelines; Management of Ischaemic Stroke. Retrieved online on 1 May 2024 from http://www.acadmed.org.my
11. Sacks, F. M., Svetkey, L. P., Vollmer, W. M., Appel, L. J., Bray, G. A., et al. (2001). Effects on blood pressure of reduced dietary sodium and the dietary approach to stop hypertension (DASH) diet. *The New England Journal of Medicine, 344*, 1. https://doi.org/10.1056/NEJM200101043440101
12. Salehi-Abargouei, A., Maghsoudi, Z., Shirani, F., & Azadbakht, L. (2013). Effects of dietary approaches to stop hypertension (DASH)-style diet on fatal or nonfatal cardiovascular dis-

eases—Incidence: A systematic review and meta-analysis on observational prospective studies. *Nutrition, 29*(4), 611–618. https://doi.org/10.1016/j.nut.2012.12.018
13. Ois, A., Gomis, M., Rodríguez-Campello, A., Cuadrado-Godia, E., Jiménez-Conde, J., Pont-Sunyer, C., Cuccurella, G., & Roquer, J. (2008). Factors associated with a high risk of recurrence in patients with transient ischemic attack or minor stroke. *Stroke, 39*, 1717–1721. https://doi.org/10.1161/STROKEAHA.107.505438
14. Mitsios, J. P., Ekinci, E. I., Mitsios, G. P., Churilov, L., & Thijs, V. (2018). Relationship between glycated hemoglobin and stroke risk: A systematic review and meta-analysis. *Journal of the American Heart Association, 7*(11), e007858.
15. Russell, R. D., Black, L. J., & Begley, A. (2022). Nutrition education programs for adults with neurological diseases are lacking: A scoping review. *Nutrients, 14*, 1577. https://doi.org/10.3390/nu14081577
16. Lijehult, J., Molsted, S., Moller, T., Overgaard, D., Adamsen, L., et al. (2020). Lifestyle counselling as secondary prevention in patients with minor stroke and transient ischemic attack: Study protocol for a randomized controlled pilot study. *Pilot and Feasibility Studies, 6*, 40. https://doi.org/10.1186/s40814-020-00583-4
17. Geib, A. (2023). Understanding barriers and facilitators of dietary change following stroke: A qualitative analysis. Preprint retrieved from https://cdr.lib.unc.edu/concern/honors_theses/ns064h02d
18. Bailey, R. R., & Ipsen, M. (2022). Facilitators and barriers to performing dietary behaviours among chronic community-dwelling stroke survivors: A qualitative secondary analysis. *Disability and Health Journal, 15*(3), 101270. https://doi.org/10.1016/j.dhjo.2022.101270
19. Hill, V. A., Vickrey, B. G., Cheng, E. M., Valle, N. P., Ayala-Rivera, M., et al. (2017). A pilot trial of a lifestyle intervention for stroke survivors: Design of healthy eating and lifestyle after stroke (HEALS). *Journal of Stroke and Cerebrovascular Diseases, 26*(12), 2806–2813. https://doi.org/10.1016/j.jstrokecerebrovasdis.2017.06.058
20. He, F. J., Tan, M., Ma, Y., & MacGregor, G. A. (2020). Salt reduction to prevent hypertension and cardiovascular disease: JACC state-of-the-art review. *Journal of the American College of Cardiology, 75*, 632–647. https://doi.org/10.1016/j.jacc.2019.11.055.97
21. Reynolds, K., Lewis, B., Nolen, J. D., Kinney, G. L., Sathya, B., He, J., & Lewis, B. L. (2003). Alcohol consumption and risk of stroke: A meta-analysis. *JAMA, 289*, 579–588. https://doi.org/10.1001/jama.289.5.579

The manufacturer's authorised representative in the EU is Springer Nature Customer Service Centre GmbH, Europaplatz 3, 69115 Heidelberg, Germany. If you have any concerns regarding our products, please contact ProductSafety@springernature.com

Printed and bound by CPI Group (UK) Ltd, Croydon, CR0 4YY
01/10/2026
02225934-0013